Lecture Notes in Bioengineering

More information about this series at http://www.springer.com/series/11564

Somsubhra Gupta · Sandip Bag
Karabi Ganguly · Indranath Sarkar
Papun Biswas
Editors

Advancements of Medical Electronics

Proceedings of the First International Conference, ICAME 2015

Springer

Editors
Somsubhra Gupta
Information Technology
JIS College of Engineering
Kalyani, West Bengal
India

Sandip Bag
Biomedical Engineering
JIS College of Engineering
Kalyani, West Bengal
India

Karabi Ganguly
Biomedical Engineering
JIS College of Engineering
Kalyani, West Bengal
India

Indranath Sarkar
Electronics and Communication Engineering
JIS College of Engineering
Kalyani, West Bengal
India

Papun Biswas
Electrical Engineering
JIS College of Engineering
Kalyani, West Bengal
India

ISSN 2195-271X ISSN 2195-2728 (electronic)
Lecture Notes in Bioengineering
ISBN 978-81-322-2255-2 ISBN 978-81-322-2256-9 (eBook)
DOI 10.1007/978-81-322-2256-9

Library of Congress Control Number: 2014958893

Springer New Delhi Heidelberg New York Dordrecht London

Printed on acid-free paper

Springer (India) Pvt. Ltd. is part of Springer Science+Business Media (www.springer.com)

Invited Talks

E. Russell Ritenour, Ph.D. Professor and Chief Medical Physicist Department of Radiology and Radiological Sciences Medical University of South Carolina 96 Jonathan Lucas St., MSC 323 Charleston, SC 29425-3230 ph: 843-792-4884	
Invited talk title	Magnetic Resonance Imaging Safety Issues at High Field Strength—E. Russell Ritenour, Ph.D.
Invited talk abstract	The increasing trend toward higher magnetic fields in Magnetic Resonance Imaging (MRI) brings with it some new challenges in safety for patients and staff in the medical environment. In particular, the stronger static magnetic field gradients have the capability of producing measurable effects in diamagnetic substances as well as paramagnetic and ferromagnetic substances. The higher peak values of static magnetic field also have the potential to produce unwanted inline forces and torques upon paramagnetic and ferromagnetic substances. In the work presented here, magnetic properties of materials will be reviewed briefly. Artifacts and safety issues will be discussed with emphasis on the increased severity of effects at higher magnetic field levels. The current status of the United States Food and Drug Administration regulation of equipment will be presented with emphasis on equipment design and labeling. The equipment discussed will include devices implanted in the patient. Practical aspects of dealing with patients and doing research with the high magnetic field large bore magnets used in MRI will be emphasized

(continued)

(continued)

Bio	E. Russell (Russ) Ritenour received his Ph.D. in Physics from the University of Virginia in 1980. He was selected for a National Institutes of Health Postdoctoral Fellowship at the University of Colorado and stayed on the faculty there as director of the Graduate Program in Medical Physics until 1989 when he moved to Minnesota to become Professor and Chief of Physics and Director of the Graduate Program in Biophysical Sciences and Medical Physics at the University of Minnesota. In 2014 he became Professor and Chief Medical Physicist of the Department of Radiology and Radiological Science of the Medical University of South Carolina. He has served the American Board of Radiology in various capacities since 1986 including membership and then chair of the physics committee for the radiology resident's exam. He has served as a consultant to the U.S. Army for resident physics training and is a founding member of the Society of Directors of Academic Medical Physics Programs. He is a fellow of the ACR and a fellow and past president of the AAPM. He also served as imaging editor of the journal Radiographics and as Board Member and Treasurer of the RSNA Research and Education Foundation and as Board Member of the International Organization of Medical Physics. He is currently the Medical Physics and Informatics Editor of the American Journal of Roentgenology. His research interests include radiation safety, efficacy of diagnostic imaging, and the use of high speed networks for medical education and clinical communication
Xiaoping Hu, Ph.D. The Wallace H. Coulter Department of Biomedical Engineering Georgia Institute of Technology and Emory University Atlanta, Georgia, USA xhu3@emory.edu	
Invited talk title	Recent Developments in Magnetic Resonance Neuroimaging and Molecular Imaging—Xiaoping Hu, Ph.D.

(continued)

(continued)

Invited talk abstract	While MRI has been around for more than 40 years, significant advances are still being made. In particular, there is remarkable progress in the methodologies and applications of MRI in the study of the brain, i.e., neuroimaging, and there are also numerous innovations in molecular imaging with MRI. There are two main directions in MRI of the brain: functional brain imaging and structural connectivity. The former was ushered into the field in the early 1990s and has generated an unprecedented interest in a wide range of disciplines; it is now used not only for mapping brain function but also for understanding brain activity and ascertaining brain connectivity. The latter was introduced at about the same time but has been exploding in the last decade with the advances in gradient technology; it is now widely used for assessing the structural connectivity of the brain. For molecular imaging, MRI has been used for cell tracking, targeted imaging of biomarkers, and reporting of gene express. In this talk, I will provide an overview of these two broad directions and highlight some of the recent in my lab. More highlights in neuroimaging will include methodological developments and applications of functional and structural connectivity of the brain and MR detection of action current. As for molecular imaging, I will highlight methods for better detection of magnetic nanoparticles and development of MR reporter genes
Bio	Dr. Hu obtained his Ph.D. in Medical Physics from the University of Chicago in 1988 and his postdoctoral training there from 1988 to 1990. From 1990 to 2002, he was on the faculty of the University of Minnesota, where he became full professor in 1998. Since 2002, he has been Professor and Georgia Research Alliance Eminent Scholar in Imaging in the Wallace H. Coulter joint department of biomedical engineering at Georgia Tech and Emory University and the director of Biomedical Imaging Technology Center in the Emory University School of Medicine. Dr. Hu has worked on the development and biomedical application of magnetic resonance imaging for three decades. Dr. Hu has authored or co-authored 245 peer-reviewed journal articles. As one of the early players, Dr. Hu has conducted extensive and pioneering work in functional MRI (fMRI), including methods for removing physiological noise, development of ultrahigh field fMRI, systematic investigation of the initial dip in the fMRI signal, Granger causality analysis of fMRI data, and, more recently, characterization of the dynamic nature of resting state fMRI data. In addition to neuroimaging, his research interest also includes MR molecular imaging. Dr. Hu was deputy editor of *Magnetic Resonance in Medicine* from 2005 to 2013, Associate Editor of *IEEE Transactions on Medical Imaging* since 1994, editor of *Brain Connectivity* since its inception, and editorial board member of *IEEE Transactions on Biomedical Engineering* since 2012. He was named a fellow of the International Society for Magnetic Resonance in Medicine in 2004. He is also a fellow of IEEE and a fellow of the American Institute of Medical and Biological Engineering

Keynote Talk

Hiro Yoshida, Ph.D. Director, 3D Imaging Research, Department of Radiology, Massachusetts General Hospital Associate Professor of Radiology, HMS 25 New Chardon St, Suite 400C, Boston, MA 02114	
Keynote talk title	Advancement of diagnostic imaging decision support systems
Keynote talk abstract	In clinical practice, there is an increasing demand for fast turnaround time in obtaining high-quality multi-dimensional medical images with quantitative analysis and diagnostic results obtained from big data. Access to such images and analysis results from anywhere at any time facilitates collaboration between clinicians and specialists as well as patients and healthcare providers, thus improves the quality and timeliness of care for the patient. However, such advanced imaging is computationally intensive and requires high-end resources such as computational servers and clusters. In today's economic climate, investment in such resources is often cost-prohibitive, thus limiting broader adoption of advanced imaging techniques and quantitative analysis

(continued)

(continued)

	tools. Cloud supercomputing, or high-performance cloud computing, is an integration of high-performance computing with today's ubiquitous cloud computing. Recent advances of cloud technology provides, for the first time in history, an affordable infrastructure that delivers the supercomputer power needed for real-time processing of these state-of-the-art diagnostic images on mobile and wearable devices with an easy, gesture-based user interface. This keynote shows advancements of diagnostic imaging decision support system that address the above clinical and technical needs, by using computer-assisted virtual colonoscopy for cancer screening as a representative example. Virtual colonoscopy, also known as computed tomographic (CT) colonography, provides a patient-friendly method for early detection of colorectal lesions, and has the potential to solve the problems of capacity and safety with conventional colorectal screening methods. Virtual colonoscopy has been endorsed as a viable option for colorectal cancer screening and shown to increase patient adherence in the United States and in Europe. Anytime, anywhere access to the VC images will facilitate the high-throughput colon cancer screening. A could-supercomputer-assisted virtual colonoscopy demonstration system will be presented. In this system, virtual colonoscopy images are processed by computationally intensive algorithms such as virtual bowel cleansing and real-time computer-aided detection for improved detection performance of colonic polyps—precursor of colon cancer. A high-resolution mobile display system is connected to the could-supercomputer-assisted virtual colonoscopy to allow for visualization of the entire colonic lumens and diagnosis of colonic lesions at anytime, anywhere. The navigation through the colonic lumen is driven by a motion-based natural user interface for easy navigation and localization of colonic lesions. The current status, challenges, and promises of the system for realizing efficient diagnostic imaging decision support system will be described.

(continued)

(continued)

Bio	Hiro (Hiroyuki) Yoshida received his B.S. and M.S. degrees in Physics and a Ph.D. degree in Information Science from the University of Tokyo, Japan. He previously held Assistant Professorship in the Department of Radiology at the University of Chicago. He was a tenured Associate Professor when he left the university and joined the Massachusetts General Hospital (MGH) and Harvard Medical School (HMS), where he is currently the Director of 3D Imaging Research in the Department of Radiology, MGH and Associate Professor of Radiology at HMS. His research interests include computer-aided diagnosis, quantitative medical imaging, and imaging informatics systems in general, and in particular, in the area of the diagnosis of colorectal cancers with CT colonography. In these research areas, he has been the principal investigator on 11 national research projects funded by the National Institutes of Health (NIH) and two projects by the American Cancer Society (ACS) with total approximate direct cost of $6.5 million, and received 9 academic awards: a Magna Cum Laude, two Cum Laude, four Certificate of Merit, and two Excellences in Design Awards from the Annual Meetings of Radiological Society of North America (RSNA), an Honorable Mention award from the Society of Photo-Optical Instrumentation Engineers (SPIE) Medical Imaging, and Gold medal of CyPos Award from the Japan Radiology Congress (JRC). He also received two industrial awards on his work on system developments: 2012 Innovation Award from Microsoft Health Users Group and Partners in Excellence award from Partners HealthCare. He is author or co-author of more than 170 journal and proceedings papers and 16 book chapters, author, co-author, or editor of 14 books, and inventor and co-inventor of seven issued patents. He was guest editor of IEEE Transaction on Medical Imaging in 2004, and currently serves on the editorial boards of the International Journal of Computer Assisted Radiology, the International Journal of Computers in Healthcare, Intelligent Decision Technology: An International Journal, and ad hoc Associate Editor of Medical Physics.

From the Convener's Desk

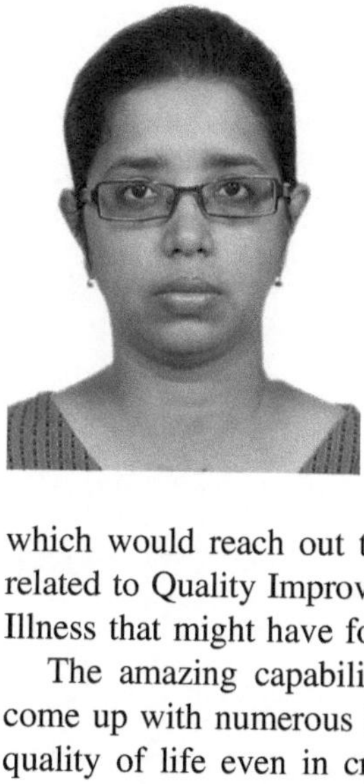

The **1st International Conference** on ***'Advancements of Medical Electronics' (ICAME 2015)*** scheduled from ***29 to 30 January 2015*** has been organized with the objective to provide a platform for exchange of research ideas, discussions on the arena of *Medical Electronics*, development of new methodologies which reach out to mankind and is beneficial for the society.

Understanding the need for interdisciplinary subjects like Biomedical Engineering, Electronics and Communication Engineering, Nanotechnology, the idea cropped up into the minds of a young group of Faculty members at JIS College of Engineering to organize a Conference which would reach out to the objective mentioned above and address the issues related to Quality Improvement of Life, Rehabilitation after Surgery or Long-term Illness that might have forced a person to compromise with a healthy lifestyle.

The amazing capabilities of Science and Technology in today's world have come up with numerous innovation and discoveries aimed towards improving the quality of life even in critical cases like quadriplegia and terminal illnesses like cancer. But often all such 'Gifts of Science and Technology' reach out to the advanced countries while the Developing Countries like India remain still in a dearth of knowledge and indigenous developments which could have catered to a hundred million population like ours.

With the zeal to serve the nation and human life, ICAME 2015 was organized. The international status of the conference will enable eminent speakers from fields like Radio Imaging, Nanotechnology, Nanomedicine, Communication, Electronics from countries outside India to amalgamate with national scientists and researchers. This would, I firmly believe, generate resources for our country and serve our people with affordable medical devices and thus increase the lifespan and quality life of our fellow Indians which are perhaps the best rewards for technologists like us.

So, I invite all of you from the international and national arena to make the most of the two days and enable us to come up with research works beneficial for mankind.

Dr. Meghamala Dutta
Head of Department of Biomedical Engineering and
Convener ICAME 2015

Program Committee

International Conference on Advancements of Medical Electronics
29–30th January 2015
Organized by
Department of Biomedical Engineering and Department of Electronics and Communication Engineering
JIS College of Engineering, Kalyani

Chief Patron

Sardar Jodh Singh, Chairman, JIS Group, India
Prof. S.M. Chatterjee, Former VC, BESU, Shibpur, Howrah

Patrons

Shri Taranjit Singh, Managing Director, JIS Group
Prof. Dr. Ajoy Roy, Vice Chancellor BESU, Shibpur, Howrah
Dr. Sajal Dasgupta, Director, Technical Education
Manpreet Kaur, CEO, JIS Group, India
Jaspreet Kaur, Trusty Member, JIS Group

Organizing Chair

Prof. Dr. Asit Guha, Advisor, JIS Group

Executive Committee Member

Mr. U.S. Mukherjee, Deputy Director, JIS Group
Prof. Sankar Ray, Principal, JISCE
Prof. Urmibrata Bandopadhyay, Former Principal, JISCE
Mrs. Shila Singh Ghosh, General Manager (Corporate Relation), JIS Group, India
Prof. Ardhendu Bhattacharya, Dean, JISCE
Prof. Debatosh Guha, Chair, IEEE Kolkata Section
Prof. S.K. Mitra, HOD-Department of Electrical Engineering, JISCE
Prof. P.K. Bardhan, HOD-Department of Mechanical Engineering, JISCE
Prof. Pranab K. Banerjee, Former Professor, Jadavpur University

Convener

Dr. Meghamala Dutta, HOD-Department of Biomedical Engineering, JISCE

Co-Convener

Mrs. Swastika Chakraborty, HOD-Department of Electronics and Communication Engineering

Dr. Sandip Bag, Department of Biomedical Engineering

International Advisory Board

- Dr. Hiro Yoshida—Professor, Harvard Medical School
- Dr. E. Russel Ritenour—Professor, University of Minnesota
- Dr. Jim Holte—Emeritus Professor, University of Minnesota
- Dr. Milind Pimprikar—Caneus Canada
- Dr. Xiaoping Hu—Professor, Georgia Tech/Emory School
- Dr. Todd Parrish—Professor, North Western University
- Dr. Rinti Banerjee—Professor, IIT, Powai
- Dr. Amit K. Roy—Professor, IT-BHU
- Prof. T. Asokan—Professor, IIT Madras
- Prof. Amit Konar—Professor, Department of Electronics and Telecommunication, JU
- Prof. Salil Sanyal—Professor, Department of Computer Science and Engineering, JU
- Prof. D. Patronobis—Former Professor, Department of Instrumentation Engineering, JU
- Prof. Amitava Gupta—Professor, Department of Power Plant Engineering, JU
- Prof. D.N. Tibarewala—Professor, School of Bioscience and Engineering, JU
- Prof. Chandan Sarkar—Department of Electronics Engineering, JU

Organizing Committee

Program Chair

Dr. Somsubhra Gupta, HOD-Department of Information Technology, JISCE

Publication Chair

Dr. Karabi Ganguly, Department of Biomedical Engineering, JISCE

Publication Committee

Dr. Sabyasachi Sen, HOD-Department of Physics

Dr. Indranath Sarkar, Department of Electronics and Communication Engineering, JISCE

Partha Roy, Department of Electrical Engineering, JISCE

Papun Biswas, Department of Electrical Engineering, JISCE
Ranjana Ray, Department of ECE, JISCE
Dr. Biswaroop Neogi, Department of Electronics and Communication Engineering, JISCE

Finance Chair

Dr. Anindya Guha, HOD-Department of Humanities, JISCE

Finance Committee

Manas Paul, Department of Computer Application, JISCE
Anirban Patra, Department of Electronics and Communication Engineering, JISCE
Santanu Mondal, Department of Mathematics, JISCE
Rupankar Mukherjee, Department of Biomedical Engineering, JISCE
S.k. Malek, Accountant, JISCE

Reception and Registration Chair

Nilotpal Manna, HOD-Department of Electronics and Instrumentation Engineering, JISCE

Reception and Registration Committee

Souvik Das, Department of Biomedical Engineering, JISCE
Bikash Dey, Department of Electronics and Communication Engineering, JISCE
Sourish Halder, Department of Electronics and Communication Engineering, JISCE
Suparna Dasgupta, Department of Information Technology, JISCE
Soumen Ghosh, Department of Electronics and Instrumentation Engineering, JISCE
Sumanta Bhattacharya, Department of Computer Application, JISCE
Moumita Pal, Department of Electronics and Communication Engineering, JISCE
Sraboni Biswas, Administrative Staff, JISCE

Travel and Accomodation

Debashis Sanki, Department of Information Technology, JISCE
Debashis Majumder, Department of Mathematics, JISCE
Mainuck Das, Department of Electronics and Instrumentation Engineering, JISCE
Basudeb Dey, Department of Electrical Engineering, JISCE
Saptarshi Nandi, Department of Civil Engineering, JISCE
Kunal Banerjee, Department of Mechanical Engineering, JISCE

Hospitality

Debashis Sanki, Department of Information Technology, JISCE
Rupak Bhattacharjee, Department of Mathematics, JISCE
Aniruddha Ghosh, Department of Electronics and Communication Engineering, JISCE
Dipankar Ghosh, Department of Mathematics, JISCE
Prolay Ghosh, Department of Information Technology, JISCE

Saikat Chakraborty, Administrative Staff
Gourango Halder, Department of Biomedical Engineering, JISCE
G.C. Sarkar, HR Manager, JISCE
Sumindar Roy, Department of Electronics and Instrumentation Engineering, JISCE

Website and e-Communication

Soumyabrata Saha, Department of Information Technology, JISCE
Nilotpal Haldar, Department of Electronics and Communication Engineering, JISCE
Sumanta Bhattacharya, Department of Computer Application, JISCE

Volunteer and Student Management

Avik Sanyal, CMS, JISCE
Subham Ghosh, Department of Electronics and Communication Engineering, JISCE
Dr. Shubhomoy Singha Roy, Department of Physics, JISCE

Contents

About the Editors

Dr. Somsubhra Gupta is presently the Head of the Department of Information Technology, JIS College of Engineering (An Autonomous Institution). He graduated from the University of Calcutta and completed his Master's from Indian Institute of Technology, Kharagpur. He received Ph.D. from University of Kalyani. His area of teaching is Algorithm and allied domains and research area is Machine Intelligence. In research, he has around 56 papers including Book Chapters so far in national/international journals/proceedings and over 40 citations. He is Principal Investigator/Project Coordinator of some Research projects (viz. RPS scheme AICTE).

Dr. Sandip Bag is Assistant Professor in the Department of Biomedical Engineering, JIS College of Engineering, Kalyani since 2005. He completed his Ph.D., Postgraduation and Graduation from Jadavpur University, Kolkata. Dr. Bag has published over 22 papers in national and international journals and proceedings and also presented papers in India and abroad. He received financial and research grant from DST and UGC, respectively. He developed various laboratories such as Biomaterials and Biomedical Instrumentation laboratory in JISCE.

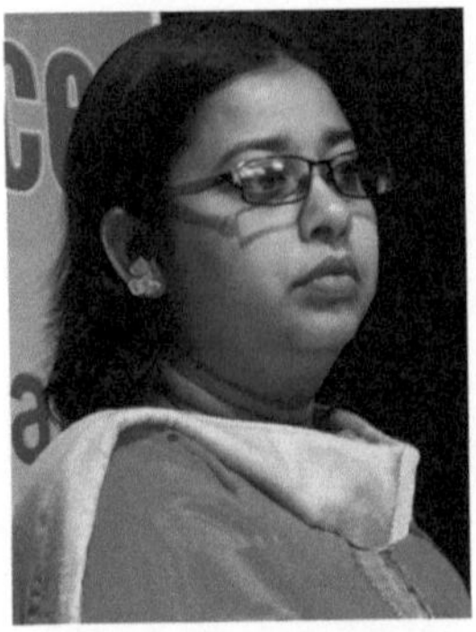

Dr. Karabi Ganguly is Assistant Professor in the Department of Biomedical Engineering, JIS College of Engineering, Kalyani. She received her Ph.D. degree from Jadavpur University, India and completed her Postrgaduation and Graduation from the University of Calcutta. She published many papers in national/international journals and proceedings. She has been invited as the reviewer in many international and national Conferences. Her research interest includes Cellular Biochemistry, Physiology, and Clinical Oncology.

Dr. Indranath Sarkar obtained his Ph.D. from University of Kalyani in the year 2013. He earned M.E. from Jadavpur University in the year 2002 and B.E. in Electronics and Communication Engineering from Regional Engineering College (Presently known as National Institute of Technology), Durgapur in the year 1999. He is presently working as Assistant Professor in the Department of Electronics and Communication Engineering, JIS College of Engineering Kalyani, India. He has published many papers in national/international journals and proceedings.

Papun Biwas is Assistant Professor in the Department of Electrical Engineering, JIS College of Engineering, Kalyani, West Bengal, India. He received his M.Tech. degree in Electrical Engineering from the University of Calcutta in 2007. Currently, he is a Research Fellow in the Department of Computer Science and Engineering, University of Kalyani. His research interests pertain to different areas of Soft and Evolutionary Computing in the area of Fuzzy Multi-objective Decision Making.

Part I
Medical Image Processing and Analysis

Proposed Intelligent System to Identify the Level of Risk of Cardiovascular Diseases Under the Framework of Bioinformatics

Somsubhra Gupta and Annwesha Banerjee

Abstract This paper proposed a method to implement an intelligent system to find out the risk of cardiovascular diseases in human being. Genetics play a direct and indirect role in increasing the risks of cardiovascular diseases. Habits and individual symptom viz. suffering from diabetes, obesity and hypertension also can influence the risk of the said diseases. Excessive energy accumulation in ones body can create fatal problem in health. In this paper, method has been proposed to the proposed to investigate three major factors i.e. family history of CVD, Other diseases and Average Energy Expenditure and find out the of level of risks of cardiovascular diseases.

Keywords Bioinformatics · Cardiovascular disease · Energy expenditure · Genetics · Intelligent systems · Production system

1 Introduction

In recent times, the cardiovascular diseases are one of the major causes of mortality in human beings. Numerous factors are there that increases the risks of CVD like obesity, diabetes, hypertensions which in turn are being caused by less energy expenditure in human being and heredity also plays a major role in the causing on CVD. However, it is also likely that people with a family history of heart disease share common environments and risk factors that increase their risk.

Work is one form of energy, often called mechanical energy. When to throw a ball or run a mile, work has been done; mechanical energy has been produced.

The sun is the ultimate source of energy. Solar energy is harnessed by plants, through photosynthesis, to produce plant carbohydrates, fats, or proteins, all forms

S. Gupta (✉) · A. Banerjee
Department of Information Technology, JIS College of Engineering, Kalyani Block A Phase III, Nadia, West Bengal, India
e-mail: gsomsubhra@gmail.com

A. Banerjee
e-mail: annwesha.banerjee@gmail.com

S. Gupta et al. (eds.), *Advancements of Medical Electronics*,
Lecture Notes in Bioengineering, DOI 10.1007/978-81-322-2256-9_1

of stored chemical energy. When humans consume plant and animal products, then carbohydrates, fats, and proteins undergo a series of metabolic changes and are utilized to develop body structure, to regulate body processes, or to provide a storage form of chemical energy. Less energy expenditure causes obesity that in turn increases the risk on CVD. By 2015, nearly one in every three people worldwide is projected to be overweight, and one in ten is expected to be obese [1].

Cardiovascular disease is one of the alarming threats. World Health Report showed that cardiovascular disease will be the major cause of death [2]. Sitting time and no exercise activity have been linked in epidemiological studies to rates of metabolic syndrome, type 2 diabetes, obesity, and CVD Regional Body fat distribution is one of the major factors that increase the CVD [3]. As a definite cause of cardiovascular morbidity and mortality [4] it is important to consider the potential impact of dietary sugar on weight gain.

Sugar intake can increase carbohydrate fuel reserves and physical performance [5]. There have been a number of studies that link sugar consumption to hypertension in animals [6]. In humans, there is one report that high dietary sugar intake enhances the risk of CHD in diabetic individuals who use diuretics [7].

Hypertension increases the risk of stroke in individuals [8]. It was been reported by Kornegay et al. [9] that there is a reasonable agreement between proband-reported family history of stroke and self-reported personal history of stroke in members of the proband's family. It is been examine the role of sedentary behaviors, especially sitting, on mortality, cardiovascular disease, type 2 diabetes, metabolic syndrome risk factors, and obesity [10].

Significant evidence for linkage heterogeneity among hypertensive sib pairs stratified by family history of stroke suggests the presence of genes influencing susceptibility to both hypertension and stroke [11].

Physical inactivity may induce negative effects on relatively fast-acting cellular processes in skeletal muscles or other tissues regulating risk factors like plasma triglycerides and HDL cholesterol [12–14].

More than 90 % of the calories expended in all forms of physical activity were due to this pattern of standing and no exercise ambulatory movements because individuals did not exercise and because the energy expenditure associated with no exercise activity thermogenesis (NEAT) while sitting was small [15]. Obviously, 6–12 h/day of no exercise activity is beyond what anyone would exercise regularly. Laboratory rats housed in standard cages without running wheels also recruit postural leg muscles for 8 h/day [16].

In this paper, how to find the influence of heredity and energy expenditure in the cardiovascular disease has been presented.

2 Genetic Pattern and Inheritance

DNA is treated as "blue print of life". It contains all the information to create life. DNA contains the information needed to create the amino acids sequences of proteins. The unit of building block of DNA Adenine (A), Cytosine (C), Guanine

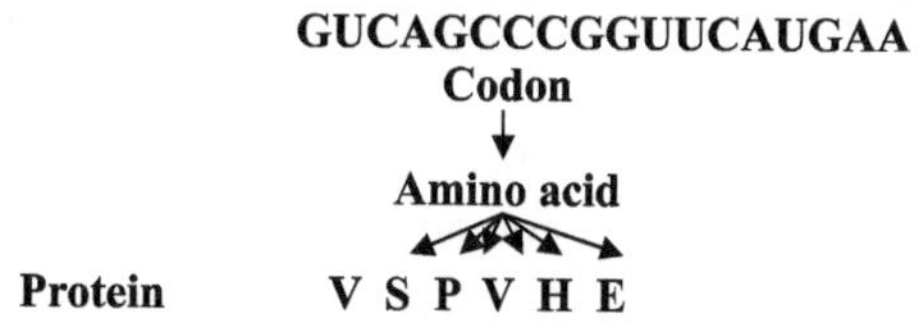

Fig. 1 mRNA to protein

(G), and Thymine (T) are the four bases in DNA. A pairs with T that is 2H bond and C pairs with G that is 3H bond [17].

Protein is a linear sequence of amino acids, shown in the Fig. 1 form a very long chain via peptide linkage. Gene is a segment of DNA. Inheritance pattern are the predictable pattern seen in the transmission of genes from one generation to the next and their expression in the organism that possesses them.

Offspring inherits genotypes from their parents. Diseases runs in family hierarchy. A team of Portugal Hospital de Criancas studied the issue of genetic susceptibility in stokes [18]. It was concluded that two genes did contribute to the development of disease.

2.1 Genetics and Cardiovascular Diseases

Life style and environment play a role in causing cardiovascular disease in individual. But it is also proved by research that heredity also has a major role in cardiovascular disease. Gene mutation and polymorphism are sometimes directly be related to cardiovascular diseases. A study of University of Texas showed that chromosome carries polymorphism for hypertension and stroke for Caucasian patient and chromosome 19 for African and American [19].

It was been reported by Kornegay et al. [9] that there is a reasonable agreement between proband-reported family history of stroke and self-reported personal history of stroke in members of the proband's family. The accuracy of reporting is high for other common diseases, such as myocardial infarction [20] coronary heart disease, diabetes, hypertension, and asthma [21]. Positive family history was defined by proband-reported history of stroke or cerebral hemorrhage diagnosed by a physician for either biologic parent or at least 1 full biologic sibling.

3 Other Diseases and CVD

Other diseases also influences the heart disease that genetic also responsible indirectly for the heart diseases. Hypertension, obesity, diabetes play a major role in increasing risk of heart diseases and at the same time behaviours like smoking habits also influences the probability of heart diseases. In humans, there is one report that high dietary sugar intake enhances the risk of CHD in diabetic individuals who use diuretics [7].

A number of studies have shown that specific topographic features of adipose tissue are associated with metabolic complications that are considered as risk factors for CVD such as insulin resistance, hyperinsulinemia, glucose intolerance and type II diabetes mellitus, hypertension, and changes in the concentration of plasma lipids and lipoproteins. Metabolic correlates the body fat distribution [3].

4 Energy Expenditure and Disease

Recent studies have suggested that respiratory diseases, such as chronic obstructive pulmonary disease (COPD) and obstructive sleep apnea syndrome (OSAS), influence energy expenditure (EE) [22]. Energy can be measured in either joules or calories. A joule (J) can be defined as the energy used when 1 kilogram (kg) is moved 1 metre (m) by the force of 1 newton (N). A calorie (cal) can be defined as the energy needed to raise the temperature of 1 g of water from 14.5 to 15.50 °C. In practice, both units are used just as different units are used to measure liquids, e.g. pints, liters. One calorie is equivalent to 4.184 J.

There are three main components of Total Energy Expenditure (TEE) in humans:

1. **Basal Metabolic Rate (BMR)**—Energy expended at complete rest in a post-absorptive state; accounts for approximately 60 % of TEE in sedentary individuals.
2. **Thermal Effect of Food (TEF)**—Increase in energy expenditure associated with digestion, absorption, and storage of food and nutrients; accounts for approximately 10 % of TEE.
3. **Energy Expenditure of Activity**—Further classified as Exercise-related Activity Thermo genesis
4. **Growth**—The energy cost of growth has two components: (1) the energy needed to synthesize growing tissues; and (2) the energy deposited in those tissues. The energy cost of growth is about 35 % of total energy requirement during the first 3 months of age, falls rapidly to about 5 % at 12 months and about 3 % in the second year, remains at 1–2 % until mid-adolescence, and is negligible in the late teens.
5. **Pregnancy**—During pregnancy, extra energy is needed for the growth of the foetus, placenta and various maternal tissues, such as in the uterus, breasts and fat stores, as well as for changes in maternal metabolism and the increase in maternal effort at rest and during physical activity.
6. **Lactation**—The energy cost of lactation has two components: (1) the energy content of the milk secreted; and (2) the energy required to produce that milk. Well-nourished lactating women can derive part of this additional requirement from body fat stores accumulated during pregnancy.

Associated with active sports or exercise and Non-Exercise Activity Thermo genesis (NEAT) associated with activities of daily living, fidgeting, spontaneous muscle contraction etc. Exercise-related Activity.

4.1 Energy Expenditure Measurement

There are few methods are available to measure the energy of human body. Following tables summaries the commonly used methods to measure energy (Table 1):

Over the last few decades, research in public health has associated inactivity with a number of ailments and chronic diseases, such as colon cancer, type II diabetes, osteoporosis and coronary heart disease.

Humans have been increasingly spending more time in sedentary behaviours involving prolonged sitting. It is found by research that most of the time people sitting physically idle i.e. with out any exercise. The amount of exercise is very limited, generally to <150 min/week [23].

Table 1 Human body energy measuring methods

Sl. No.	Methods	Description
1.	Accelerometer	Techniques for measuring energy expenditure involve either measuring heat loss directly (direct calorimetry) or measuring a proxy of heat loss (indirect calorimetry) such as oxygen (O_2) consumption or carbon dioxide (CO_2) production [27]
2.	Direct calorimeter	Done under control-led laboratory conditions in insulated chambers that measure changes in air temperature associated with the heat being released by a subject
3.	Indirect calorimeter	Measures inspired and expired oxygen and carbon dioxide in order to calculate resting energy expenditure (REE) and respiratory quotient (RQ) using the abbreviated Weir equation, where: REE (Kcal/day) = $[(VO_2 \times 3.94) + (VCO_2 \times 1.11)] \times 1{,}440$ min/day VCO_2 (carbon dioxide production) VO_2 (oxygen consumption)
4.	DLW	Technique in which stable isotopes of hydrogen and oxygen in water ($2H_2\ {}_{18}O$) are ingested
5.	First beat beat-by-beat based heart rate method	Probably the most frequently used indirect parameter in EE-assessment, mainly due to following reasons: • HR is easy to measure • EE is easily accessible from HR data • HR-based EE-estimates are relatively accurate in steady exercise conditions [28]

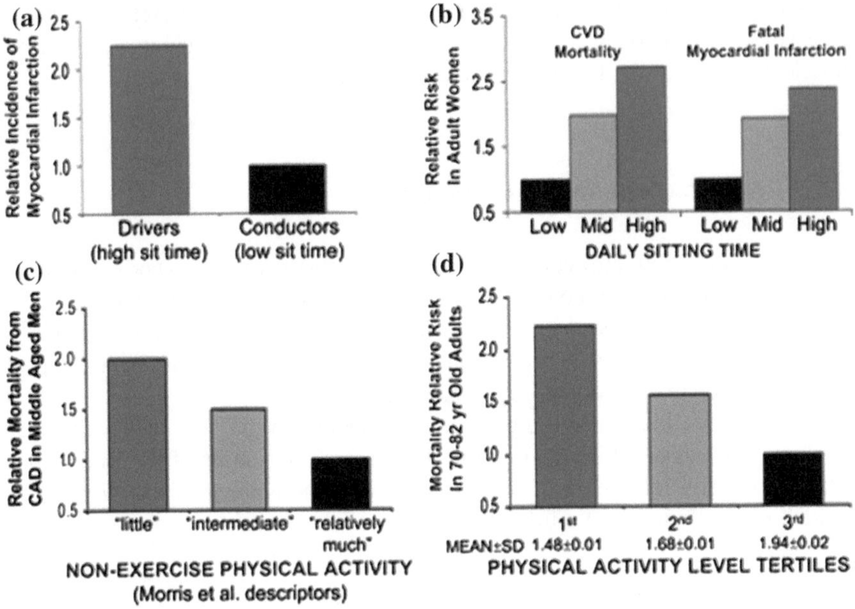

Fig. 2 Physical activity rate versus mortality

Figure 2 shows that the less activity tends to the risk of premature myocardial infarction (*A*) and mortality from coronary artery disease (*C*) [24]. These general findings were subsequently confirmed in studies in middle-aged women (*B*) [25] and an elderly group (*D*) [26].

5 Proposed Method

Through our paper we proposed a intelligent system that working on the prediction of probability Cardiovascular diseases. Some Genetics test to identify the gene which influences the CVD but theses types of tests are costly from most the countries. So we proposed the methods depending upon the following parameters. Figure 3 shows the block diagram of the proposed method.

Family History of Cardiovascular disease: We proposed to investigate up to two level of hierarchy in parents.

Here H is the set of data values assigned for different level of CVD associated in family hierarchy.

H = {0, 0.25, 0.50, 0.75, 1}
0: no family history of CVD
0.25: CVD history in one level of ancestor in parents

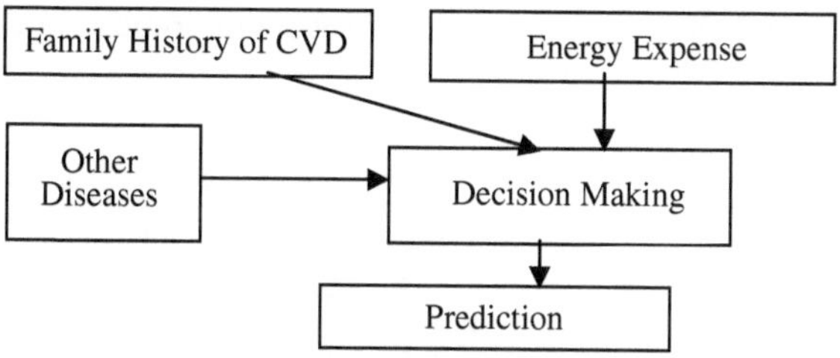

Fig. 3 Proposed intelligent systems to predict: block diagram

0.50: CVD history in one level of ancestors in either of the parents
0.75: CVD history in two levels of ancestors in either of the parents
1: CVD history in two level ancestors in parents

Energy Expenditure (EE)

Average weekly energy expenditure will be measured for individuals.
EE is the set of data values for energy expenditure amount.

EE = {0,0.20, 0.40, 0.60, 0.80, 1}
0: 0 % of EE
0.20: (>0 % to ≤25 % of EE)
0.60: (>25 % to ≤50 % of EE)
0.80: (>50 % to ≤75 % of EE)
1: (>75 % to ≤100 % of EE)

Other diseases: Hypertension, Obesity, diabetes:

OD is the set of data values for other disease that influence the risk of CVD

OD = {0, 0.25, 0.50, 1}
0: Having none of the diseases
0.25: Having one diseases
0.50: Having two diseases
1: Having all three diseases

The production system of the proposed Intelligent System has been presented below diagrammatically in the Fig. 3.

Probability of CVD:

$$P\ (CVD) = w1\ (V1) + w2\ (V2) + w3\ (V3)$$

where w1, w2 and w3 are the weight values for the considered three factor that influence the risk of CVD i.e. w1 is the weight for family history, w2 is weight for energy expenditure and w3 is the weight for having other diseases.

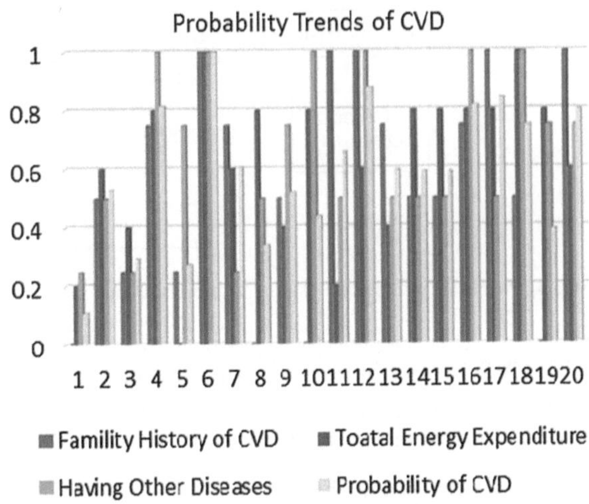

Fig. 4 Probability trends of CVD considering three factor of our proposed methods

We considered the value of weight as follows:

w: {0.5, 0.3, 0.2}

v1, v2 and v3 are the data value for the three factors being considered for our proposed method i.e. v1 is for family history, v2 is for EE and v3 is for having other diseases.

Figure 4 shows the Probability trend of CVD among individuals considering the Family History of CVD, Total Energy expenditure and Effects of other diseases.

6 Implication and Future Scope

From the investigation of our proposed method it is being proved that low energy expenditure, previous family history of CVD and having hypertension. Obesity and diabetes increases the risk of CVD. Previous family history of CVD and having other diseases are directly proportional in causing of CVD where Energy Expenditure is inversely proportional to the causing of CVD.

1. Family History of CVD ∞ CVD
2. Other Diseases ∞ CVD
3. Energy expenditure ∞ 1/CVD

Our future scope of study is to gene wide scan to find out the probability of CVD and other diseases among family history and to find out the relationship among them.

7 Conclusions

In this paper, the heredity of CVD, Other diseases and Energy expenditure of human body has been investigated and then implement an intelligent machine to find out the probability of CVD in individuals has been proposed. The proposed method is deterministic and generalized one but in practical the establishment of the factor is trivial.

References

1. World Health Organization (2011) Obesity and Overweight. Fact sheet No. 311. http://www.who.int/mediacentre/factsheets/fs311/en/index.html. Accessed 20 Jan 2011
2. World Health Organization (2002) World health report 2002: reducing risks, Promoting healthy life. WHO, Geneva, 2002
3. Despres JP, Moorjani S, Lupien PJ, Tremblay A, Adeau A, Bouchard C (2014) Obesity favors apolipoprotein E- and C-III-containing high density lipoprotein subfractions associated with risk of heart disease. J Lipid Res 55(10):2167–2177
4. Eckel RH, Krauss RM (1998) American heart association call to action obesity as a major risk factor for coronary heart disease. AHA Nutr Committee Circ 97:2099–2100
5. Hill JO, Prentice AM (1995) Sugar and body weight regulation. Am J Clin Nutr 62(suppl 1):264S–273S
6. Preuss HG, Zein M, MacArthy P et al (1998) Sugar-induced blood pressure elevations over the lifespan of three substrains of Wistar rats. J Am Coll Nutr 17:36–47
7. Sherman WM (1995) Metabolism of sugars and physical performance. Am J Clin Nutr 62(suppl):228S–241S
8. Klungel O, Stricker B, Paes A, Seidell J, Bakker A, Voko Z, Breteler M, Boer A (1999) Excess stroke among hypertensive men and women attributable to undertreatment of hypertension. Stroke 30:1312–1318
9. Kornegay C, Liao D, Bensen J, Province M, Folsom A, Ellison C (1997) The accuracy of proband-reported family history of stroke: the FHS Study. Am J Epidemiol 145:S82(Abstract)
10. Hamilton T, Deborah G. Hamilton T, Theodore W. Zderic1 (2007) Role of low energy expenditure and sitting in obesity, metabolic syndrome, type 2 diabetes, and cardiovascular disease. Marc Diabetes 56(11):2655−2667
11. Morrison AC, Brown A, Kardia SLR, Turner ST, Boerwinkle E (2003) Evaluating the context-dependent effect of family history of stroke in a genome scan for hypertension. Stroke 34:1170–1175. doi:10.1161/01.STR.0000068780.47411.16
12. Bey L, Hamilton MT (2003) Suppression of skeletal muscle lipoprotein lipase activity during physical inactivity: a molecular reason to maintain daily low-intensity activity. J Physiol 551:673–682
13. Hamilton MT, Hamilton DG, Zderic TW (2004) Exercise physiology versus inactivity physiology: an essential concept for understanding lipoprotein lipase regulation. Exerc Sport Sci Rev 32:161–166
14. Zderic TW, Hamilton MT (2006) Physical inactivity amplifies the sensitivity of skeletal muscle to the lipid-induced downregulation of lipoprotein lipase activity. J Appl Physiol 100:249–257
15. Levine JA, Lanningham-Foster LM, McCrady SK, Krizan AC, Olson LR, Kane PH, Jensen MD, Clark MM (2005) Interindividual variation in posture allocation: possible role in human obesity. Sci 307:584–586
16. Hennig R, Lømo T (1985) Firing patterns of motor units in normal rats. Nat 314:164–166

17. Wong L (2011) Some new results and tools for protein function prediction, RNA target site prediction, genotype calling, environmental genomics, and more. J Bioinform Comput Biol 9 (6):5–7
18. Wang JG, Staessen Ja (2000) Genetic polymorphism in the rennin-angiotenisn system: relevance for susceptibility to cardio vascular diseases. Eur J Phamacol 410(2–3):289–302
19. Morrison AC, Brown A, Kardia SL, Turner ST Boerwinkle E Genetic Epidemiology Network of Arteriopathy (GENOA) Study
20. Kee F, Tiret L, Robo J, Nicaud V, McCrum E, Evans A, Cambien F (1993) Reliability of reported family history of myocardial infarction. BMJ 307:1528–1530
21. Bensen J, Liese A, Rushing J, Province M, Folsom A, Rich S, Higgins M (1999) Accuracy of proband reported family history: the NHLBI Family Heart Study (FHS). Genet Epidemiol 17:141–150
22. Gregersen NT, Chaput JP, Astrup A, Tremblay A (2008) Human Energy Energy expenditure and respiratory diseases: is there a link. Expert Rev Respir Med 2(4):495–503. doi:10.1586/17476348.2.4.495
23. Morrison AC, Brown A, Kardia SL (2003) Evaluating the context-dependent effect of family history of stroke in a genome scan for hypertension. Stroke 34(5):1170–1175 (Epub 2003 Apr 24)
24. Morris JN, Heady JA, Raffle PA, Roberts CG, Parks JW (1953) Coronary heart-disease and physical activity of work. Lancet 265:1053–1057
25. Weller I, Corey P (1998) The impact of excluding non-leisure energy expenditure on the relation between physical activity and mortality in women. Epidemiol 9:632–635
26. Manini TM, Everhart JE, Patel KV, Schoeller DA, Colbert LH, Visser M, Tylavsky F, Bauer DC, Goodpaster BH, Harris TB (2006) Daily activity energy expenditure and mortality among older adults. JAMA 296:171–179
27. An Energy Expenditure Estimation Method Based on Heart Rate, Measurement Firstbeat Technologies Ltd
28. Kruger J, Yore MM, Kohl HW 3rd (2007) 3rd Leisure-time physical activity patterns by weight control status: 1999–2002 NHANES. Med Sci Sports Exerc 39:788–795

Real Time Eye Detection and Tracking Method for Driver Assistance System

Sayani Ghosh, Tanaya Nandy and Nilotpal Manna

Abstract Drowsiness and fatigue of automobile drivers reduce the drivers' abilities of vehicle control, natural reflex, recognition and perception. Such diminished vigilance level of drivers is observed at night driving or overdriving, causing accident and pose severe threat to mankind and society. Therefore it is very much necessary in this recent trend in automobile industry to incorporate driver assistance system that can detect drowsiness and fatigue of the drivers. This paper presents a nonintrusive prototype computer vision system for monitoring a driver's vigilance in realtime. Eye tracking is one of the key technologies for future driver assistance systems since human eyes contain much information about the driver's condition such as gaze, attention level, and fatigue level. One problem common to many eye tracking methods proposed so far is their sensitivity to lighting condition change. This tends to significantly limit their scope for automotive applications. This paper describes real time eye detection and tracking method that works under variable and realistic lighting conditions. It is based on a hardware system for the real-time acquisition of a driver's images using IR illuminator and the software implementation for monitoring eye that can avoid the accidents.

Keywords Vigilance level · Eye tracking · Deformable template · Edge detection · Template-based correlation · IR illuminator

S. Ghosh (✉) · T. Nandy · N. Manna
Department of Electronics and Instrumentation Engineering,
JIS College of Engineering, Kalyani, Nadia 741235, India
e-mail: sayani.rimi@gmail.com

T. Nandy
e-mail: tanayanandy@gmail.com

N. Manna
e-mail: nmanna324@gmail.com

S. Gupta et al. (eds.), *Advancements of Medical Electronics*,
Lecture Notes in Bioengineering, DOI 10.1007/978-81-322-2256-9_2

1 Introduction

The increasing number of traffic accidents due to a drivers' diminished vigilance level is a serious problem for the society. Drivers' abilities of vehicle control, natural reflex, recognition and perception decline due to drowsiness and fatigue, reducing the drivers' vigilance level. These pose serious danger to their own lives as well as lives of other people. According to the U.S. National Highway Traffic Safety Administration (NHTSA), drowsiness and falling asleep while driving are responsible for at least 100,000 automobile crashes annually [1, 2]. An annual average of roughly 40,000 nonfatal injuries and 1,550 fatalities result from these crashes. These figures only present the casualties happening during midnight to early morning, and underestimate the true level of the involvement of drowsiness because they do not include crashes during daytime hours. Vehicles with driver intelligence system that can detect drowsiness of the driver and send alarm may avert fatal accidents.

Several efforts to develop active safety systems have been reported in the literature [3–11] for reducing the number of automobile accidents due to reduced vigilance. Among deferent techniques, measurement of physiological conditions like brain waves, heart rate, and pulse rate [10, 12] yields maximum detection accuracy. However these techniques are intrusive as the sensing elements (electrodes) for measurement require physical contact with drivers causing annoyance. Less intrusive techniques like eyelid movement or gaze or head movement monitoring techniques [13] with head-mounted devices as eye tracker or special contact lens also deliver good results. These techniques, though less intrusive, are still not practically acceptable. A driver's state of vigilance can also be characterized by the behaviors of the vehicle he/she operates. Vehicle behaviors including speed, lateral position, turning angle, and moving course are good indicators of a driver's alertness level. While these techniques may be implemented non-intrusively, they are, nevertheless, subject to several limitations including the vehicle type, driver experiences, and driving conditions [4].

Fatigue in people can be easily observed by certain visual behaviors and changes in their facial features like the eyes, head, and face. The image of a person with reduced alertness level exhibits some typical visual characteristics that include slow eyelid movement [14, 15], smaller degree of eye openness (or even closed), frequent nodding [16], yawning, gaze (narrowness in the line of sight), sluggish in facial expression, and sagging posture. To make use of these visual cues, increasingly popular and non-invasive approach for monitoring fatigue is to assess a driver's vigilance level through visual observation of his/her physical conditions using a camera and state-of-the-art technologies in computer vision. Techniques using computer vision are aimed at extracting visual characteristics that typically characterize a driver's vigilance level from his/her video images. In a recent workshop [17] sponsored by the Department of Transportation (DOT) on driver's vigilance, it is concluded that computer vision represents the most promising non-invasive technology to monitor driver's vigilance.

There are many literatures reporting the fatigue monitoring systems based on active real-time image processing technique [3–7, 9, 10, 15, 18–21]. Detection of driver fatigue is primarily focused in efforts. Characterization of a driver's mental state from his facial expression is discussed by Ishii et al. [9]. A vision system from line of sight (gaze) to detect a driver's physical and mental conditions is proposed by Saito et al. [3]. A system for monitoring driving vigilance by studying the eyelid movement is described by Boverie et al. [5] and results are revealed to be very promising. A system for detection of drowsiness is explained at Ueno et al. [4] by recognizing the openness or closeness of driver's eyes and computing the degree of openness. Qiang et al. [18] describes a real-time prototype computer vision system for monitoring driver vigilance, consisting of a remotely located video CCD camera, a specially designed hardware system for real-time image acquisition and various computer vision algorithms for simultaneously, real-time and non-intrusively monitoring various visual bio-behaviors typically characterizing a driver's level of vigilance. The performance of these systems is reported to be promising and comparable to the techniques using physiological signals. This paper focuses an effort to develop a low cost hardware system that may be incorporated at dashboard of vehicles to monitor eye images movements pertaining to drowsiness of driver. The paper is organized with background theory describing various processes of eye detection, followed by proposed scheme and implementation. Finally experimental observations and results are tabulated and discussed.

2 Background Theory

To analyze the eye movement and tracking, image processing technique is employed which treats images as two dimensional signals while applying already set signal processing methods. It is a type of signal dispensation in which input is image, like video frame or photograph and output may be image or characteristics associated with that image. The images from camera are converted into digital form that are thereby enhanced and performed some filtering and logic operations on them, to extract some useful and desired information.

Methods used for eye detection and tracking rely on various prior assumptions of the image data and in general two classes of approaches exist. One common and effective approach is to exploit active illumination from infrared (IR) light emitters. Through appropriate synchronization schemes and by using the reflective properties of the pupil when exposed to near infrared light (dark or bright pupil effects) the eye can be detected and tracked effectively. In addition to controlling the light conditions, IR also plays an important role for some gaze estimation methods. Other approaches avoid the use of active illumination, and rely solely on natural light. But eye detection much difficult in absence of active illumination as some assumptions on the image data is to be made. A method based on active light is the most predominate in both research and in commercial systems. Ebisawa and Satoh [22] use a novel synchronization scheme in which the difference between images obtained from on axis and off axis light emitters are used for tracking. Kalman filtering, the Mean shift algorithm, and combinations of Kalman and mean shift filtering are applied for eye tracking. The success of these approaches is highly dependent on external light sources and the apparent size of the pupil. Efforts are made to focus on improving eye tracking under various light conditions. Sun light and glasses can seriously disturb the reflective properties of IR light.

Eye tracking and detection methods fall broadly within three categories, namely deformable templates, appearance-based, and feature-based methods. Deformable template and appearance-based methods are based on building models directly on the appearance of the eye region while the feature-based method is based on extraction of local features of the region. In general appearance models detect and track eyes based on the photometry of the eye region. A simple way of tracking eyes is through template-based correlation. Tracking is performed by correlation maximization of the target model in a search region. Grauman et al. [23] uses background subtraction and anthropomorphic constraints to initialize a correlation-based tracker. Matsumoto and Zelinsky [24] present trackers based on template matching and stereo cameras. Excellent tracking performance is reported, but the method requires a fully calibrated stereo setup and a full facial model for each user. The appearance of eye regions share commonalities across race, illumination and viewing angle. Rather than relying on a single instance of the eye region, the eye model can be constructed from a large set of training examples with varying pose and light conditions. Based on the statistics of the training set a classifier can be constructed for detection purposes over a larger set of subjects. Eye region localization by Eigen images uses a subset of the principal components of the training data to construct a low-dimensional object subspace to represent the image data. Recognition is performed by measuring distances to the object subspace. The limitations of the methods, which are purely based on detection of eyes in individual frames, are that they do not make use of prior information from previous frames which can be avoided by temporal filtering.

Deformable template-based method relies on a generic template which is matched to the image. In particular deformable templates, an eye model is constructed in which the eye is located through energy minimization. The system is to be robust to the variations of the template and the actual image for analytical approximations. Yuille et al. [25] uses statistical measures in a deformable template approach to account for statistical variations. The method uses an idealized eye consisting of two regions with uniform intensity. One region corresponds to the iris region and the other the area of the sclera. Ivins and Porrill [26] describe a method of tracking the three-dimensional motion of the iris in a video sequence. A five-parameter scalable and deformable model is developed to relate translations, rotation, scaling due to changes in eye-camera distance, and partial scaling due to expansion and contraction. The approximate positions of the eyes are then found by anthropomorphic averages. Detected eye corners are used to reduce the number of iterations of the optimization of a deformable template. This model consists of parabolas for the eyelids and a subset of a circle for iris outline. A speedup is obtained compared to Yuille et al. [25] by exploiting the positions of the corners of the eye. This method requires the presence of four corners on the eye, which, in turn, only occur if the iris is partially occluded by the upper eyelid. When the eyes are wide open, the method fails as these corners do not exist. Combination of deformable template and edge detection is used for an extended iris mask to select edges of iris through an edge image. The template is initialized by manually locating the eye region along with its parameters. Once this is done the template is allowed to deform in an energy minimization manner. The position of the template in an initial frame is used as a starting point for deformations that are carried out in successive frames. The faces must be nearly frontal-view and the image of the eyes should be large enough to be described by the template. The deformable template-based methods seem logical and are generally accurate. They are also computationally demanding, require high contrast images and usually needs to be initialized close to the eye. While the shape and boundaries of the eye are important to model so is the texture within the regions. For example the sclera is usually white while the region of the iris is darker. Larger movements can be handled using Active Appearance Models for local optimization and a mean shift color tracker. These effectively combine pure template-based methods with appearance methods. This model shares some of the problems with template-based methods; theoretically it should be able to handle changes in light due to its statistical nature. In practice they are quite sensitive to these changes and especially light coming from the side can have a significant influence on their convergence.

Feature-based methods extract particular features such as skin-color, color distribution of the eye region. Kawato et al. [27] use a circle frequency filter and background subtraction to track the in-between eyes area and then recursively binaries a search area to locate the eyes. Sommer et al. [28] utilize Gabor filters to locate and track the features of eyes. They construct a model-based approach which controls steerable Gabor filters: The method initially locates particular edge (i.e. left corner of the iris) then use steerable Gabor filters to track the edge of the iris or the corners of the eyes. Nixon demonstrates the effectiveness of the Hough transform modeled for circles for extracting iris measurements, while the eye boundaries are modeled using an exponential function. Young et al. [29] show that using a head mounted camera and after some calibration, an ellipse model of the iris has only two degrees of freedom (corresponding to pan and tilt). They use this to build a Hough transform and active contour method for iris tracking using head mounted cameras. propose the Fast Radial Symmetry Transform for detecting eyes in which they exploit the symmetrical properties of the face. Explicit feature detection (such as edges) in eye tracking methods relies on thresholds. In general defining thresholds can be difficult since light conditions and image focus change. Therefore, methods on explicit feature detection may be vulnerable to these changes.

In this paper real time eye detection and tracking method is presented that works under variable and realistic lighting conditions which is applicable to driver assistance systems. Eye tracking is one of the key technologies for future driver assistance systems since human eyes contain much information about the driver's condition such as gaze, attention level, and fatigue level. Thus, non-intrusive methods for eye detection and tracking are important for many applications of vision-based driver-automotive interaction. One common problem to many eye tracking methods is their sensitivity due to lighting condition change. This tends to significantly limit the scope for automotive applications. By combining image processing and IR light the proposed method can robustly track eyes.

3 Proposed Scheme

To detect and track eye images with complex background, distinctive features of user eye are used. Generally, an eye-tracking and detection system can be divided into four steps: (i) Face detection, (ii) Eye region detection, (iii) Pupil detection and (iv) Eye tracking.

Image processing technique is incorporated for detection of these. Figure 1 illustrates the scheme. Camera is incorporated in the dashboard of vehicle which takes the images of the driver regularly at certain interval. From the images first the face portion is recognized from the complex background. It is followed by eye region detection and thereafter the pupil or eyelid detection. The detection algorithm finally detects the eyelid movement or closeness and openness of eyes. In the proposed method, eye detection and tracking are applied on testing sets, gathered from different images of face data with complex backgrounds. This method combines the location and detection algorithm with the grey prediction for eye tracking.

The accuracy and robustness of the system depends on consistency of image acquisition of the driver face in real time under variable and complex background. For this purpose the driver's face is illuminated using a near-infrared (NIR) illuminator. It serves three purposes:

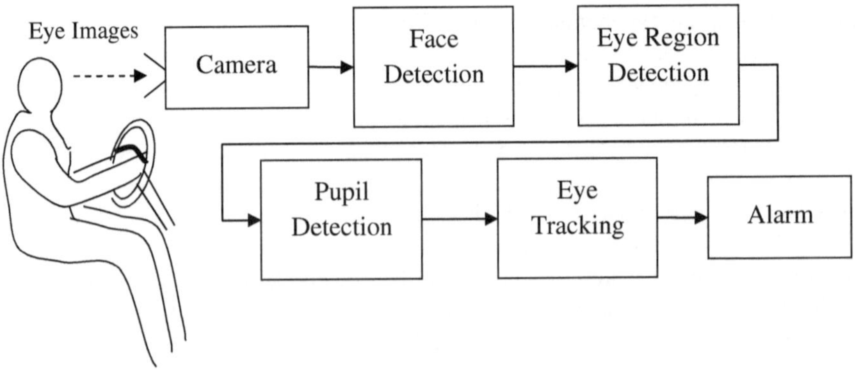

Fig. 1 Image acquisition scheme

- It minimizes the impact of different ambient light conditions, and hence the image quality is ensured under varying real-world conditions including poor illumination, day, and night;
- It allows producing the bright pupil effect, which constitutes the foundation for detection and tracking the visual cues.
- As the near-infrared illuminator is barely visible to the driver, any interference with the driver's driving will be minimized.

If the eyes are illuminated with a NIR illuminator at certain wavelength beaming light along the camera optical axis, a bright pupil can be obtained. At the NIR wavelength, almost all IR light is reflected from the pupils along the path back to the camera. Thus bright pupil effect is produced which is very much similar to the red eye effect in photography. The pupils appear dark if illuminated off the camera optical axis, since the reflected light will not enter the camera lens which is called dark pupil effect. It is physically difficult to place IR light-emitting diodes (LEDs) as illuminators along the optical axis since it may block the view of the camera, limiting the camera's operational field of view. Therefore quite a few numbers of IR illuminator LEDs are placed evenly and symmetrically along the circumference of two coplanar concentric rings, the center of both rings coincides with the camera optical axis as shown at Fig. 2. In the proposed scheme, the camera acquires the images of face of the driver at certain interval. Every time the image is analyzed and bright pupil effect is detected. Whenever dark pupil effect is detected i.e., eyelid is at closed condition at prolonged time, it may be assumed that driver's vigilance level has been diminished. Subsequently alarm is activated to draw the attention of the driver.

Fig. 2 IR illuminator with camera

4 Implementation

A laboratory model has been developed to implement above scheme. A web camera with IR illuminators has been employed focusing the face region of a person (driver) to acquire the images of face. The acquired image signal is fed to Data Acquisition Card and subsequently to a microcontroller. The microcontroller analyses the images and detects the pupil characteristics. If the eyelid is closed for several seconds, it may be assumed that the drowsiness occurred to the person and alarm is activated by the microcontroller. The circuit scheme is shown at Fig. 3. Microcontroller ATMEGA 8 is employed here in association with voltage regulator IC 7805 and driver IC L2930 for buzzer.

To find the position of pupil, first, face region must be separated from the rest of the image using *boundaries* function, which is a process of segmentation. This will cause the images background to be non-effective. *Region prop* technique is used to separate a region from total face and the region containing eyes and eyebrow. This will result in decreasing the computational complexity. Finally, in proposed method points with the highest values are selected as the eye candidate's using *centroid* function. The eye region is well detected among these points. If eye is detected then the next frame is simulated, and if the eye is not detected then a signal is passed to the micro controller for raising the alarm. The indicator also turns red. And when the eye is detected the indicator turns green with raising no alarm. This is how the system can awake the driver in long drive or in fatigue condition. The implementation flow chart is given at Fig. 4.

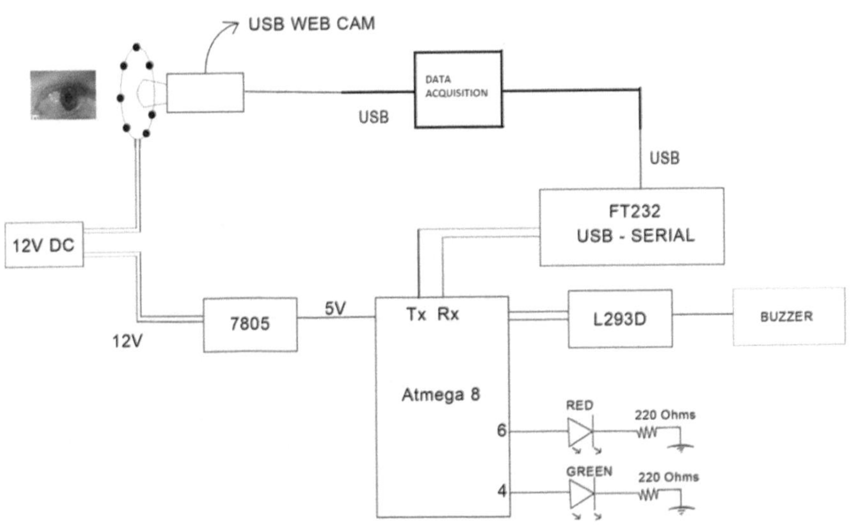

Fig. 3 Circuit scheme

Fig. 4 Flow chart

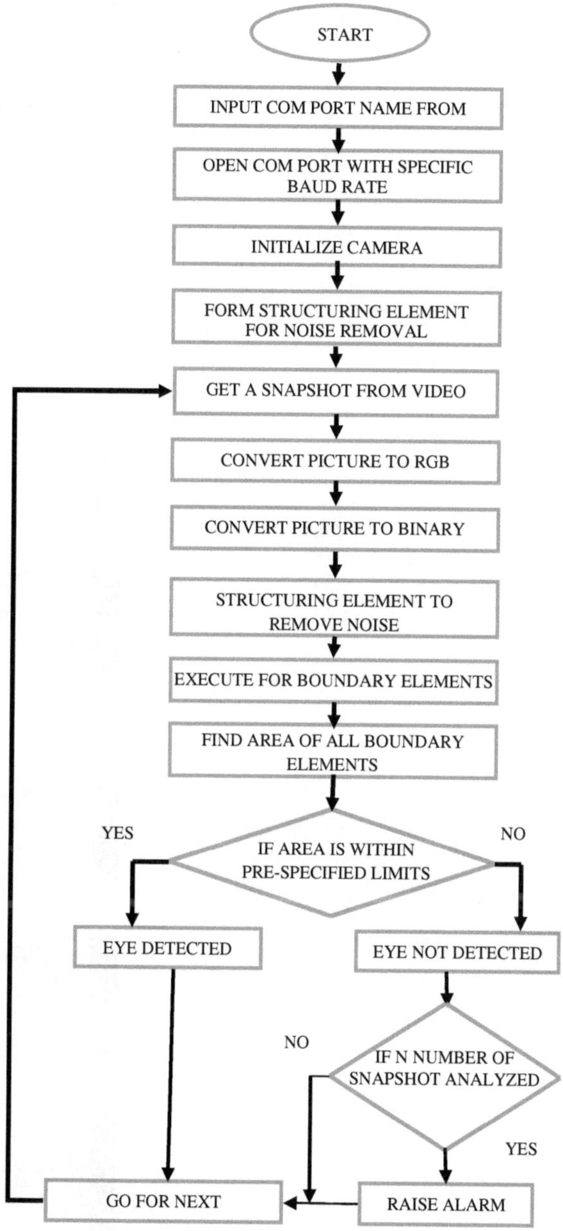

Fig. 5 Experimental results

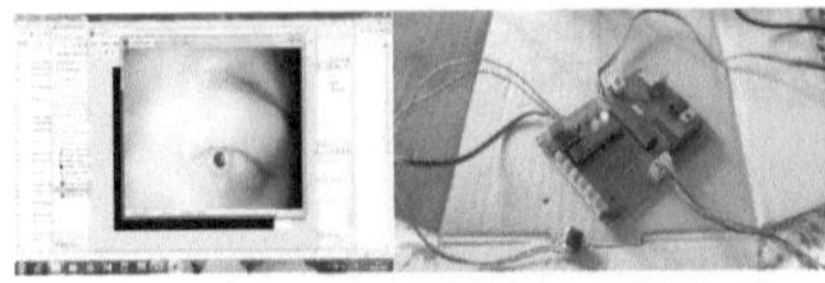

OBSERVATION 1

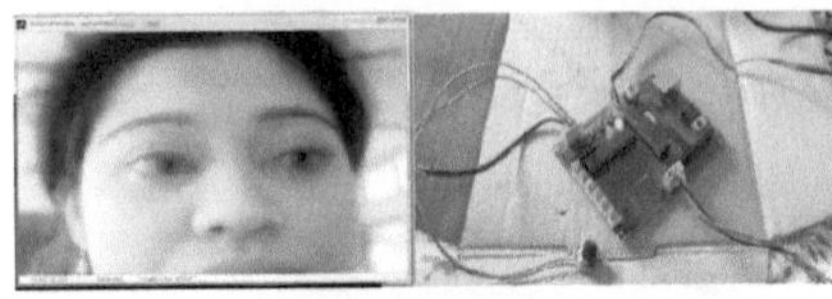

OBSERVATION 2

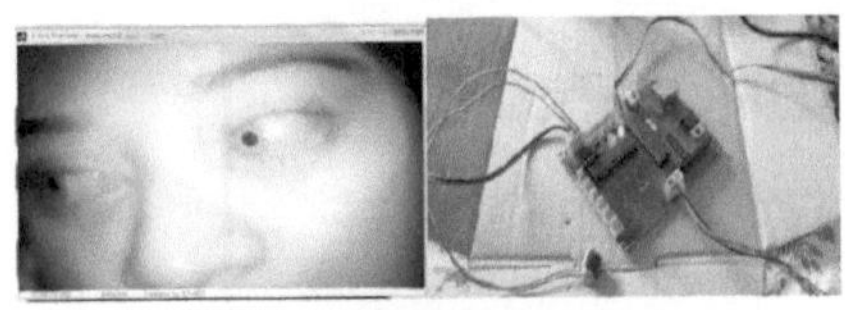

OBSERVATION 3

OBSERVATION 4

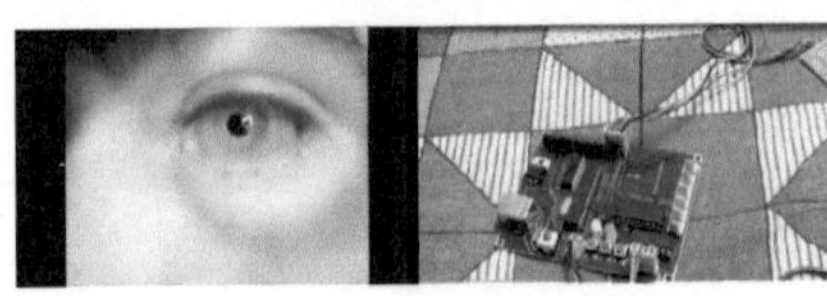

OBSERVATION 5

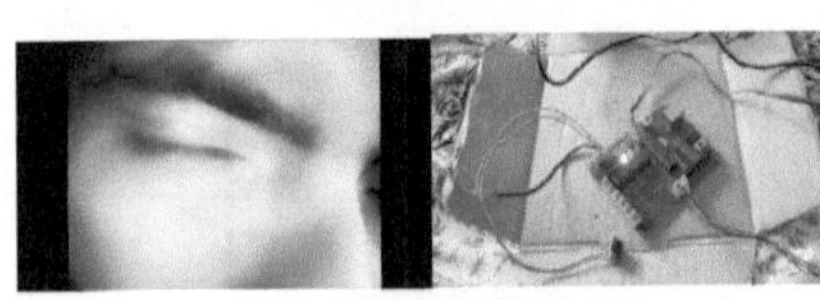

OBSERVATION 6

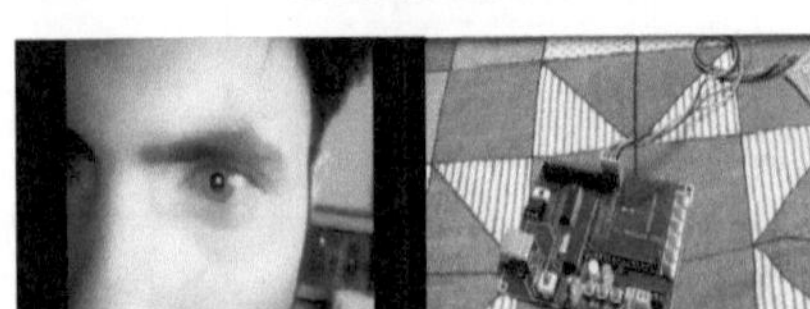

OBSERVATION 7

Table 1 Observations on alarm conditions with respect to the eye condition

	Eye condition	Alarm condition
Observation 1	Open	Green LED is ON. No buzzer
Observation 2	Open	Green LED is ON. No buzzer
Observation 3	Open (at night)	Green LED is ON. No buzzer
Observation 4	Closed	Red LED is ON. Buzzer raise the alarm
Observation 5	Open	Green LED is ON. No buzzer
Observation 6	Closed	Red LED is ON. Buzzer raise the alarm
Observation 7	Open	Green LED is ON. No buzzer

5 Observations

Experiments have been carried out on different person and different time. These indicate the high correct detection rate which is indicative of the method's superiority and high robustness. In the experimental set up two different colors of LEDs —Red and Green are used to indicate fatigue condition (closed eyes) and normal condition (open eyes) respectively. A buzzer is also incorporated whenever fatigue condition is detected. The experimental results for image sequence of eye tracking are given at Fig. 5, and observations on alarm conditions with respect to the eye condition are also tabulated at Table 1. It may be noticed that at closed eye condition the Red LED glows as well as buzzer is activated. These observations show that the model can track eye region robustly and correctly and can avoid the accident as well.

6 Discussions

The experimental model as developed in the laboratory is of minimum complexities. Experiments and observations have been carried out at different time, person and environmental condition that prove its good robust performance. It is helpful in driver vigilance and accident avoidance system. However the performance and effectiveness of the project depends on the quality of the camera and finding out the threshold while removing noise from the acquired picture.

In this model the alarming system is only to alert the driver. The system can be integrated with brake and accelerator system of the vehicle. Also this alarming system may be attached with the front end and the back end light indicator with good audible sound to alert other drivers and passer-by on that road to minimize the fatal rate.

References

1. Elzohairy Y (2008) Fatal and injury fatigue-related crashes on ontario's roads: a 5-year review. In: Working together to understand driver fatigue: report on symposium proceedings, february 2008
2. Dingus TA, Jahns SK, Horowitz AD, Knipling R (1998) Human factors design issues for crash avoidance systems. In: Barfield W, Dingus TA (eds) Human factors in intelligent transportation systems. Lawrence Associates, Mahwah, pp 55–93
3. Saito H, Ishiwaka T, Sakata M, Okabayashi S (1994) Applications of driver's line of sight to automobiles—what can driver's eye tell. In: Proceedings of vehicle navigation and information systems conference, Yokohama, Japan, pp 21–26
4. Ueno H, Kaneda M, Tsukino M (1994) Development of drowsiness detection system. In: Proceedings of vehicle navigation and information systems conference, Yokohama, Japan, pp 15–20
5. Boverie S, Leqellec JM, Hirl A (1998) Intelligent systems for video monitoring of vehicle cockpit. In: International Congress and exposition ITS: advanced controls and vehicle navigation systems, pp 1–5
6. Kaneda M et al (1994) Development of a drowsiness warning system. In: The 11th international conference on enhanced safety of vehicle, Munich
7. Onken R (1994) Daisy, an adaptive knowledge-based driver monitoring and warning system. In: Proceedings of vehicle navigation and information systems conference, Yokohama, Japan, pp 3–10
8. Feraric J, Kopf M, Onken R (1992) Statistical versus neural bet approach for driver behaviour description and adaptive warning. The 11th European annual manual, pp 429–436
9. Ishii T, Hirose M, Iwata H (1987) Automatic recognition of driver's facial expression by image analysis. J Soc Automot Eng Jap 41:1398–1403
10. Yammamoto K, Higuchi S (1992) Development of a drowsiness warning system. J Soc Automot Eng Jap 46:127–133
11. Smith P, Shah M, da Vitoria Lobo N (2000) Monitoring head/eye motion for driver alertness with one camera. In: The 15th international conference on pattern recognition, vol 4, pp 636–642
12. Saito S (1992) Does fatigue exist in a quantitative of eye movement? Ergonomics 35:607–615
13. Anon (1999) Perclos and eye tracking: challenge and opportunity. Technical Report Applied Science Laboratories, Bedford
14. Wierville WW (1994) Overview of research on driver drowsiness definition and driver drowsiness detection. ESV, Munich
15. Dinges DF, Mallis M, Maislin G, Powell JW (1998) Evaluation of techniques for ocular measurement as an index of fatigue and the basis for alertness management. Dept Transp Highw Saf Publ 808:762
16. Anon (1998) Proximity array sensing system: head position monitor/metric. Advanced safety concepts, Inc., Sante Fe, NM87504
17. Anon (1999) Conference on ocular measures of driver alertness, Washington DC, April 1999
18. Qiang J, Xiaojie Y (2002) Real-Time Eye, Gaze, and face pose tracking for monitoring driver vigilance. Real-Time Imag 8:357–377
19. D'Orazio T, Leo M, Guaragnella C, Distante A (2007) A visual approach for driver inattention detection. Pattern Recogn 40(8):2341–2355
20. Boyraz P, Acar M, Kerr D (2008) Multi-sensor driver drowsiness monitoring. Proceedings of the institution of mechanical engineers, Part D: J Automobile Eng 222(11):2041–2062
21. Ebisawa Y (1989) Unconstrained pupil detection technique using two light sources and the image difference method. Vis Intell Des Eng, pp 79–89
22. Grauman K, Betke M, Gips J, Bradski GR (2001) Communication via eye blinks: detection and duration analysis in real time. In: Proceedings of IEEE conference on computer vision and pattern recognition, WIT Press, pp 1010–1017

23. Matsumoto Y, Zelinsky A (2000) An algorithm for real-time stereo vision Implementation of Head pose and gaze direction measurements. In: Proceedings of IEEE 4th international conference on face and gesture recognition, pp 499–505
24. Yuille AL, Hallinan PW, Cohen DS (1992) Feature extraction from faces using deformable templates. Int J Comput Vis 8(2):99–111
25. Ivins JP, Porrill J (1998) A deformable model of the human iris for measuring small 3-dimensional eye movements. Mach Vis Appl 11(1):42–51
26. Kawato S, Tetsutani N (2002) Real-time detection of between-the-eyes with a circle frequency filter. In: Asian conference on computer vision
27. Sommer G, Michaelis M, Herpers R (1998) The SVD approach for steerable filter design. In: Proceedings of international symposium on circuits and systems 1998, Monterey, California, vol 5, pp 349–353
28. Yang G, Waibel A (1996) A real-time face tracker. In: Workshop on applications of computer vision, pp 142–147
29. Loy G, Zelinsky A (2003) Fast radial symmetry transform for detecting points of interest. IEEE Trans Pattern Anal Mach Intell 25(8):959–973

Preprocessing in Early Stage Detection of Diabetic Retinopathy Using Fundus Images

Vijay M. Mane, D.V. Jadhav and Ramish B. Kawadiwale

Abstract Automated retinal image processing is becoming a primary important screening tool for early detection of diabetic retinopathy (DR). An automated system reduces human errors and also reduces the burden on the ophthalmologists. The accurate detection of microaneurysms (MAs) is an important step for early detection of DR. This paper present some methods to improve the quality of input retinal image and extraction of blood vessels, as a preprocessing step in automatic early stage detection of DR. Experimental results are performed for preprocessing and blood vessel extraction techniques using standard fundus image database.

Keywords Preprocessing · Contrast enhancement · Segmentation · Fundus images · Blood vessels

1 Introduction

Diabetes mellitus (DM) is the name of a systemic and serious disease [1]. It occurs when the pancreas does not produce an adequate amount of insulin or the body is unable to process it properly. This results in an abnormal increase of the glucose level in the blood. Eventually this high level of glucose causes damage to blood vessels. This damage affects almost all organs like eyes, nervous system, heart, kidneys etc. Diabetes mellitus commonly results in diabetic retinopathy (DR),

V.M. Mane (✉)
JSPM'S RSCOE Tathwade, Savitribai Phule Pune University, Pune, India
e-mail: vijay.mane@vit.edu

D.V. Jadhav
TSSM BSCOE&R, Narhe, Pune, India
e-mail: dvjadhao@yahoo.com

R.B. Kawadiwale
Department of Electronics, Vishwakarma Institute of Technology, Pune, India
e-mail: ramish.kawadiwale.vit@gmail.com

S. Gupta et al. (eds.), *Advancements of Medical Electronics*,
Lecture Notes in Bioengineering, DOI 10.1007/978-81-322-2256-9_3

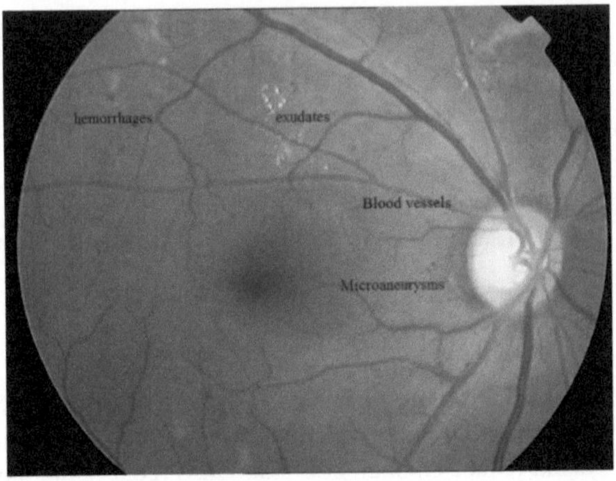

Fig. 1 Fundus image showing features of diabetic retinopathy

which is caused by pathological alteration of the blood vessels which nourish the retina. As a result of this damage, the capillaries leak blood and fluid on the retina [2]. DR is the main cause of new cases of blindness among adults aged 20–74 years [3–7]. Microaneurysms, hemorrhages, exudates, cotton wool spots or venous loops etc. can be seen as visual features on retinal images as shown in Fig. 1. Microaneurysms (MAs) are a common and often early appearance of DR. So the MA detector is an attractive candidate for an automatic screening system able to detect early findings of DR. The main processing components for detection of MAs using retinal fundus images include: preprocessing, selection of a candidate MA, feature extraction and classification as shown in Fig. 2. The performance of lesion detection algorithms solely depends on quality of retinal images that are captured by fundus camera.

Preprocessing step is used to minimize image variations and improve image quality. MAs do not appear in a vessel but many MA candidates can be detected in retinal vessels. This false detection may be due to the dark red dots in the blood vessels. To reduce false MA detection, blood vessels have to be removed to prevent

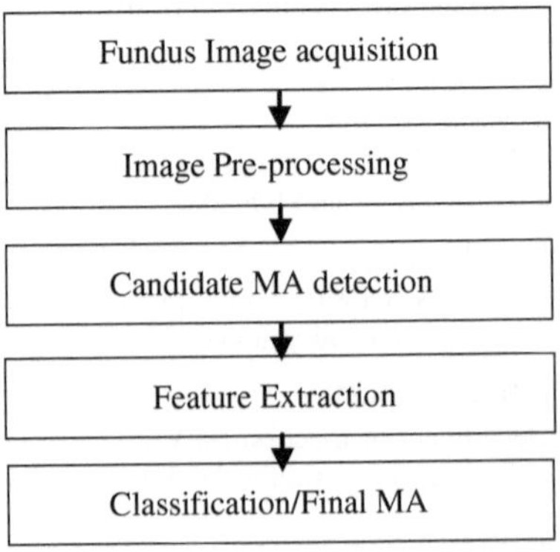

Fig. 2 Detection of MAs

a misclassification. In this paper, we present comparison of retinal image preprocessing techniques that enhance the fundus images by extracting Green plane of color fundus image. Contrast enhancement using Histogram stretching, histogram equalization and adaptive histogram equalization are implemented to enhance the input fundus image. Blood vessels are extracted using simple thresholding, top hat transform, K means clustering and fuzzy C means clustering methods. Paper is organized in five sections. Sections 2 and 3 present the techniques for colored retinal image enhancement and vessels segmentation respectively. Experimental results of the tests on the images of the DRIVE database and their analysis are given in Sect. 4 followed by conclusion in Sect. 5.

2 Contrast Enhancement

The main objective of pre-processing technique is to attenuate image variation by normalizing the original retinal image against a reference model or data set for subsequent viewing, processing or analysis. Variations typically arise within the same image (intra-image variability) as well as between images (inter-image variability). Intra-image variation occurs due to differences in light diffusion, presence of abnormalities, variations in fundus reflectivity and fundus thickness. Inter-image variations arise due to differences in cameras, illumination, acquisition angle and retinal pigmentation.

2.1 Green Plane Selection

It is observed that the green channel of color fundus images is commonly used by unsupervised methods to detect MAs from fundus images, as it has the best vessel—background contrast as observed in middle column of Fig. 3.

2.2 Histogram Stretching [8]

The contrast of an image is the distribution of intensity of its pixels. A low contrast image show small differences between its light and dark pixel values. Human eyes are sensitive to contrast rather than absolute pixel intensities. a perceptually better image could be obtained by stretching the histogram of an image, in which dynamic range of the image is filled. Figure 4a shows the original and Fig. 4b, c are log and power law contrast enhanced images.

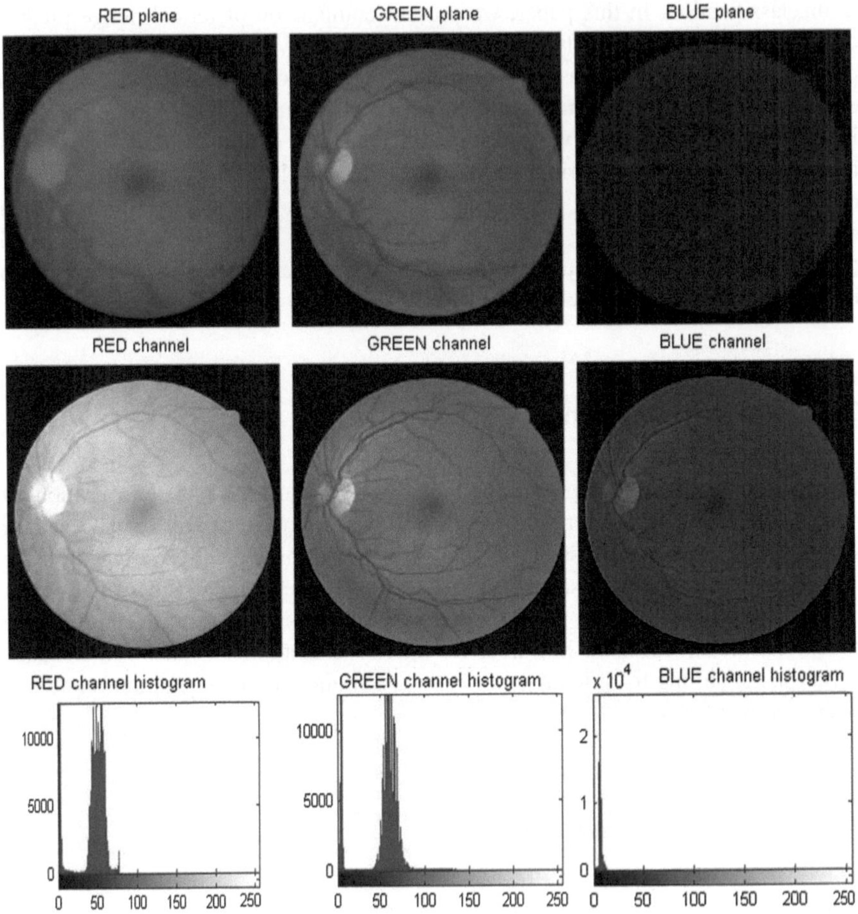

Fig. 3 RGB colour band with respective histograms

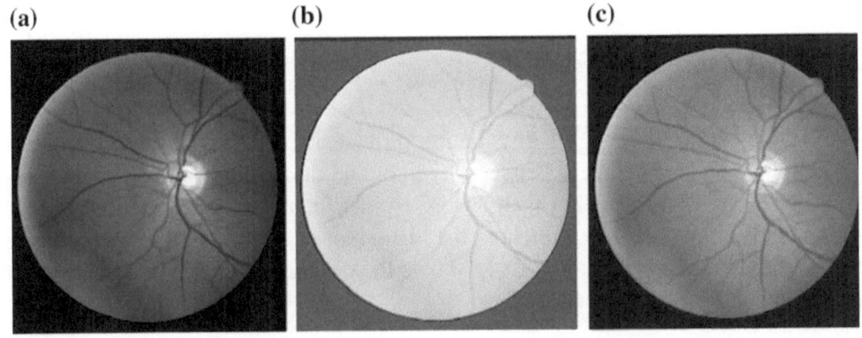

Fig. 4 Contrast stretched images. **a** Input image. **b** LOG transformed image. **c** Power law corrected

- Take input fundus image, RGB image
- Extract Green channel component from this image
- Calculate:

  1. L = C * log(1 + intensity of pixel)
  2. P = C * (intensity of pixel)$^{\gamma}$

2.3 *Histogram Equalization [8, 9]*

A histogram equalized image is obtained by mapping each pixel in the input image to a corresponding pixel in the output image using an equation based on the cumulative distribution function. Figure 5a, b are input image and its histogram whereas Fig. 5c, d are histogram equalized image and its histogram.

- Take input fundus image.
- Extract Green channel component from this image.
- Get the histogram of the image.
- Find probability density function and cumulative distribution function for each intensity level.

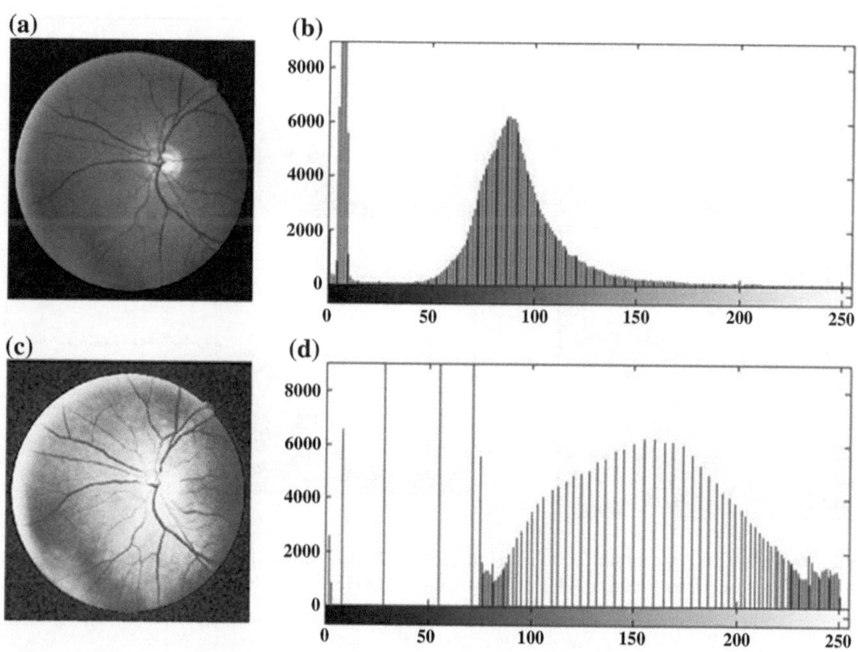

Fig. 5 Histogram equalized images and respective histograms. **a** Input image. **b** Histogram of input image. **c** Histogram equalized image. **d** Histogram of equalized image

- Calculate new intensity values using equalization equations.
- Build new image by replacing original gray values with the new gray values.

2.4 Adaptive Histogram Equalization (AHE) [9]

Adaptive method computes several histograms, each corresponding to a distinct section of the image, and uses them to redistribute the lightness values of the image. Figure 6a, b shows the original image and its histogram and Fig. 6c, d are image after adaptive histogram equalization and its histogram respectively.

- Take input fundus image.
- Extract Green channel component from this image.
- Select appropriate window size, centered at a pixel called grid point.
- For each grid point calculate the histogram of the region around it, having area equal to window size and centered at the grid point.
- Enhance the contrast of window selected and map the grid point to new image.

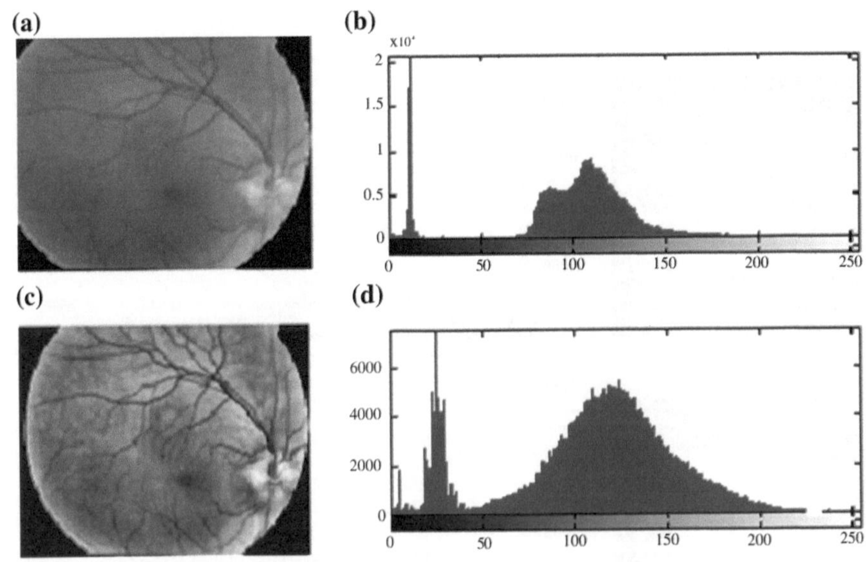

Fig. 6 Adaptive histogram equalized images and respective histograms. **a** Green plane. **b** Histogram of green plane. **c** AHE. **d** Histogram of AHE.

3 Blood Vessels Extraction

3.1 Shade Correction Followed by Thresholding [10]

A simple manually selected thresholding is applied to segment blood vessels from background as shown in Fig. 7.

- Apply adaptive histogram equalization algorithm to improve the contrast.
- Apply median filter to contrast enhanced image with filter size greater than blood vessel width.
- Subtract adaptive histogram equalized image from median filtered image.
- Threshold this shade corrected image to get blood vessels.

3.2 Top Hat Transform [9, 11, 12]

Top-hat transform is an operation that extracts small elements and details from given images. The top-hat transformation applied on the filtered image with a disk structure element with a size large enough to fill all the holes in blood vessels. The top hat transformation is then performed by a closing operation, as shown in Fig. 8a input image and Fig. 8b the output.

- Take shade corrected image as input.
- Define suitable structuring element.
- Apply morphological opening with structuring element on shade corrected image.
- Manually threshold the opened image.
- Top hat transform—Subtract the threshold opened image from shade corrected image.

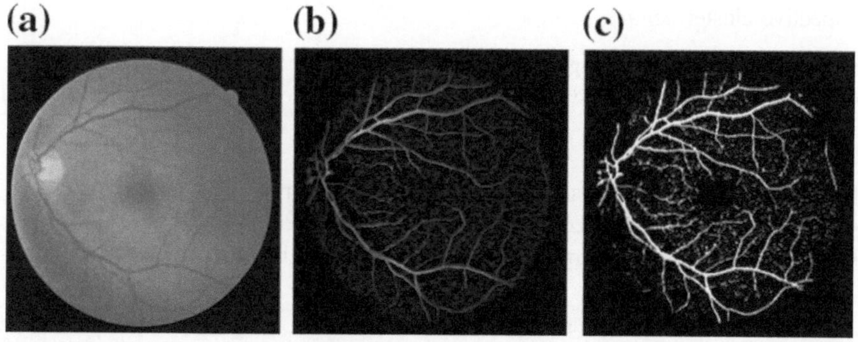

Fig. 7 Output of shade correction algorithm. **a** Input image. **b** Shade corrected. **c** Thresholding output

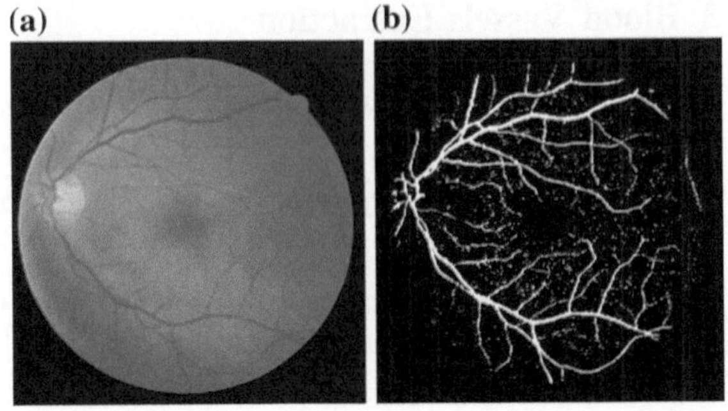

Fig. 8 Top hat transform. **a** Input image. **b** Output of top hat transform

- $T(f) = f - (f \ o \ b)$ Where $T(f)$ is the top-hat transformation, (o) is the opening operation, b is the structuring element and f is the shade corrected image.
- Subtract $T(f)$ pixels from shade corrected image.
- Threshold this image to get blood vessels.

3.3 K-Means Clustering [13]

K-means clustering is one of the simplest unsupervised learning algorithms. The procedure follows a simple and easy way to classify a given data set through a certain number of clusters fixed a priori. This algorithm aims at minimizing an objective function:

$$J = \sum_{j=1}^{k} \sum_{i=0}^{n} \left\| \mathrm{X}_{\mathrm{i}}^{\mathrm{j}} - \mathrm{C}_{\mathrm{j}} \right\|^2$$

where $||\mathrm{X}_{\mathrm{i}}^{\mathrm{j}} - \mathrm{C}_{\mathrm{j}}||^2$ is a chosen distance measure between a data point $\mathrm{X}_{\mathrm{i}}^{\mathrm{j}}$ and the cluster centre C_{j} is an indicator of the distance of the n data points from their respective cluster centres.

- Take pre-processed image as input.
- Empirically set threshold to divide the image into blood vessels and background clusters.
- Find cluster centers.
- Find distance of each data point from cluster center.
- Create an array whose columns are two distance vectors.
- According to the minimum distance, assign original image pixel to corresponding cluster.

3.4 Fuzzy C-Means Clustering [14]

Fuzzy clustering technique allows each of data object to belong to more than one cluster. Membership value, which specifies degree of belongingness to each cluster, is assigned to each data object. The membership value of each data object is updated in each iteration. Data point may partially or fully belongs to a cluster.

- Take pre-processed image as input of size [m n] and convert into single vector of m*n data objects.
- Set threshold empirically, so as to divide image into 2 clusters, blood vessels and background.
- Find cluster centers "C_i" by using equation:

$$C_i = \frac{\sum_{j=1}^{n} u_{ij}^m x_j}{\sum_{j=1}^{n} u_{ij}^m}$$

- Calculate objective function "J" to be minimized:

$$J(U, c_1, c_2, \ldots, c_c) = \sum_{i=1}^{c} J_i = \sum_{i=1}^{c} \sum_{j=1}^{n} u_{ij}^m d_{ij}^2$$

- Calculate membership function "U_{ij}":

$$u_{ij} = \frac{1}{\sum_{k=1}^{c} \left(\frac{d_{ij}}{d_{kj}} \right)^{2/(m-1)}}$$

$$\sum_{i=1}^{c} u_{ij} = 1, \quad \forall j = 1, 2, \ldots, n$$

d_{ij} = Euclidean distance between ith cluster center and jth data point
m = weighting exponent (m > 1)

- De-fuzzify the single vector to get data clusters (Fig. 9).

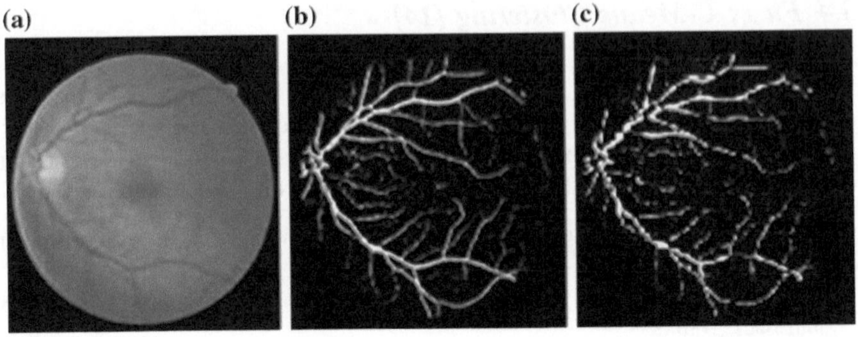

Fig. 9 Results of clustering algorithm. **a** Input image. **b** K-means clustering. **c** Fuzzy C-means clustering

4 Experimental Results

The performance of preprocessing methods is evaluated using publicly available DRIVE database [15, 16]. The DRIVE database consists of 40 RGB color images of size 768 × 584 pixels, eight bits per color channel. The image set is divided into a test and training set and each one contains 20 images. There are two hand-labelling available for the 20 images of test set made by two different human observers.

4.1 Evaluation of Contrast Enhancement Algorithms

The performance of preprocessing algorithms is evaluated using peak signal to noise ratio (PSNR) and mean square error (MSE). Ideally, PSNR should be high and MSE should be low. The presented algorithms for contrast enhancement of fundus images shows promising results. Adaptive histogram equalization shows a good contrasted fundus image as seen from Table 1.

Table 1 PSNR and MSE of drive images

Algorithm	TEST_10 image		TEST_19 image	
	MSE	PSNR	MSE	PSNR
Contrast stretching	58.5081	30.4586	58.801	30.436
Histogram equalization	42.374	31.859	42.371	31.8600
Adaptive histogram equalization	38.561	32.269	38.259	32.3041

Table 2 Comparison of blood vessel extraction algorithms

Algorithm		TEST_10 image (%)	TEST_19 image (%)
Shade correction	Sensitivity	65.9	78.5
	Specificity	98.33	97.75
	Accuracy	95.77	96.15
Top hat transform	Sensitivity	63.45	76.83
	Specificity	98.3	98.31
	Accuracy	95.3	96.53
K-means clustering	Sensitivity	70.04	83.02
	Specificity	97.38	97.67
	Accuracy	94.9	96.45
FCM clustering	Sensitivity	54.77	71.5
	Specificity	99.4	99.47
	Accuracy	95.72	97.15

4.2 Evaluation of Blood Vessels Detection Algorithm

Blood vessels detection algorithms are evaluated in terms of their sensitivity, specificity and accuracy. It is observed that given a contrast enhanced fundus image to different blood vessel extraction methods, each gives a promising output in terms of accuracy. These parameters are evaluated on two images from database as shown in Table 2. K means clustering and FCM clustering shows good results.

5 Conclusions

While capturing the retinal image using fundus camera, the retina is not illuminated uniformly because of circular shape of retina. The preprocessing stage in early stage detection of MAs is necessary to correct the non-uniform illumination and to enhance the contrast. This paper presents preprocessing step for fundus image contrast enhancement and blood vessel extraction. Operations like Histogram stretching, histogram equalization and adaptive histogram equalization are implemented to enhance the contrast of input fundus images. Adaptive histogram equalization gives better contrast enhanced image which is evaluated in terms of MSE and PSNR. Blood vessels are extracted using simple thresholding, top hat transform, K means clustering and fuzzy C means clustering algorithms on retinal images of publicly available DRIVE database. Blood vessels detection algorithms are evaluated in terms of their sensitivity, specificity and accuracy. K means clustering and FCM clustering outperform in terms of accuracy than other methods presented in this paper. Segmentation of retinal blood vessels is a challenging task mainly because of the presence of a wide variety of vessel widths, low contrast and the vessel color is close to that of the background.

References

1. Chiulla TA, Amador AG, Zinman B (2003) Diabetic retinopathy and diabetic macular edema: pathophysiology, screening, and novel therapies. Diabetes Care 26(9):2653–2664
2. Frank RN (1995) Diabetic retinopathy. Prog Retin Eye Res 14(2):361–392
3. Klein R, Klein BEK, Moss SE (1994) Visual impairment in diabetes. Ophthalmology 91:1–9
4. Klonoff DC, Schwartz DM (2000) An economic analysis of interventions for diabetes. Diabetes Care 23(3):390–404
5. Center for Disease Control and Prevention (2011) National diabetes fact sheet: technical report, U.S.
6. Bresnick GH, Mukamel DB, Dickinson JC, Cole DR (2000) A screening approach to the surveillance of patients with diabetes for the presence of vision-threatening retinopathy. Opthalmology 107(1):19–24
7. Susman EJ, Tsiaras WJ, Soper KA (1982) Diagnosis of diabetic eye disease. J Am Med Assoc 247(23):3231–3234
8. Hatanaka Y, Inoue T, Okumura S, Muramatsu C, Fujita S (2012) Automated microaneurysm detection method based on double-ring filter and feature analysis in retinal fundus images. In: Proceedings of 25th IEEE international symposium on computer-based medical systems, paper-150
9. Saleh MD, Eswaran C (2012) An automated decision-support system for non-proliferative diabetic retinopathy disease based on MAs and HAs detection. Elsevier—Comput Meth Programs Biomed 108:186–196
10. Marín D, Aquino A, Gegúndez-Arias ME, Bravo JM (2011) A new supervised method for blood vessel segmentation in retinal images by using gray-level and moment invariants-based features. IEEE Trans Med Imaging 30(1):146–158
11. El Abbadi NK, Al Saadi EH (2013) Blood vessels extraction using mathematical morphology. J Comput Sci 9(10):1389–1395
12. Ram K, Joshi GD, Sivaswamy J (2011) A successive clutter-rejection-based approach for early detection of diabetic retinopathy. IEEE Trans Biomed Eng 58(3)
13. Masroor AM, Mohammad DB (2008) Segmentation of brain MR images for tumor extraction by combining K means clustering and Perona-Malik anisotropic diffusion model. Int J Image Proc 2(1)
14. Dey N, Roy AB, Pal M, Das A (2012) FCM based blood vessel segmentation method for retinal images. Int J Comput Sci Netw (IJCSN) 1(3). ISSN:2277-5420
15. Staal JJ, Abramoff MD, Niemeijer M, Viergever MA, van Ginneken B (2004) Ridge based vessel segmentation in color images of the retina. IEEE Trans Med Imaging 23:501–509
16. Image Sciences Institute (2001) DRIVE: digital retinal images for vessel extraction. http://www.isi.uu.nl/Research/Databases/DRIVE

Magnetic Resonance Image Quality Enhancement Using Transform Based Hybrid Filtering

Manas K. Nag, Subhranil Koley, Chandan Chakraborty and Anup Kumar Sadhu

Abstract This paper proposes a novel methodology for improving the quality of magnetic resonance image (MRI). The presence of noise affects the image analysis task by degrading the visual contents of image. The proposed methodology integrates transform domain method, discrete wavelet transform with spatial domain filter, Non local means to smoothed out noisy interferences leading to the improvement of visual characteristics of MRI insights. The quantitative validation of the proposed technique has been done and experimental result shows the effectiveness of this algorithm over anisotropic diffusion, bilateral, trilateral and wavelet shrinkage filters.

Keywords MRI · Wavelet transform · Non-local means filter · Quality metrics

1 Introduction

MRI is a non invasive radiation free imaging technology for capturing the visual information of internal body tissues, which provides aid in diagnosing variety of abnormalities to clinicians. It has been well appreciated in doctor's community that magnetic resonance (MR) images are well enough to produce necessary features in brain tumor diagnosis. It uses a powerful magnetic field, radio frequency waves (RF waves) and a computer to produce image of internal structure of body. It has been observed that MR signal fluctuates in random manner because of the presence of thermal noise. Moreover, impulse noise is very common in medical images, which appears during the acquisition or transmission of image through channel.

M.K. Nag · S. Koley · C. Chakraborty (✉)
School of Medical Science and Technology, Indian Institute of Technology Kharagpur, Kharagpur 721302, India
e-mail: chandanc@smst.iitkgp.ernet.in

A.K. Sadhu
EKO CT & MRI Scan Centre Medical College and Hospitals Campus, Kolkata 700073, India

S. Gupta et al. (eds.), *Advancements of Medical Electronics*,
Lecture Notes in Bioengineering, DOI 10.1007/978-81-322-2256-9_4

The appearance of noise severely affects the visual features which are the key markers of disease recognition from the image. In reality, radiologist does the analysis on the basis of his ability in investigating visual imaging features apart from case study. Therefore, the noisy artifact often misleads them to reach the accurate diagnosis. In continuation, it also makes the next level image processing tasks i.e. segmentation, feature extraction etc. difficult. It is well known that in case of teleradiology; where MRI data can be transmitted from one place to another require high end network set up especially for transmitting images in DICOM format. The establishment of this kind of infrastructure is costly and not available in remote area. Under such situation transmission of JPEG images becomes easy and does not require any extra facility other than internet and hence cost effective. In such case, noise may appear. After concerning all the issues, this study focuses in reducing the noisy impact from MR images at the preprocessing stage of an automated computer assistive brain tumor screening system.

In this section, we address the previous works in reducing the effect of noise from MRI. Several denoising methods have been developed to enhance the quality of image. Perona and Malik proposed a new filtering approach based on heat equation to overcome the drawbacks of spatial filtering by emphasizing on preserving the edge information and this filter is quiet effective for homogeneous region [1]. Later on Krissian and Aja-Fernandez modified anisotropic diffusion filter (ADF) and tested over MR images to eliminate rician noise [2]. On the other hand Tomasi and Manduchi extended the concept of domain filtering through introducing additional range information. Hence bilateral filter (BLF) [3] adds value in smoothing along with keeping contour intact. Moreover its non-iterative policy gives privilege over anisotropic diffusion. Later on Wong et al. [4] introduced a novel methodology, trilateral filtering (TLF), as an extension of BLF for suppressing noise from medical images. Apart from photometric and geometric similarity of BLF, it makes an addition of structural information. Low pass filter is employed in homogeneous region, whereas pixel belongs to heterogeneous region is replaced by the weighted average of those three similarity indexes. They showed the improved performance over BLF through preserving edges while smoothing. Transform based technique such as wavelet transform (WT) is popular approach in image denoising through keeping the characteristics intact [5, 6]. Nowak [7] proposed the wavelet domain filter for denoising MR images where noise follows rician distribution properties. Baudes et al. [8] developed non local means (NLM) filter and showed its ability in preserving image structure. After some years, Manjón et al. [9] introduced the application of NLM filter and also proposed unbiased NLM for dealing with the rician noise distribution in the magnitude MR image. They performed the parameterization task of filter for various noise levels. Later on, Manjon et al. [10] have also developed adaptive NLM for denoising MR images when noise level in image is spatially varying. In the next year, Erturk et al. [11] proposed a novel denoising technique based on spectral subtraction for improving signal-to-noise ratio (SNR).

In this paper, we have introduced a hybrid filtering technique as a combination of discrete WT and NLM filter.

The transform coefficients obtained from discrete WT would have been undergone through NLM filter for modification and finally filtered image can be reconstructed from inverse transformation. This experiment has been performed without addition (assume presence of noise) and with addition of noise. The proposed algorithm shows its robustness and effectiveness over ADF, BLF, TLF and wavelet shrinkage (WS) in both cases for enhancing the quality of MR images.

2 Materials and Method

The step wise schematic of proposed filter is shown in Fig. 1. In this section we discuss the type of MR images and the algorithms employed in this study.

2.1 MR Imaging

The MR imaging were done with 1.5 Tesla MRI scanner and acquired in Digital Imaging and Communications in Medicine (DICOM) format. The resolution, slice thickness and flip angle were kept as 512 × 512, 5 mm and 90° respectively. All total 50 axial slices of MRI from 9 cases of brain tumor have been considered in this study. The DICOM images are compressed in .jpg (Joint photographic experts group) for processing. Moreover, the background i.e. non brain part of each individual has been reduced through cropping and the resulting image would be considered for filtering.

2.2 Wavelet Decomposition

Discrete WT is a multiresolution decomposition framework of 1D or 2D signal i.e. image into low (approximation coefficients) and high frequency components (detail coefficients) [5]. The high frequency components are precious in nature as they

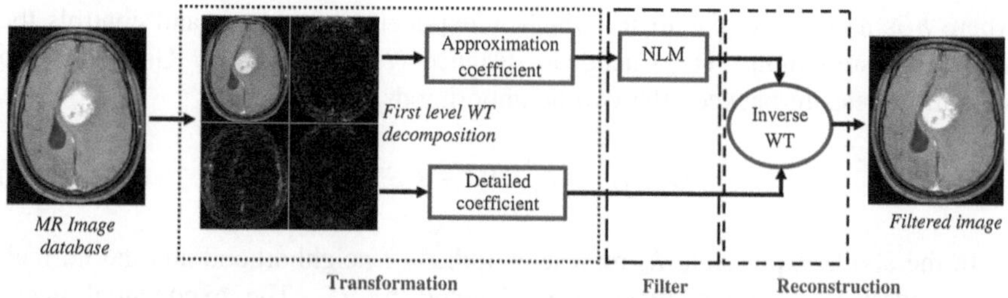

Fig. 1 The step by step representation of proposed methodology for filtering brain MR images

carry the edge information. WT is capable of representing edge information in several directions viz. horizontal, vertical and diagonal. The major advantage of WT over fourier transform is that this is localised in spatial as well as frequency domain, whereas fourier is restricted in frequency domain only. The coefficients of wavelet decomposition provide information which are independent of original image resolution. In the context of this research, single level wavelet decomposition has been adopted to generate the coefficients from which approximation coefficients will be considered for filtering task, whereas detail coefficients will be kept intact.

2.3 Non Local Means

NLM filter replaces the pixel value being filtered with the weighted average of pixels within a specified region. The weights are computed from the region based comparison instead of pixel based comparison and this makes it different and effective from bilateral filtering and also capable of removing thermal noise efficiently by reducing the variation of signal intensity in a region [9].

According to this approach, the filtered value (F) of a pixel i is obtained from

$$F(I(i)) = \sum_{\forall j \in I} w(i,j)I(j) \quad j = 1, 2, \ldots, n. \tag{1}$$

Here, I is an input MR image containing n number of pixels and $w(i, j)$ denotes the weight between *ith* and *jth* pixel, which has following properties as mentioned below

$$0 \le w(i,j) \le 1 \quad and \quad \sum_{\forall j \in I} w(i,j) = 1$$

The two square neighborhoods S_i and S_j centered at *ith* and *jth* pixel respectively are employed to measure the neighborhood based similarity between these two pixels from which the normalized weight between them can be computed as defined below

$$w(i,j) = \exp\left[-\frac{D(i,j)}{h^2}\right] / \sum_{\forall j} \exp\left[-\frac{D(i,j)}{h^2}\right], \tag{2}$$

where h is a decaying parameter which is exponential in nature and controls the amount of smoothing. The Euclidean distance (D) weighted by Gaussian as a similarity measure between the two neighborhoods is defined as

$$D(i,j) = G_i \left\| I(S_i) - I(S_j) \right\|_{R_s}^2. \tag{3}$$

In the above expression, R_s represents radius of neighborhood and the normalized Gaussian weighting function is denoted by G_i. The exponential decay parameter can be computed from the standard deviation σ of the noise which can be estimated from the background of the image as

$$\sigma = \sqrt{\frac{\mu}{2}} \tag{4}$$

where μ is the mean of background extracted by Otsu thresholding [9, 12]. A special condition in weight calculation is appeared when $i = j$ and under such situation it is computed as shown below

$$w(i, i) = \max(w(i, j)\forall j \neq i). \tag{5}$$

The actual process of computing weights by considering all pixels in an image increases time complexity and makes the technique inefficient. Therefore, a search window with radius R_{search} can be employed for weight calculation. As per the suggestion of Manjón et al. [9], a search window of dimension 11 × 11 and a 5 × 5 window for similarity measure along with $h = 1.2 \times \sigma$ have been considered in this study to suppress the effect of noise from MR signals.

2.4 Performance Evaluation Metrics

In this study, Mean squared error (MSE) and Peak signal to noise ratio (PSNR) are considered as quality metrics for the quantitative analysis of filter performance in terms of image quality [10, 13]. MSE is the accumulative squared error between the input (I) and filtered image (F) of dimension $M \times N$ as expressed below

$$MSE = \frac{1}{MN}\sum_{y=1}^{M}\sum_{x=1}^{N}[I(x,y) - F(x,y)]^2. \tag{6}$$

PSNR is the ratio of the maximum power of the original image to the filtered image, which is derived from MSE as

$$PSNR = 20 \times \log_{10}\left(\frac{\max(\max(\mathrm{I}), \max(\mathrm{F}))}{\sqrt{MSE}}\right). \tag{7}$$

3 Results and Discussion

The proposed algorithm has been designed in such a way that the major characteristics or features of MR image of brain tumor should be kept intact. The boundary that separates tumor from neighboring tissues holds the most significant characteristics. Blurring due to the smoothing effect of filter reduces the significance of contour of tumor and affects the analysis task. Hence edge preserving should be taken into account during the design of filter process. Spatial domain filters are inefficient of preserving the original characteristics of edges due to the smoothing effect. Therefore in the proposed framework discrete WT is employed and the low frequency components fed to NLM for updating through the weighted average of coefficients under a certain region, while preserving the high frequencies. The computation of weights plays the pivotal role in NLM. The weights between two coefficients are derived from the similarity measure and this task is accomplished by employing two neighbourhoods centered at both coefficients. In the last stage, the inverse wavelet transformation is applied on modified approximation and preserved detail coefficients to obtain the final outcome.

We have tested the filtering techniques directly on 50 axial MR images in two ways; without adding noise and adding external noise on original images. Figure 2 represents the results of proposed methodology along with ADF, BLF, TLF and WS (Haar wavelet is used with second level decomposition) [14] filter for three different images. In this case filters have been directly applied to original MR images. On the other hand, Fig. 3 shows the results of filters when Gaussian noise with variance 0.01 has been added. The two different quality measures are employed for quantitative validation of filtering techniques over 50 images. The lower value of MSE is desirable because higher value signifies large difference between input and filter image leading to the changes in image characteristics. In contrast, large PSNR is required as it signifies the comprehensive signal-to-noise

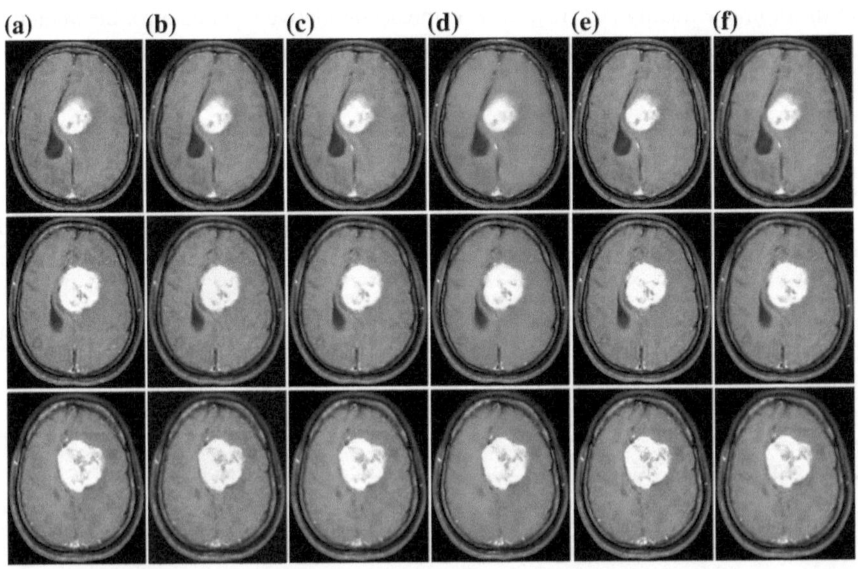

Fig. 2 Filtering outputs of axial MR images of brain tumor, **a** original MR image, **b** output of ADF, **c** BLF, **d** TLF, **e** WS and **f** proposed method

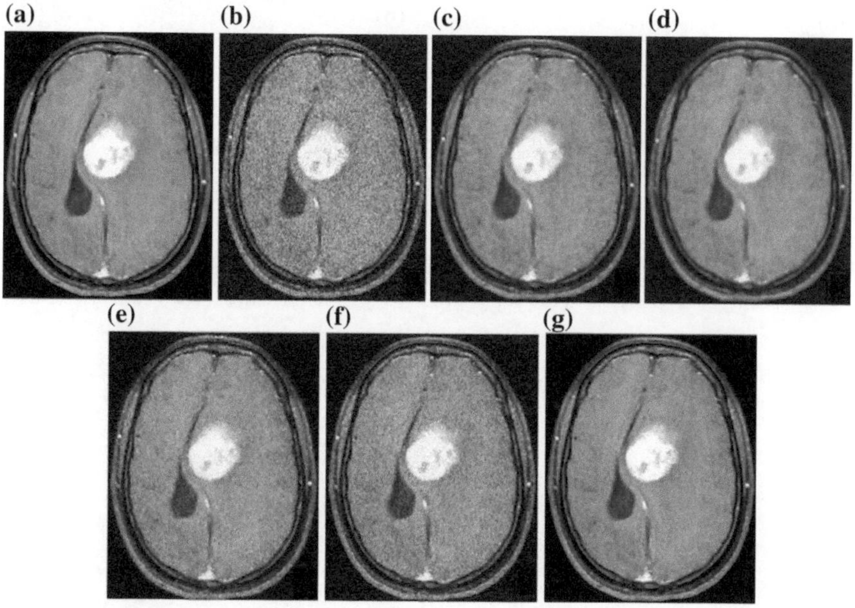

Fig. 3 Filtered outputs of MR image when noise is externally added, **a** original MR image, **b** image after noise addition, **c** output of ADF, **d** BLF, **e** TLF, **f** WS and **g** proposed methodology

ratio. The quantitative assessment of different filtering techniques is presented in Table 1 in terms of mean and standard deviation. From this table we clearly understand that proposed filtering methodology provides minimum MSE and maximum PSNR for both cases i.e. without and with adding noise as compare to ADF, BLF, TLF and WS methods. This evaluation justifies the superiority of proposed method.

We have plotted the mean and standard deviation of both MSE and PSNR for proposed method along with four other techniques as presented in Table 1 through the error bar representation. This graphical presentation has been executed for both cases i.e. with and without noise for better understanding and analysis of the results as shown in Fig. 4.

Table 1 Performance evaluation of proposed methodology, ADF, BLF, TLF and WS for reducing noise from MR images of brain tumor

Filter	With noise		Without noise	
	MSE	PSNR	MSE	PSNR
ADF	299.23 ± 12.28	23.36 ± 0.19	35.79 ± 2.47	32.61 ± 0.28
BLF	299.23 ± 14.83	23.36 ± 0.21	69.19 ± 8.32	30.00 ± 0.55
TLF	415.21 ± 13.68	21.94 ± 0.14	48.81 ± 7.3	31.37 ± 0.56
WS	480.54 ± 0.46	20.40 ± 0.08	36.25 ± 0.32	32.16 ± 0.22
Proposed	97.30 ± 0.96	28.24 ± 0.04	27.15 ± 2.25	34.06 ± 0.36

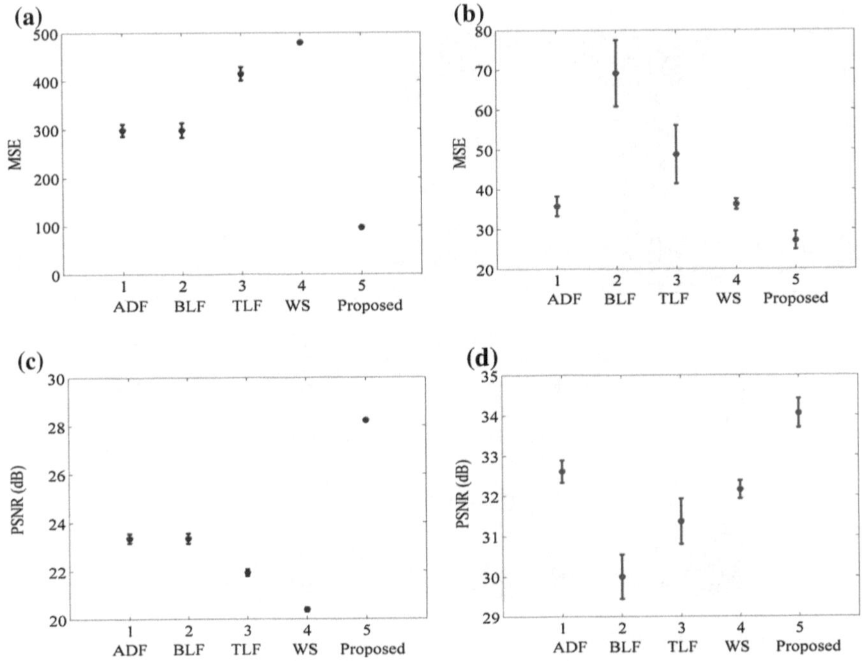

Fig. 4 *Error bar* representation of ADF, BLF, TLF, WS and proposed method with respect to **a** MSE when external noise is added, **b** MSE with no external noise, **c** PSNR when external noise is added and **d** PSNR with no external noise

To justify the superiority of proposed algorithm over the existing filters, the performance (in terms of MSE and PSNR) of all five filtering methods have been statistically analysed by Fisher's-F statistic. The performance of this statistical analysis has been reported in Table 2 by showing F value and *p*-value [15]. In Table 2, it is observed that the *p*-value is less than 0.05 for MSE and PSNR in both cases (with and without noise). Therefore the results obtained from this analysis signify that the mean MSE and PSNR of all five filters are different (in both cases) and hence the quantitative assessment of filter performances with respect to the quality metrics is statistically significant. After observing the results of proposed and other four techniques, it is well understood that our methodology reduces the impact of noise through preserving edges and avoiding the smoothing effect. Therefore, the internal characteristics are being intact. The overall performance

Table 2 Analysis of variance for MSE and PSNR of all five filters for both the cases (with and without noise)

One way ANOVA	With noise		Without noise	
	MSE	PSNR	MSE	PSNR
F value	188.900	40.053	67.898	36.274
p-value	0.000	0.000	0.000	0.000

p-value < 0.05 indicates statistical significance

analysis also supports this statement. But one thing we observe that, when filtering is performed on externally added noisy image, the image characteristics change in output, although proposed method provides efficient result than others. From the view point of visual interpretation and quantitative evaluation, the WT-NLM based hybrid filtering approach outperforms ADF, BLF, TLF and WS for improving the image quality of brain MR images of tumor.

4 Conclusion

A novel noise reduction technique combining transform domain method with spatial domain filter to enhance the image quality of MRI is presented in this paper. The proposed method proofs its effectiveness in suppressing noise in both approaches i.e. with and without adding external noise over four different filters. From the analysis of visual perception and performance evaluation, altogether we conclude that our methodology is a reasonable approach in reducing noise from MR images of brain tumor.

Acknowledgments The authors would like to acknowledge EKO CT & MRI Scan Centre at Medical College and Hospitals Campus, Kolkata-700073 for providing brain MR images. Authors would like to acknowledge Board of Research in Nuclear Sciences (BRNS), Dept. of Atomic Energy for financially supporting the research work under the grant number 2013/36/38-BRNS/ 2350 dt. 25-11-2013.

References

1. Perona P, Malik J (1990) Scale-space and edge detection using anisotropic diffusion. IEEE Trans Pattern Anal Mach Intell 12(7):629–639
2. Krissian K, Aja-Fernández S (2009) Noise-driven anisotropic diffusion filtering of MRI. IEEE Trans Image Process 18(10):2265–2274
3. Tomasi C, Manduchi R (1998) Bilateral filtering for gray and color images. In: Proceedings of the sixth international conference on computer vision, pp 839–846
4. Wong WCK, Chung ACS, Yu SCH (2004) Trilateral filtering for biomedical images. IEEE Int Symp Biomed Imaging: Nano to Macro 1:820–823
5. Xu Y, Weaver JB, Healy DM, Lu J (1994) Wavelet transform domain filters: a spatially selective noise filtration technique. IEEE Trans Image Process 3(6):747–758
6. Wood JC, Johnson KM (1999) Wavelet packet denoising of magnetic resonance images: importance of Rician noise at low SNR. Magn Reson Med 41(3):631–635
7. Nowak RD (1999) Wavelet-based Rician noise removal for magnetic resonance imaging. IEEE Trans Image Process 8(10):1408–1419
8. Buades A, Coll B, Morel J-M (2005) A review of image denoising algorithms, with a new one. Multiscale Model Simul 4(2):490–530
9. Manjón JV, Caballero JC, Lull JJ, Marti GG, Bonmati LM, Robles M (2008) MRI denoising using non-local means. Med Image Anal 12(4):514–523

10. Manjon JV, Coupe P, Bonmati LM, Collins DL, Robles M (2010) Adaptive non-local means denoising of MR images with spatially varying noise levels. J Magn Reson Imaging 31 (1):192–203
11. Erturk MA, Bottomley PA, El-Sharkawy AM (2013) Denoising MRI using spectral subtraction. IEEE Trans Biomed Eng 60(6):1556–1562
12. Ng H-F (2006) Automatic thresholding for defect detection. Pattern Recogn Lett 27(14):1644–1649
13. Starck JL, Candès EJ, Donoho DL (2002) The curvelet transform for image denoising. IEEE Trans Image Process 11(6):670–684
14. Balster EJ, Zheng YF, Ewing RL (2005) Feature-based wavelet shrinkage algorithm for image denoising. IEEE Trans Image Process 14(12):2024–2039
15. Zijdenbos A, Forghani R, Evans AC (2002) Automatic "pipeline" analysis of 3-D MRI data for clinical trials: application to multiple sclerosis. IEEE Trans Med Imaging 21(10):1280–1291

Histogram Based Thresholding for Automated Nucleus Segmentation Using Breast Imprint Cytology

Monjoy Saha, Sanjit Agarwal, Indu Arun, Rosina Ahmed, Sanjoy Chatterjee, Pabitra Mitra and Chandan Chakraborty

Abstract Breast imprint cytology is a well-recognized technique and provides a magnificent cytological clarity. For imprint cytology slide preparation, tissue samples from the needle taken out to touch and rolled over glass slide and finally stained by hematoxilin and eosin (H&E). The aim of this research is to segment breast imprint cytology nucleus. Images from imprint cytology slides were grabbed by an optical microscope. The histogram based threshold technique has been used to segment nucleus. The proposed technique includes pre-processing, segmentation, post-processing, and final output stage. In pre-processing first image colors was normalized by white balance technique. Then the green channel was extracted from the normalized image. In segmentation stage target nucleus was segmented by pixel intensities. Post-processing stage refers to the clear border nucleus or sharpening the edges. Finally, the three channels were concatenated to get RGB image. The proposed technique performs best in imprint cytology nucleus segmentation, and capable of distinguishing nucleus and non-nucleus objects. Performance of our proposed algorithm is quite high and much more useful for further analysis.

Keywords Hematoxilin · Eosin · Histopathology · Imprint cytology · Morphology · Intensity · Nucleus

M. Saha · C. Chakraborty (✉)
School of Medical Science and Technology, Indian Institute of Technology, Kharagpur 721 302, India
e-mail: chandanc@smst.iitkgp.ernet.in

S. Agarwal · I. Arun · R. Ahmed · S. Chatterjee
Tata Medical Center, New Town, Rajarhat, Kolkata 700 156, India

P. Mitra
Computer Science and Engineering, Indian Institute of Technology, Kharagpur 721 302, India

S. Gupta et al. (eds.), *Advancements of Medical Electronics*, Lecture Notes in Bioengineering, DOI 10.1007/978-81-322-2256-9_5

1 Introduction

Microscopic observation for cancer detection is one of the most preferred techniques. By microscopic visual inspection pathologist conclude grading and severity of cancer. The most prominent and effective example of microscopic image analysis is in the breast cancer grading system. Breast cancer considered a very common cancer among women [1–4].

Core biopsy confirmation of cancers is often time-consuming and entails biopsy, fixation, staining with hematoxilin and eosin (H & E) before visual examination under the microscope to report the diagnosis and characterize (grading, type) the cancer. This leads to increase patient discomfort apart from adding to the diagnostic delay. So, it is required to develop automated early screening protocol especially in rural areas to provide an efficient diagnostic prediction.

In view of this, breast imprint cytology is a well-recognized simple technique and provides an excellent cytological clarity [5, 6]. During the last few years, researchers investigated imprint cytology to screen cancer cells. This cytological approach involves taking an imprint of the core biopsy specimen, based on such imprints a result can be predicted instantly before detailed characterization in the core biopsy report. Trained technician takes breast imprint easily, and it does not cause distortion of the biopsy specimen architecture. This process could allow one needle biopsy that gives an instant report of cancer versus no-cancer as well as tumor characterization (as reported on core biopsy). The main challenge here lies in developing robust and efficient image processing algorithms for automated characterization of breast cancer nucleus cells from imprint cytological images (grabbed by optical microscope); because, there is a higher chance of having imprecision and ambiguities in these cytology images. Nuclei are the important component in breast imprint cytology. Here, we proposed to develop automated breast imprint cytology nucleus detection by intensity based image segmentation technique.

This paper is divided into four sections. First, section presents an introduction of breast imprint cytology and cancer. The second section describes materials and methods. Third section describes experimental results. Fourth section presents the conclusion.

2 Materials and Methods

2.1 Breast Imprint Cytology Slide Preparation and Image Acquisition

Ethical rules and patient's consent has been taken from every patient in our research. All the breast imprint cytology slides were prepared and maintained at Tata Medical Center, Kolkata by the approved cytologists and cytopathologists. Local anesthesia and 0.5 % epinephrine are injected around the lesions. Lidocaine hydrochloride is mixed along with 0.5 % epinephrine during anesthesia [9]. Tissue samples from the needle taken out to touch and rolled over glass slide. Then the tissue samples are fixed with 95 % ethanol [9]. Finally, H&E stain is used to stain the slides.

Breast imprints cytology images were grabbed in Leica DM750 microscope with Leica ICC50 HD camera, from Tata Medical Center, Kolkata and BMI Lab, School of Medical Science and Technology, Indian Institute of Technology, Kharagpur, West Bengal.

2.2 Proposed Methodology

The proposed methodology divided into four parts i.e., image pre-processing, segmentation, post-processing and final output. The method is summarized in Fig. 1 using the flow diagram, and Fig. 2 describes resulted image of each step of Fig. 1.

All the images were grabbed at a constant brightness, and contrast. For proper visibility and enhancement of the target (nucleus) region pre-processing step is applied. The preprocessing step is sub-divided into white balance adjustment, G-channel extraction, and image intensity adjustment.

Before going to further analysis of breast imprint cytology images, it is necessary to reduce the different color response error in the image. These types of color response error may occur due to microscopic light. Even some digital microscopic cameras sometimes give inconsistent outputs. So, white balance adjustment is necessary to normalize the unrealistic colors present in the image. Color correction model of white balance can be represented by the following matrix equation [7].

$$\begin{bmatrix} R_{co} \\ G_{co} \\ B_{co} \end{bmatrix} = \begin{bmatrix} R_{ref}/R_{meg} & 0 & 0 \\ 0 & G_{ref}/G_{meg} & 0 \\ 0 & 0 & B_{ref}/B_{meg} \end{bmatrix} \cdot \begin{bmatrix} R_o \\ G_o \\ B_o \end{bmatrix}$$

Here, R_o, G_o, and B_o corrected color coordinates R_{ref}, G_{ref}, and B_{ref} reference color coordinates, R_{meg}, G_{meg}, and B_{meg} measured coordinates and R_{co}, G_{co}, B_{co} corrected color output to a standard white illuminant. Breast imprint cytology white balance adjusted images were then separated into red (R), green (G), and blue (B) intensity levels. Our proposed methodology has been used for analyzing the images.

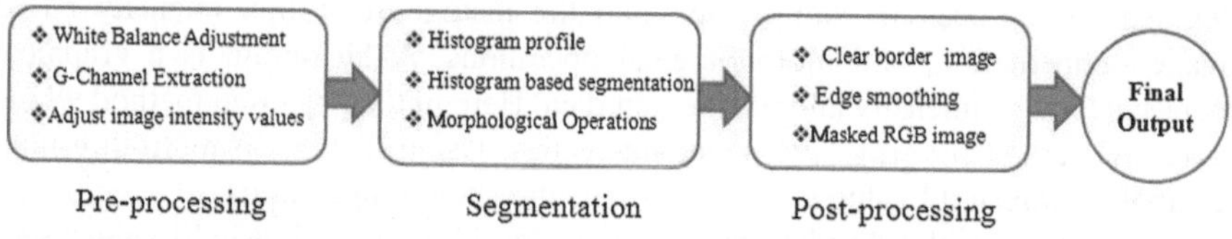

Fig. 1 Schematic of the proposed breast imprint cytology malignant nucleus segmentation technique

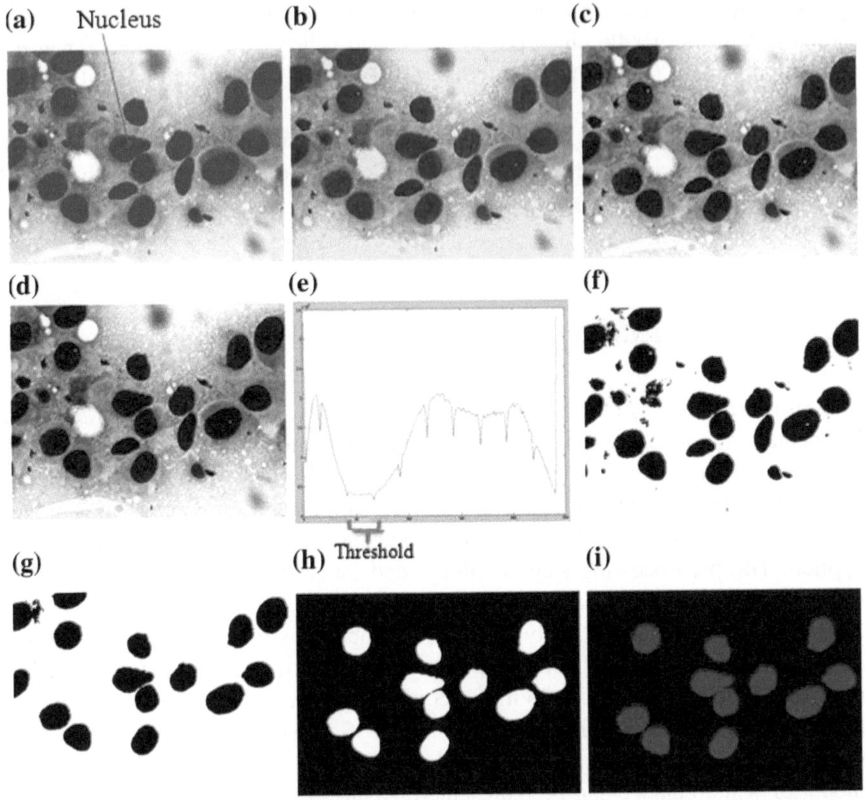

Fig. 2 Steps of Breast imprint cytology malignant nucleus segmentation. **a** Original breast imprint cytology image at 100X; **b** white balance adjustment; **c** G-Channel image; **d** image intensity adjustment; **e** histogram; **f** segmented nucleus by proposed method; **g** morphological operation image; **h** clear border image; **i** masked RGB image

We tested all images in RGB, HSV, L*a*b etc. color levels. Only green (G) channel images showed good results. G-channel extracted images have high contrast between nucleus and background. The next step is an image intensity adjustment. Image intensity adjustment maps the image intensity into 0–255 range and increases the contrast of the image.

Furthermore, intensity adjusted images go to the next segmentation process. Segmentation process is sub-divided into the histogram profile, intensity based image segmentation, and morphological operations. A histogram is a graphical representation of intensity and pixel count data. Here in the proposed method whole image is marked by gray level color intensities. Using histogram multi-threshold technique a threshold value is generated. The threshold range varies between 0–255 and it is automatically detected through our algorithm. This threshold values separate nucleus from the background. Mathematically, we can define the threshold as:

If $f(i, j) \le T_a$ then $f(i, j) = 255$
else if $T_a < f(i, j) < T_b$ then $f(i, j) = 128$
else $f(i, j) = 0$

Here, T_a and T_b are threshold values, and 'i' and 'j' are co-ordinates. In the morphological operation image erode and dilate concept used. Erosion and dilation do with a structuring element. The grayscale erosion is [8]:

$$(p \Theta q)(r, s) = \min\{p(r + r', s + s') - q(r', s')|(r', s') \in V_q\}$$

Here, $(p \Theta q)$ denotes binary erosion of p and q. Structuring element q's domain is V_q. $p(r, s) = +\alpha$.

If, $q(r, s) = 0$, we'll get flat structuring element.

Now the gray scale erosion will be [8]:

$$(p \Theta q)(r, s) = \min\{p(r + r', s + s') - |(r', s') \in V_q\}$$

For dilation equation will be [8]:

$$q(r, s) = 0;$$

$$(p \oplus q)(r, s) = \min\{p(r + r', s + s') - |(r', s') \in V_q\}$$

In post-processing stage border image has been cleared. This stage is required because if a nucleus is connected at the border, it's very tough to get full morphological information of nucleus. It may give incorrect information in nucleus counting. Nucleus edges have been shaped by image sharpening algorithm. Finally, three channels have been concatenated to get final RGB nucleus.

3 Results and Discussion

The proposed algorithm is tested on breast imprint cytology images. Figure 3, shows the result of three R, G, B channels. From the output images clearly R and G channel images are much more visible than B channel image. Again, between R and G images, G channel image has much more contrast and brightness than R channel image. So, G channel image extraction is very much suitable for our analysis.

Figure 4, shows the original image and ground truth image where the nucleus has been marked by specialists at Tata Medical center using red color.

In Fig. 5, pre-processing, segmentation, and final output results have shown in three different breast imprint cytology images. Table 1, describes the comparison table of manual and automated nucleus count of total 9 frames.

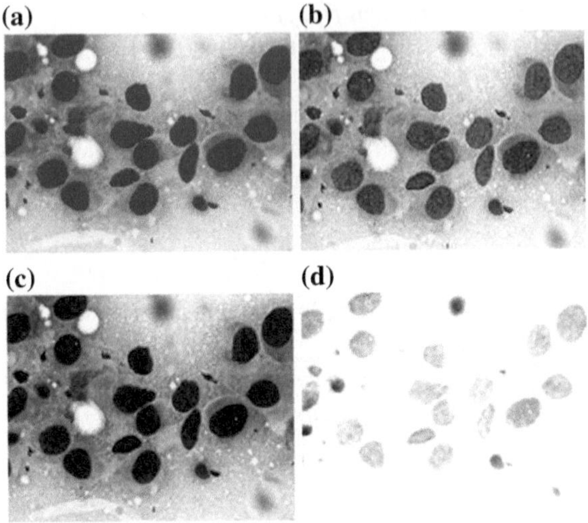

Fig. 3 Contrast visualization in RGB space. **a** Original breast imprint cytology image at 100X; **b** R-channel image; **c** G-channel image; **d** B-channel image

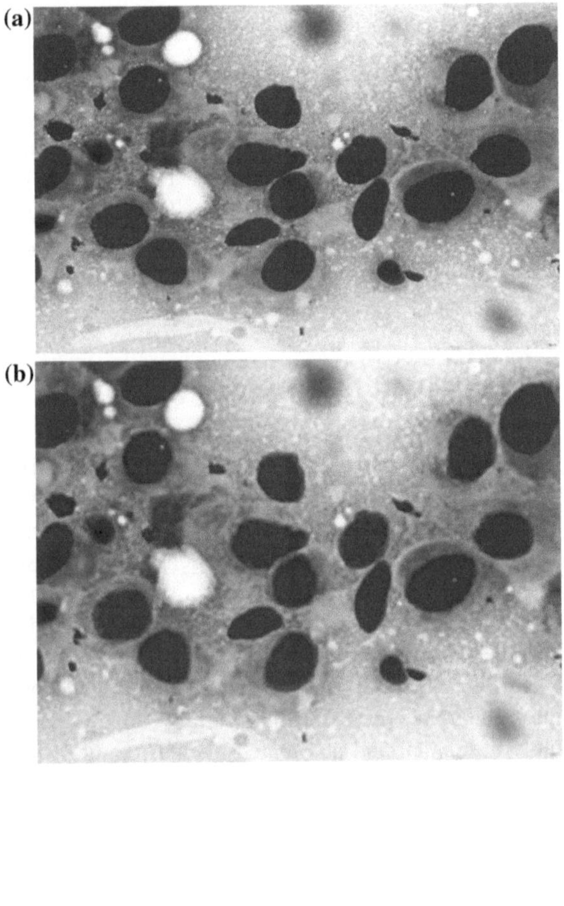

Fig. 4 Ground truth RGB image. **a** Original breast imprint cytology image at 100X; **b** ground truth RGB image (*red circle*)

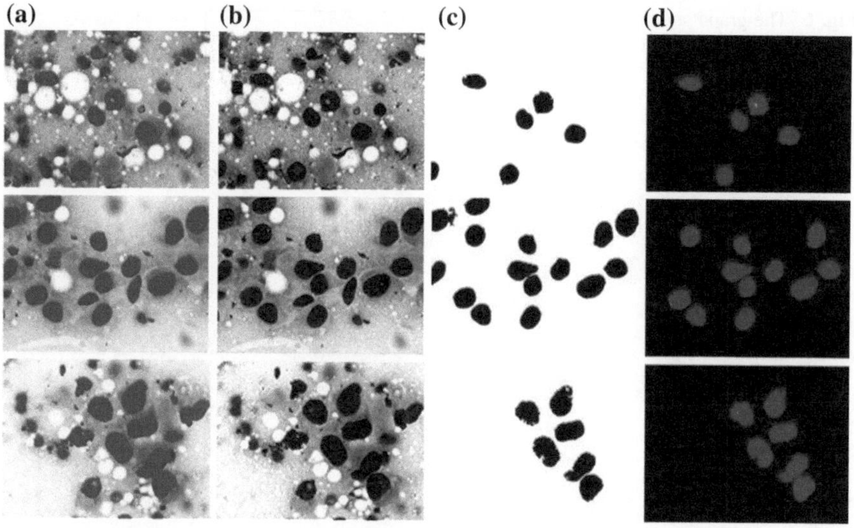

Fig. 5 Results of the proposed methodology. **a** Original cytology image at 100X; **b** pre-processed image; **c** segmented binary image; **d** final segmented nucleus

Table 1 Comparison table of nucleus count

Image index	Manual nucleus count	Automated nucleus count
1	5	5
2	11	11
3	6	6
4	6	7
5	4	4
6	10	10
7	9	9
8	5	5
9	6	6

Figure 6, shows a regression curve between manual and automated nucleus count. Here, almost all nucleus segmented normally.

The graph clearly shows that the proposed method is perfectly segmenting nucleus as machine generated nucleus count provides exact equivalent results in parallel to the ground truth, counted manually.

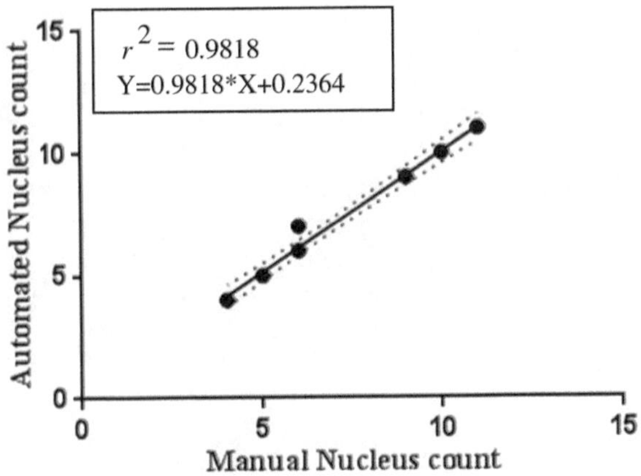

Fig. 6 The graph shows the strength of the proposed method for automated nucleus segmentation in comparison with ground truth

4 Conclusion

Nucleus identification is important in breast cancer detection in breast imprint cytology. It is concluded that this algorithm performs very well on breast imprint cytology images for nucleus segmentation. We have successfully eliminated other background objects except nucleus from the images. Our proposed method gives higher count accuracy (97 %) result in nucleus detection. Further research will be focused on the breast imprint overlapped or connected nucleus.

Acknowledgments First author acknowledges the Department of Science and Technology (DST), Govt. of India, under ***INSPIRE* fellowship**.

All other authors acknowledge the Ministry of Human Resource Development (MHRD), Govt. of India, for financial support under grant no: 4-23/2014 -T.S.I. date: 14-02-2014.

References

1. Niwas SI, Palanisamy P, Sujathan K, Bengtsson E (2013) Analysis of nuclei textures of fine needle aspirated cytology images for breast cancer diagnosis using complex Daubechies wavelets. Sig Process 93:2828–2837
2. Kowal M, Korbicz J (2010) Segmentation of breast cancer fine needle biopsy cytological images using Fuzzy clustering. In Koronacki J, Raś Z, Wierzchoń S, Kacprzyk J (eds) Advances in machine learning I, vol 262. Springer, Berlin, pp 405–417
3. Kamangar F, Dores GM, Anderson WF (2006) Patterns of cancer incidence, mortality, and prevalence across five continents: defining priorities to reduce cancer disparities in different geographic regions of the world. J Clin Oncol 24:2137–2150
4. Niwas SI, Palanisamy P, Sujathan K (2010) Wavelet based feature extraction method for breast cancer cytology images. In: IEEE symposium on industrial electronics & applications (ISIEA), 2010. pp 686–690
5. Suen K, Wood W, Syed A, Quenville N, Clement P (1978) Role of imprint cytology in intraoperative diagnosis: value and limitations. J Clin Pathol 31:328–337

6. Bell Z, Cameron I, Dace JS (2010) Imprint cytology predicts axillary node status in sentinel lymph node biopsy. Ulster Med J 79:119–122
7. Wannous H, Lucas Y, Treuillet S, Mansouri A, Voisin Y (2012) Improving color correction across camera and illumination changes by contextual sample selection. J Electron Imaging 21:023015-1–023015-14
8. Gonzalez RC, Woods RE, Eddins SL (2004) Digital image processing using MATLAB. Prentice Hall, Upper Saddle River
9. Kashiwagi S, Onoda N, Asano Y, Noda S, Kawajiri H, Takashima T et al (2013) Adjunctive imprint cytology of core needle biopsy specimens improved diagnostic accuracy for breast cancer. SpringerPlus 2:1–7

Separation of Touching and Overlapped Human Chromosome Images

V. Sri Balaji and S. Vidhya

Abstract Chromosomes are generally thread-like structures present in the nucleus of each living cells. There are twenty three pairs of chromosome in human beings. The additional in chromosome number or missing chromosome will cause chromosome abnormality i.e. chromosome anomaly in human beings. This is mainly occurring due to accident or error while sperm or egg progress in growing. The abnormality in chromosome will cause birth defects, genetic disorders and cancer to the human beings. The twenty three pairs of chromosome can be classified into twenty four different classes by using karyotyping process. In this paper, the main objective is to give idea about how to diagnose the genetic disorder. The genetic disorder can lead to cancer in human. It may happen due to touching or overlapped chromosomes in human beings. In order to overcome these genetic disorders and cancer in human, the touching and overlapped chromosomes were separated. This separation process will be easier to undergo karyotyping analysis and to be handled by a cytogeneticist.

Keywords Nucleus · Chromosome abnormality · Birth defects · Cancer · Karyotyping

1 Introduction

Chromosomes are generally thread-like structures found in the nucleus of each living cells. It carries genetic information [1]. They cannot be viewed by naked eye, but can be viewed through a microscope. There are several bands available in the chromosome. To view the image clearly under microscope, G-band metaphase or metaspread images is shown in Fig. 1 are taken for analysis [2].

V.S. Balaji
Biomedical Engineering, VIT University, Vellore 632014, Tamil Nadu, India
e-mail: sribalaji1006@gmail.com

S. Vidhya (✉)
Biomedical Engineering, VIT University, Vellore 632014, Tamil Nadu, India
e-mail: svidhyavalentina@vit.ac.in

S. Gupta et al. (eds.), *Advancements of Medical Electronics*,
Lecture Notes in Bioengineering, DOI 10.1007/978-81-322-2256-9_6

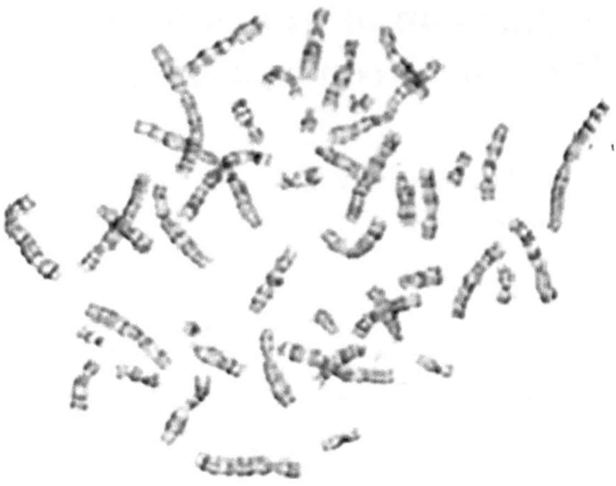

Fig. 1 The G-band metaphase human chromosomes image before karyotyping

There are twenty three pairs of chromosome in human beings. The first twenty two pairs are called as autosomes and twenty third pair is called as sex chromosome i.e. two X in female and one X and one Y in male [3]. The additional in chromosome number i.e. forty seven instead of forty six or missing chromosome will cause chromosome abnormality in human beings [4].

The anomaly in chromosome will cause birth defects which affect new born babies and lead to mental or physical disabilities, improper body function and can even be fatal sometimes [5, 6]. The leukemia, aneuploidy, deletion, duplication, inversion and translocation will occur due to genetic disorders or any defects in chromosome [7]. The study of cancer and diagnosing the defects in chromosome images i.e. genetic disorders are important process. After separating the chromosome, the classification is followed. For karyotyping process as shown in Fig. 2 is the human chromosomes are classified into twenty four different classes [8]. The karyotyping process will include segmentation and classification [4].

2 Methodology, Results and Discussion

2.1 The Steps to be Considered While Segmenting an Image

(i) Get an overlapped or touching Chromosome images as input image.
(ii) Apply Binary approximation to input image.
(iii) Obtain Binary and Thresholding output.
(iv) Apply Contour to the threshold image.
(v) Apply a suitable algorithm to obtain the segmented output.
(vi) After segmentation, classification is followed.

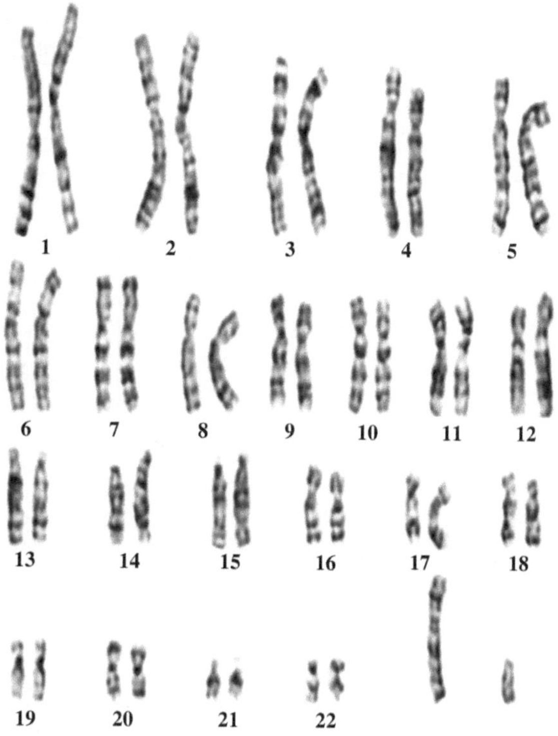

Fig. 2 The G-band metaphase human chromosomes image after karyotyping

2.2 Description and Results

The abnormal chromosomes will cause chromosome anomaly in human begins this can be identified by analysis G-band metaphase images. To overcome this problem, the suitable segmentation algorithm should be followed for the approach [5, 6]. Before segmenting, the pre-processing steps to be considered (i, ii, iii, iv) is shown in Fig. 3. In some cases the pre-processing steps will be included in algorithm itself. The first four steps contribute to the pre-processing of the image. Initially, an overlapped or touching chromosome as an input image was taken and binary approximation was applied to the input image. To get the complete and fulfilled information about an image Otsu thresholding was applied, followed by applying contour [5, 6, 9–12].

By finding either the intersecting points (concave points) [5, 6] or the cut points [10–12], or by the Voronoi diagrams and Delaunay triangulations [13] or Semi-automatic Segmentation [14] the touching and overlapped chromosome can be separated easily. After drawing a circle in Fig. 1, the image was investigated to see that the number 1 in Fig. 4 represents the overlapped chromosomes and number 2 in Fig. 4 represents the touching chromosomes in G-band image respectively.

Fig. 3 The block diagram

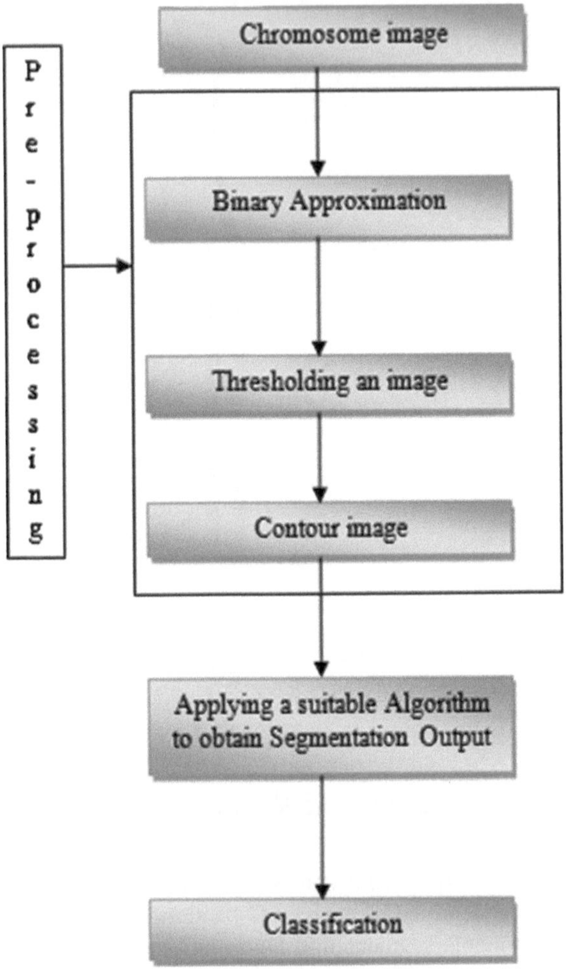

The abnormal chromosomes in Fig. 5 will cause cancer, birth defects, fatal etc. in human beings. This can be overcome by early diagnosis using appropriate segmentation, which follows the steps in Fig. 3.

The separation of touching chromosomes in Fig. 7 is easier than the separation of overlapped chromosomes in Fig. 6. This segmentation was done by semi-automatic technique and currently by automatic technique also to be followed.

After segmentation, the chromosomes are classified into twenty four different classes. This segmentation and classification of chromosome together follow the karyotyping process [4–6]. This process is lengthy, quite difficult and repetitive [15]. So the automatic segmentation of chromosome was followed. By segmenting a chromosome image, it would be helpful for karyotyping analysis. The genetic disorders in human can be diagnosed using segmentation method in Figs. 6 and 7 followed by karyotyping process in Fig. 2 [10–12].

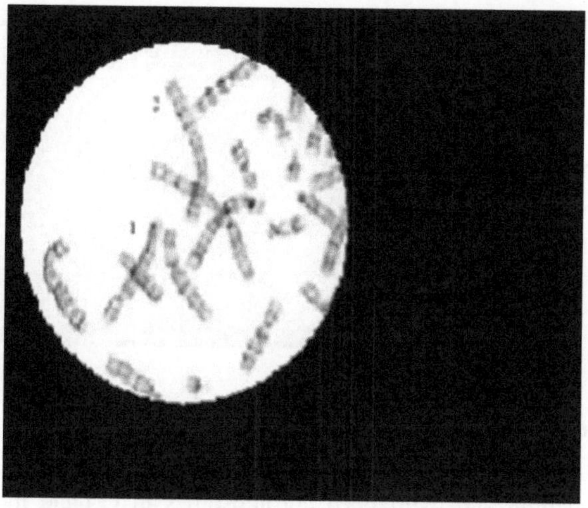

Fig. 4 The overlapped and touching chromosomes in G-band image shown in Fig. 1

Fig. 5 The sample overlapped and touching chromosomes (abnormal chromosomes) in Fig. 4

Fig. 6 The separation of overlapped chromosomes image

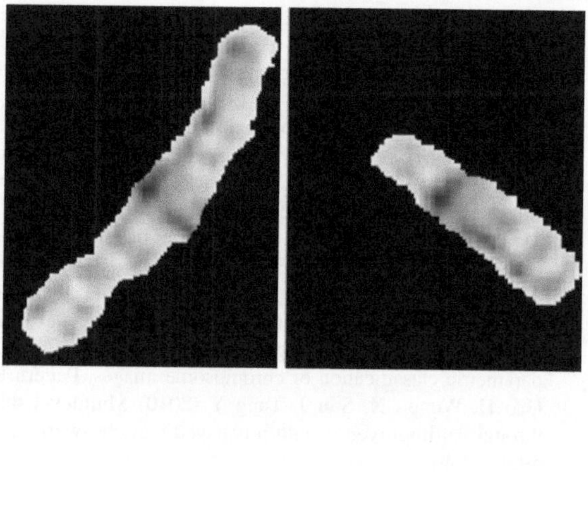

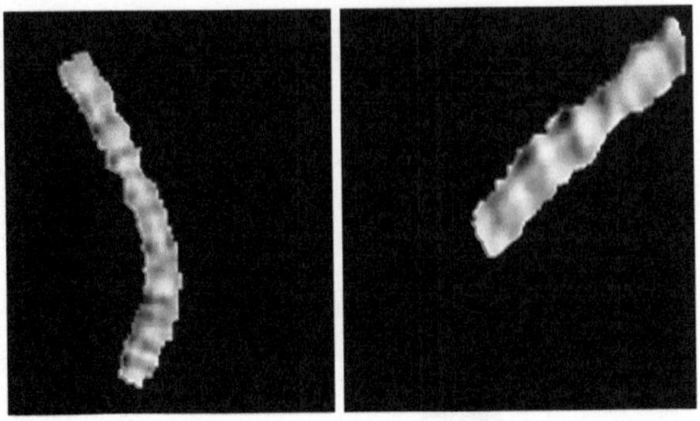

Fig. 7 The separation of touching chromosomes image

3 Conclusion

By identify touching and overlapped chromosomes in G-band metaphase image under light microscope; one can separate the touching and overlapped chromosomes. This separation method is helpful for diagnosing the genetic disorders using karyotyping method. Thus we may overcome cancer, genetic disease in human by diagnosing this disorder.

Acknowledgments The authors wish to thank Ms. Nirmala Madian working as an Assistant Professor and Phd research scholar in K.S.R College of Technology and Dr. Suresh, Director, Birth Registry of India for providing the G-band chromosome images for the work.

References

1. Legrand B, Chang CS, Ong SH, Soek-Ying N, Nallasivam P (2008) Chromosome classification using dynamic time warping. Pattern Recogn Lett 29:215–222
2. Cai N, Hu K, Xiong H, Li S, Su W, Zhu F (2004) Image segmentation of G bands of Triticum monococcum chromosomes based on the model-based neural network. Pattern Recogn Lett 25:319–329
3. Piper J (1995) Genetic algorithm for applying constraints in chromosome classification. Pattern Recogn Lett 16:857–864
4. Loganathan E, Anuja MR, Nirmala M (2013) Analysis of human chromosome images for the identification of centromere position and length. IEEE Point Care Healthc Technol, pp 314–317
5. Madian N, Jayanthi KB (2002) Overlapped chromosome segmentation and separation of touching chromosome for automated chromosome classification. In: IEEE EMBS, pp 5392–5395
6. Madian N, Jayanthi KB (2014) Analysis of human chromosome classification using centromere position. Measurement 47:287–295
7. Roshtkhari MJ, Setarehdan SK (2008) A novel algorithm for straightening highly curved images of human chromosome. Pattern Recogn Lett 29:1208–1217
8. Sampat MP, Bovik AC, Aggarwal JK, Castleman KR (2005) Supervised parametric and non-parametric classification of chromosome images. Pattern Recognit 38:1209–1223
9. Gao H, Wenbo X, Sun J, Tang Y (2010) Multilevel thresholding for image segmentation through an improved quantum-behaved particle swarm algorithm. IEEE Trans Instrum Meas 59:934–946

10. Somasundaram D, Vijay Kumar VR (2014) Separation of overlapped chromosomes and pairing of similar chromosomes for karyotyping analysis. Measurement 48:274–281
11. Somasundaram D, Vijay Kumar VR (2014) Straightening of highly curved human chromosome for cytogenetic analysis. Measurement 47:880–892
12. Somasundaram D, Palaniswami S, Vijayabhasker R, Venkatesakumar V (2014) G-Band chromosome segmentation, overlapped chromosome separation and visible band calculation. Int J Hum Genet 14:73–81
13. Wacharapong S, Krisanadej J, Mullica J (2006) Segmentation of overlapping chromosome images using computational geometry. Walailak J Sci Tech 3:181–194
14. Munot MV, Joshi MA, Mandhawkar P (2012) Semi automated segmentation of chromosomes in metaphase cells. IET Conference on Image Processing, pp 1–6
15. Grisan E, Poletti E, Ruggeri A (2009) Automatic segmentation and disentangling of chromosomes in Q-Band prometaphase images. IEEE Trans Inf Technol Biomed 13:575–581

Combination of CT Scan and Radioimmunoscintigraphy in Diagnosis and Prognosis of Colorectal Cancer

Sutapa Biswas Majee, Narayan Chandra Majee and Gopa Roy Biswas

Abstract Staging of colorectal cancer constitutes an important part of its diagnosis and prognosis. However, both invasive and non-invasive techniques prevail with their own advantages and disadvantages. The present review focuses on the existence of complementarity between the information obtained from computerized tomography and radioimmunoscintigraphy in the study of hepatic and extra-hepatic lesions of significance and relevant to colorectal cancer. The latter technique utilizes different monoclonal antibodies which are tagged with radioisotopes and imaging done by gamma-camera. For complete diagnosis of recurrent carcinoma or metastases, knowledge form both the pre-surgical procedures is an absolute necessity to choose the correct therapeutic modality.

1 Introduction

Colorectal cancer is the commonest and the most preventable form of cancer where survival rate can be improved markedly by selection of proper diagnosis of its stages. Progression to adeno-carcinoma from adenomatous polyps takes place over a span of time and through various stages as proposed by American Joint Committee on Cancer/Union Internationale Contre le Cancer (AJCC/UICC). The system is referred to as TNM system which is used for both clinical and pathological classification, where T denotes local extent of untreated primary tumor, N indicates tumor involvement of regional lymph nodes and lymphatic system and finally M refers to metastatic disease.

S.B. Majee (✉) · G.R. Biswas
NSHM College of Pharmaceutical Technology, NSHM Knowledge Campus, Kolkata-Group of Institution, 124 B L Saha Road, Kolkata 700053, India
e-mail: sutapabiswas2001@yahoo.co.in

N.C. Majee
Department of Mathematics, Jadavpur University, Kolkata 700053, India

S. Gupta et al. (eds.), *Advancements of Medical Electronics*,
Lecture Notes in Bioengineering, DOI 10.1007/978-81-322-2256-9_7

The prognosis of colorectal cancer depends on the stage at which examination is being carried out which helps in determining the extent of tumor as well as detecting regional and distant metastases. Invasive methods of staging colorectal cancers like sigmoidoscopy, optical and screening computerized tomography (CT) enabled colonoscopy involve surgical operations and have several drawbacks including the failure to identify tumor deposits in lymph nodes, soft tissues in the abdomen and retro peritoneal areas. However, any less invasive method, e.g. CT colonography or virtual colonoscopy, capable of diagnosing and staging cancer prior to surgery would always be more preferable. This method is suitable mainly for screening of patients with polyps of larger size (>1 cm in diameter) which may grow into tumors. Recent development in non-invasive pre-surgical cancer staging is related to targeting of monoclonal antibodies to the antigenic markers of colorectal cancer cells and radiolabeling with gamma-isotopes. Once administered, the distribution of the radioisotope which is studied by gamma-immunoscintigraphy gives a precise indication about the location, locoregional and metastatic spread, size and severity of the tumor. Radioimmunoscintigraphy with ^{123}I has been routinely used by various scientists with different fragments of IMMU-4[F(ab′)2 and F (ab′)]in confirming suspected tumors and disclosing occult lesions along with CT. Similarly, immunoscintigraphy with anti-CEA monoclonal antibody coupled with single photon emission computerized tomography (SPECT) reduces the time delay between diagnosis and treatment since it permits early diagnosis of recurrent carcinoma or metastasis of colorectal cancer [1–3].

2 Role of Monoclonal Antibodies (mAbs) in Radioimmunoscintigraphy (RIS)

Monoclonal antibodies with high affinity are designed for specific tumor markers or antigens expressed on the cell surface due to alterations in cell DNA. Ideally, antigen should be produced in abundance (5,000 epitopes per cell) only from tumor cells and not from normal cells or during other pathological condition. Moreover, it should be expressed from tumors at various stages of differentiation. An ideal monoclonal antibody should recognize only tumor cells and should possess limited reactivity with the non-malignant cells. The most widely used monoclonal antibodies used in diagnosis of colorectal cancer are IMMU-4 and PRIA-3, both of which are targeted against carcinoembryonic antigen (CEA), an onco-fetal antigen arising from gastrointestinal epithelium. Monoclonal murine antibody B72.3 can be exploited for radioimmunoscintigraphy as TAG-72, its tumor-associated cell-surface glycoprotein target can interact specifically with majority of the mucin-producing colon adenocarcinomas. After isolation and purification, monoclonal antibody can be site specifically conjugated with satumomab pendetide to form the immunoconjugate which can then be labeled with either radioisotopes of Technetium, Iodine, Indium or Rhenium. These isotopes are used because of the ease of labeling with them, physical characteristics, high percentage of intake per gram of tumor tissue. ^{99m}Tc emits optimal energy gamma rays detectable by gamma camera possesses a short half-life enabling imaging to be completed within the same day. Although, uptake by liver and marrow is less compared to those of ^{111}Indium, higher percentage of urinary excretion leads to accumulation in bladder thereby causing a hindrance to pelvis imaging. Moreover, due to its long half-life, imaging should be continued for 48–72 h after administration. Radioisotope of Indium also fails to provide good image of hepatic tissues, where colorectal cancer is known to spread commonly during metastasis. ^{125}I produces low radiation energy and long half-life. Labeling of antibody fragment of anti-CEA, IMMU-4 with Technetium has proved satisfactory in diagnosis of occult metastatic cancer which could not be detected by abdominal and pelvic CT scans in patients with elevated CEA levels. Stability of rhenium-labeled antibodies can be improved by chelating with tetrafluorophenyl-activated ester derivative of triamide thiolate. Intravenous administration of the B 72.3 conjugate followed by imaging with gamma camera between second and 7 day and separated by an interval of not less than 24 h, has been able to detect cases of both primary and local recurrence of colorectal cancer including occult lesions and incidences of liver metastases successfully and with high sensitivity. However, there are reports of non-specific uptake in spleen and bone marrow as well as gastrointestinal and genitourinary systems. Monoclonal antibody PRIA-3 possesses high selectivity for CEA and could detect recurrent colorectal cancers with high degree of accuracy [4–9].

3 Role of Monoclonal Antibody Fragments (Fabs) in Radioimmunoscintigraphy (RIS)

Antibodies are characterized by two heavy chains and two light chains linked together by disulphide bonds to form a Y-shaped configuration. The stalk of the Y is the Fc portion constitutes the stalk of the Y and Fab portions represent the arms. The tip of the Fab portion is responsible for reaction with antigen. Whole murine monoclonal antibodies may induce an immune response in the form of allergic reaction in 5–40 % of patients, due to the formation of human anti-murine antibodies (HAMA) which are targeted against the Fc portion of the antibody. This can be potentially avoided by use of Fab portion. There are also other reasons which favour the use of antibody fragments instead of whole antibodies in radioimmunoscintigraphy. Use of antibody fragments accelerates the blood clearance compared to intact form of monoclonal antibodies thereby reducing high background noises. Moreover, cocktail of antibody fragments helps in recognition of different epitopes, otherwise not recognizable by individual fractions due to heterogeneity of tumor. An important monoclonal antibody which found use in detection of occult metastatic colorectal cancer was radiolabeled and stabilized F(ab′)2 fragments of ZCE-025, an anti-CEA monoclonal antibody and investigated by single photon emission CT (SPECT). It exhibited higher sensitivity at smaller protein doses than the intact immunoglobulin antibody. Moreover, cancer could be detected in patients who showed negative results with laparotomies and presence of tumor as observed from positive scans could be confirmed by exploratory surgery or by diagnostic biopsy. Indium-labelled ZCE-025 has been found to be quite useful in pre-operative staging, distinguishing recurrent tumors from post-operative or post-radiation alterations manifested on CT scans or any other radiographic images. Radioimmunoscintigraphy and detection of colorectal cancer was also conducted with radiolabelled monoclonal antibody fragments [F(ab′)2 BW 431/31] against CEA. However, efforts should be taken to improve and optimize tumor affinity and specificity. IMACIS 1 is a cocktail of mAb19–9F (ab′)2 and mAb anti-CEA F (ab′) 2 which is labeled with ^{131}I. Another related antibody is Indimacis 19–9 containing 19–9F(ab′)2/DTPA fragments of mAbs.

There are several factors governing targeting and imaging modality with monoclonal antibodies like stability of the monoclonal antibody fragment, the labeling chemistry, modification of critical residues, the number of antigens expressed on the cell surface, recycling rate of the target and ability of the target to internalize upon binding [10–13].

4 CT Scan and RIS: Comparative Assessment in Diagnosis of Colorectal Cancer

CT is an important imaging method for detection of lymph node metastases, where an idea about size, number and sometimes morphology can be obtained. A significant drawback of the method is that it possesses sensitivity varying from 22–73 % and may provide incorrect information about malignancy if the size of the tumor is less than 1 cm. It has been reported that clusters of drainage nodes of smaller size located near the primary tumor may be considered as malignant although it diminishes the extent of specificity. On the other hand, radioimmunoscintigraphy has been found to locate tumor growth in lymph nodes with size less than 1 cm and also strands of tumor within the peritoneum. However, the technique seems to fail in detecting extrahepatic metastases which present themselves as localized areas of increased uptake within the abdomen. The main disadvantage with RIS is its low degree of spatial resolution. The type of the radionuclide used alters the appearance of hepatic metastases which appear as areas of increased isotope activity or photopenic areas. This problem seems more acute with Indium-labelled antibodies. In scintiscans using technetium-labelled antibodies against CEA-derived antigens hepatic metastases are observed as either areas of increased radioactivity with a high uptake boundary, most probably due to central necrosis in the core. In CT scan, accuracy in detection of hepatic metastases is comparatively greater.

CT alone does not provide accurate information in the early stage of recurrent carcinoma as the local anatomy is distorted after surgery. It has been observed that if recurrence were suspected from CT and RIS scans, the patients actually demonstrated recurrence and reduced the need for biopsy. Any false positive interpretation due to presence of fecal matter and bladder activity can be confirmed by correlating the observations from both CT and RIS investigations. In some other instances, Technetium-labeled anti-CEA scan could detect local disease and distant metastases with improved sensitivity. Elevated levels of CEA in blood of post-operative patients indicated signs of recurrence which could only be confirmed by combining the favorable features of both RIS and CT study.

Moreover, following therapy with radiation, it becomes difficult to differentiate fibrotic areas from the viable tissue. Introduction of multi-detector CT scanning (MDCT) and improved processing software has enhanced the accuracy rates of stage detection primarily for T patients but not so much for N patients. Therefore, it must be realized that information from both the pre-surgical diagnostic procedures is required for complete evaluation of different anatomic regions of human gastro-intestinal tract and thus they provide complementary tools in the diagnosis and prognosis of colorectal cancer, especially in those cases where it is known or suspected to extend beyond the bowel. CT colonography shows promise in assessment of synchronous lesions and metastasis [14, 15].

5 Radioimmunoscintigraphy Adds a New Milestone

Radioimmunoscintigraphy has paved the way for radioimmunoguided surgery (RIGS) as a surgical intervention in management of colorectal cancer. In the procedure, an antitumor antibody is injected intravenously before surgery which is followed during the operation with the help of a hand-held gamma camera probe, in order to locate tumor selectively by counting of radioactivity in the operative field. But prolonged blood clearance time of whole antibodies causes a delay prior to surgical resection of tumor. Therefore, here also, single chain Fv antibodies have been used with better tumor penetration and faster blood clearance enabling identification of small areas of increased uptake which would otherwise be invisible and therefore difficult to localize. The success rate of the technique depends highly on the availability of good and specific antibodies and of appropriate nuclides [16–18].

6 Conclusion

Tumor imaging with radiolabeled antibodies can be considered useful as only neoplastic cells are specifically targeted without being localized in non-cancerous cells. However, for complete anatomical profile of the patient with proven or suspected incidence of colonic adenocarcinoma, combination of both CT scan and radioimmunoscintigraphy is mandatory to avoid any biased-ness or false-negative results and finally for better therapeutic management.

References

1. Artiko V, Petrovic M, Jankovic Z, Jaukovic L, Sobic-Saranovic D, Grozdic I, Odalovic S, Pavlovic S, Jaksic E, Zuvela M, Ajdinovic B, Matic S, Obradovic V (2012) Scintigraphic detection of colon carcinomas with iodinated monoclonal antibodies. J BUON 17:695–699
2. Artiko V, Koljevic Marković A, Šobić-Šaranović D, Petrović M, Antić A, Stojković M, Žuvela M, Šaranović D, Stojković M, Radovanović N, Galun D, Milovanović A, Milovanović J, Bobić-Radovanović A, Krivokapic Z, Obradović V (2011) Monoclonal immunoscintigraphy for detection of metastasis and recurrence of colorectal cancer. World J Gastroenterol 17:2424–2430
3. Doerr RJ, Abdel-Nabi H, Krag D, Mitchell E (1991) Radiolabeled antibody imaging in the management of colorectal cancer results of a multicenter clinical study. Ann Surg 214:118–124
4. Rodriguez-Bigas MA, Bakshi S, Stomper P, Blumenson LE, Petrelli NJ (1992) ^{99m}Tc—IMMU-4 monoclonal antibody scan in colorectal cancer: a prospective study. Arch Surg 127:1321–1324
5. Patt YZ, Podoloff DA, Curley S, Kasi L, Smith R, Bhadkamkar V, Charnsangavej C (1994) Technetium 99m-labeled IMMU-4, a monoclonal antibody against carcinoembryonic antigen, for imaging of occult recurrent colorectal cancer in patients with rising serum carcinoembryonic antigen levels. J Clin Oncol 12:489–495
6. Durbin H, Young S, Stewart LM, Wrba F, Rowan AJ, Snary D, Bodmer WF (1994) An epitope on carcinoembryonic antigen defined by the clinically relevant antibody PR1A3. PNAS 91:4313–4317
7. Hosono MN, Hosono M, Zamara PO, Guhlke S, Haberberger T, Bender H, Russ Knapp FF, Biersack HJ (1998) Localisation of colorectal carcinoma by rhenium-188-labelled B 72.3 antibody in xenografted mice. Ann Nucl Med 12:83–88
8. Doerr RJ, Abdel-Nabi H, Merchant B (1990) Indium 111 ZCE-025 Immunoscintigraphy in occult recurrent colorectal cancer with elevated carcinoembryonic antigen level. Arch Surg 125:226–229
9. Lamki LM, Patt YZ, Rosenblum MG, Shanken LJ, Thompson LB, Schweighardt SA, Frincke JM, Murray JL (1990) Metastatic colorectal cancer: radioimmunoscintigraphy with a stabilized In-111-labeled F(ab′)2 fragment of an anti-CEA monoclonal antibody. Radiol 174:147–151
10. Patt YZ, Lamki LM, Shanken J, Jessup JM, Charnsangavej C, Levin B, Merchant C, Halverson C, Murray JL (1990) Imaging with indium111-labeled anticarcinoembryonic antigen monoclonal antibody ZCE-025 of recurrent colorectal or carcinoembryonic antigen-producing cancer in patients with rising serum carcinoembryonic antigen levels and occult metastases. J Clin Oncol 8:1246–1254
11. Abdel-Nabi HH, Schwartz AN, Higano CS, Wechter DG, Unger MW (1987) Colorectal carcinoma: detection with indium-111 anticarcinoembryonic-antigen monoclonal antibody ZCE-025. Radiol 164:617–621
12. Bares R, Fass J, Truong S, Buell U, Schumpelick V (1989) Radioimmunoscintigraphy with 111In labelled monoclonal antibody fragments (F(ab′)2 BW 431/31) against CEA: radiolabelling, antibody kinetics and distribution, findings in tumour and non-tumour patients. Nucl Med Commun 10:627–641
13. Oymada H, Uchida I, Nomura E, Yamada Y, Ohta H, Kasumi F, Yoshomoto M (1992) Clinical trials of radioimmunoimaging with ^{111}Indium-ZCE-025 and IMACIS-1. Radioisotopes 41:14–22
14. Saunders TH, Mendes Ribeiro HK, Gleeson FV (2002) New techniques for imaging colorectal cancer: the use of MRI, PET and radioimmunoscintigraphy for primary staging and follow-up. Brit Med Bull 64:81–99
15. Corbisiero RM, Yamauchi DM, Williams LE et al (1991) Comparison of immunoscintigraphy and computerized tomography in identifying colorectal cancer: individual lesion analysis. Cancer Res 51:5704–5711

16. Kim JC, Kim WS, Ryu JS et al (2000) Applicability of carcinoembryonic antigen-specific monoclonal antibodies to radioimmunoguided surgery for human colorectal carcinoma. Cancer Res 60:4825–4829
17. Mayer A, Tsiompanou E, O'Malley D et al (2000) Radioimmunoguided surgery in colorectal cancer using a genetically engineered anti-CEA single-chain Fv antibody. Clin Cancer Res 6:1711–1719
18. Arnold MW, Schneebaum S, Martin EW Jr (1999) Radioimmunoguided surgery in the treatment and evaluation of rectal cancer patients. Cancer Cont:JMCC 3:42–45

Enhanced Color Image Segmentation by Graph Cut Method in General and Medical Images

B. Basavaprasad and M. Ravi

Abstract Segmentation of color images is a tricky task. Regaining the segments in an image using the content image is a difficult and significant problem. Medical imaging is the most alive research topic from the past two decades. From the medical diagnosis of patients suffering from various diseases, abnormal regions in the organs can be easily identified, which is a greatest achievement. It is experimentally proved that graph based segmentation methods are better than the other segmentation techniques, especially when combined with statistical methods. We have proposed color image segmentation by adaptive graph cut method in this paper. It consists of two important stages. During the first phase we enhance the input color image using transformation technique as the image may contain noise, may be of low contrast and missing some color statistics. Then this enhanced color image is processed under graph cut technique to get better results, especially for the analysis of medical and general images. The proposed method contributes to medical imaging by means of image segmentation and also to other general image analysis. Our experimental results are found to very good in segmenting color images.

Keywords Medical · Image enhancement · Fast fourier transform · Image segmentation · Graph · Maximum flow · Minimum cut

1 Introduction

In computer visualization, image segmentation is the process of dividing an image into several segments which consists of sets of pixels. There set of pixels are also called as super pixels. The objective of image segmentation is to make image

B. Basavaprasad (✉) · M. Ravi
Research and Development Centre, Bharathiar University, Coimbatore 641046, India
e-mail: b.basavaprasad@gmail.com

M. Ravi
e-mail: ravimsk27@gmail.com

S. Gupta et al. (eds.), *Advancements of Medical Electronics*,
Lecture Notes in Bioengineering, DOI 10.1007/978-81-322-2256-9_8

simpler and modify the presentation into further meaningful to understand and easier to analyze [1–3]. The outcome of image segmentation is a collection of segments that together cover the whole image, or a collection of regions taken out from the image [4]. Every pixel in a segment is identical by means of to some futuristic or calculated property, such as intensity, color or texture. Neighboring regions also called as segments are considerably dissimilar by means of the same properties or characteristics.

Graphs theoretical techniques can effectively be used for image segmentation. Though many segmentation techniques exist but graph theoretical methods are proven more efficient and accurate [5]. Generally a pixel or a collection of pixels are vertices and edges define the dissimilarity among the neighborhood pixels. Graph cut is one among the segmentation algorithms. Mincut/Maxflow [6] is one of the techniques used to get cut on the graph. In this paper we have proposed enhanced graph cut method for image segmentation consists of two main phases. During the first phase the input color image is enhanced using the frequency domain methods. The resultant enhanced image is then segmented using the conventional graph cut technique. The proposed method can be used in various applications of image processing and pattern recognition systems such as medical imaging, in particular analysis of endoscopic images, tumor images, defected organs etc. It can be used in other applications such as face recognition, biometric, finger print recognition, detection of satellite images and much more. The paper is organized as follows. Proposed method is described in Sect. 2. Results and discussion are explained in Sect. 3. And finally experimental results and comparison with other segmentation methods are discussed in Sects. 3 and 4.

2 Proposed Method

A method which is a combination of two techniques is presented in this paper and hence contains two phases. During the first phase we are enhancing the input color image using transform technique [7]. During the second and final phase the enhanced image is processed under graph cut to get the quality segmented image [8].

2.1 Image Enhancement

The main goal of this technique is to get an enhanced image to protect the naturalness of an image. This method can be applied for all kinds of image; also a dark image can be brightening to visualize the dark image information. This method analyses the Bilog Transform, Fast Fourier Transform and NTSC space color enhancement [7] as shown in the following Fig. 1.

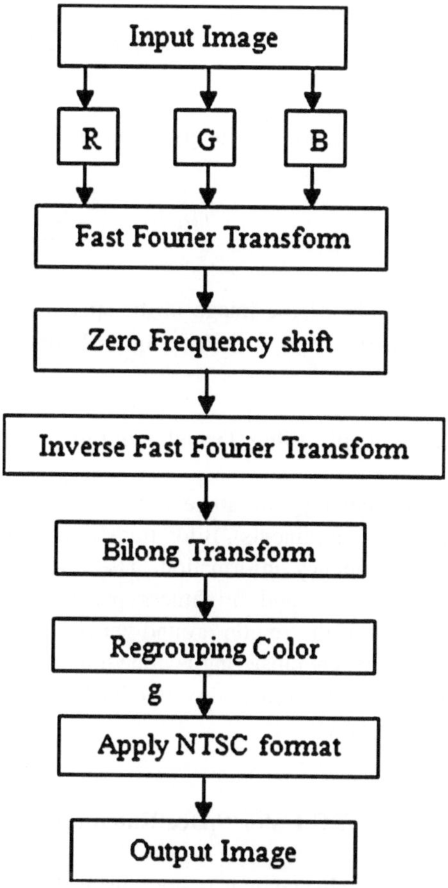

Fig. 1 Image enhancement using transform technique

2.1.1 Apply Fast Fourier Transform

The input image is a color image and this color image consists of three primary spectral components of red, green and blue i.e., RGB color panels. This RGB color Panels has to be segregated to grayscale images i.e., binary images. HSV. FFT filter is used to filter the pixels based on intensity. Then the low frequency components representing zeros are to be shifted to the center of array matrix in frequency domain. Inverse operation is performed to reconstruct the image in frequency domain to its spatial domain so as to picturize the image whose dimension is $N * N$. The equation for FFT is given by

$$F(k,l) = \frac{1}{N}\sum_{b=0}^{N-1} P(k,b)e^{-l2\pi\frac{lb}{N}} \tag{1}$$

$$P(k,b) = \frac{1}{N}\sum_{a=0}^{N-1} f(a,b)e^{-l2\pi\frac{ka}{N}} \tag{2}$$

The equation for inverse Fourier Transform is given by

$$f(a,b) = \frac{1}{N^2}\sum_{k=0}^{N-1}\sum_{l=0}^{N-1} F(k,l)e^{l2\pi\left(\frac{ka}{n}+\frac{lb}{N}\right)} \tag{3}$$

where $f(a,b)$ the image in the spatial domain is, $\frac{1}{N^2}$ is the normalization term in inverse FT, $F(k,l)$ is the point in Fourier space.

2.1.2 Apply Bilog Transformation

Still there may be some presence of negative frequency components (zero frequency components). Bilog transformation is made use of here to perform action on low frequency information. The region near zeros is to be highlighted for the enhancement and brightness preservation. Hence, after the application of this transform, the region around zeros is enhanced. This is followed by grouping of pixels, where clustering is done to increase the high resolution pixels. At this stage, the image pixels are converted back to RGB color model and pixels highlighted to a certain level.

2.1.3 NTSC Color Space Enhancement

Further enhancement is done using NTSC format. The NTSC stands for national television system committee. This color space is used in televisions in the United States. In this format the RGB color panels are converted to YIQ and converting back YIQ panels to RGB color model to process the gray scale and color information present in the image hence, the resulting image is an enhanced image.

The algorithm steps are as follows:

- Read the input image from the file and output the image. Transform RGB color panels to binary images i.e., to HSV images.
- Image matrix is initialized as KLAP and other variables. Then rewrite the values of matrix by comparing it with the original or actual image matrix.
- Use FFT filter to perform FAST FOURIER TRANSFORM, as FFT filter is used to filter the pixels based on intensity i.e., low resolution pixels and high resolution pixels.
- Use Envelope function to convert low resolution pixels into high resolution pixels. Perform Inverse FFT wherever needed to reconstruct the image.

- Perform BILOG TRANSFORM, as image does not contain same pixels some might be very large or small and regroup Pixels using envelope check function to increase high resolution pixels in image.
- Further enhance the image using NTSC format. Cost function performs the NTSC color space enhancement.
- Enhance the image by Performing $L. * R$ and Convert back HSV into RGB image to view image in RGB color model.

2.2 Image Segmentation Using Graph Cuts

Let $G = (V, E)$ be graph with V as, set containing all the vertices of a graph G and E is an edge set which contains all the edges of a graph G. A cut is a set of edges $C \subset E$ such that the two terminals become separated on the induced graph $G' = (V, E \backslash C)$ Denoting a source terminal as s and a sink terminal as t, a cut (S, T) of is a partition of V into S and $T = V \backslash S$, such that $t \in T$ and $s \in S$. A flow network is defined as a directed graph where an edge has a nonnegative capacity. A flow in G is a real-valued (often integer) function that satisfies the following three properties:

$$\text{Capacity Constraint: } \forall u, v \in V, f(u, v) \leq c(u, v) \tag{4}$$

$$\text{Skew Symmetry: } \forall u, v \in V, f(u, v) = -f(u, v) \tag{5}$$

$$\text{Flow Conservation: } u \in \left(V\{s, t\}, \sum_{v \in V} f(u, v) \right) = 0 \tag{6}$$

- If f is a flow, then the net flow across the cut (S, T) is defined to be $f(S, T)$, which is the sum of all edge capacities from S to T subtracted by the sum of all edge capacities from T to S.
- The capacity of the cut (S, T) is $C(S, T)$, which is the sum of the capacities of all edge from S to T.
- A minimum cut is a cut whose capacity is the minimum over all cuts of G.

2.2.1 Finding the Min-Cut

After the max-flow is found, the minimum cut is determined by S $= \{\text{All vertices } \textit{reachable from s}\}$.

Finding the cut with minimal cost is solvable in polynomial time as shown in the Fig. 2. A directed graph shown below in Fig. 3a with positive edge weights and two special vertices: A source s with only outgoing edges and a sink T with only

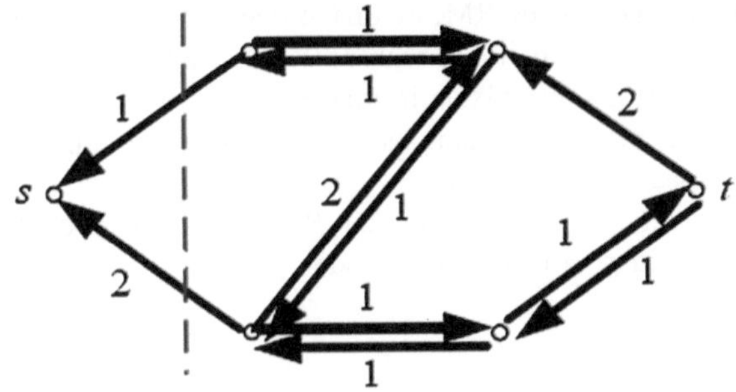

Fig. 2 Finding the min-cut of a graph

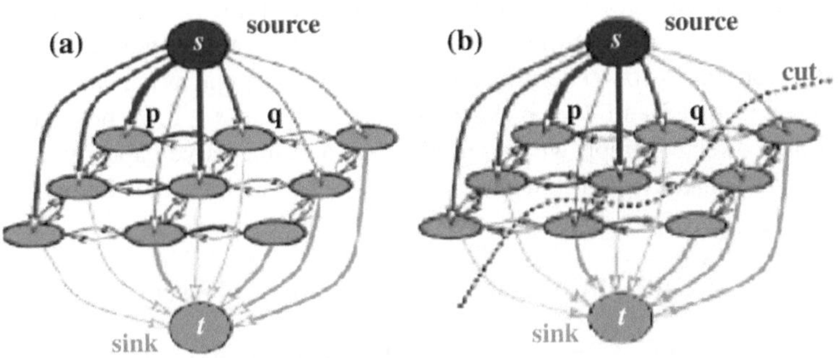

Fig. 3 A graph G and its corresponding cut

incoming edges. On this graph a cut (shown Fig. 3b) is a binary partition of the vertices into a set *S* around the source and a set *T* around the sink. The cost of the cut is the sum of the weights of all the edges inducing flow from source to sink. Cut edges that induce flow in the opposite direction. Binary labeling is equivalent to partitioning, so construct a directed graph. All edges in the graph are assigned some weight or cost. A cost of a directed edge (p, q) may differ from the cost of the reverse edge (q, p). In fact, ability to assign different edge weights for (p, q) and (q, p) is important for many graph-based applications in vision. Normally, there are two types of edges in the graph: *N-links* and *T-links*. *N-links* connect pairs of neighboring pixels or voxels. Thus, they represent a neighborhood system in the image. Cost of *N-links* corresponds to a penalty for discontinuity between the pixels. *T-links* connect pixels with terminals (labels). The cost of a *T-link* connecting a pixel and a terminal corresponds to a penalty for assigning the corresponding label to the pixel.

2.2.2 Maxflow Algorithm

The max-flow algorithm presented here belongs to the group of algorithms. Figure 4 is an example of the search trees S (red nodes) and T (blue nodes) at the end of the growth stage when a path (yellow line) from the source s to the sink t is found. Active and passive nodes are labeled by letters A and P, correspondingly. Free nodes appear in black on augmenting paths. It builds search trees for detecting augmenting paths. Two trees search is constructed; first tree is called source and the second is called as the sink. The other difference is that reuse of these trees is done and never start building them from scratch. This method has one drawback i.e., paths of the augmentation created are not essentially need to be shortest; thus the time complexity of the shortest augmenting path is no longer valid. The trivial upper bound on the number of augmentations for our algorithm is the cost of the minimum cut |C|, which results in the worst case complexity $O(mn2|C|)$.

$$S \subset V, s \in S, T \subset V, t \in T, S \cap T = \emptyset \quad (7)$$

Figure 4 illustrates the basic terminology. Two non-overlapping search trees '*S*' and '*T*' with roots at the source *s* and the sink *t*, correspondingly. The nodes which do not exist in *S* or *T* are called as free nodes. These nodes can be either active or passive. The nodes which are active stand for the external border in every tree at the same time the nodes which are passive represents internal. One thing can be notices that the nodes which are active permits the trees to grow by attaining children which are new beside non-saturated edges from a group of nodes which are free. The nodes which are passive are totally blocked by remaining nodes from the tree and hence cannot grow. It is also vital that active nodes are able to come in touch by way of the nodes from the outside tree.

There are three important stages of this algorithm which iteratively repeats.

- Growth: Look for trees *S* and *T* grow until they touch giving a path $s \rightarrow t$.
- Augmentation: In this phase the path is augmented and the tree searches will split into forests.
- Adoption: Here the both the *S* and *T* trees are regained.

During the stage of growth the trees get bigger. The active nodes travel around the neighboring edges which are non-saturated and obtain fresh children from a group of nodes which are free. The recently acquired nodes would be the active members of the consequent trees. Every adjacent active node which would travel around these

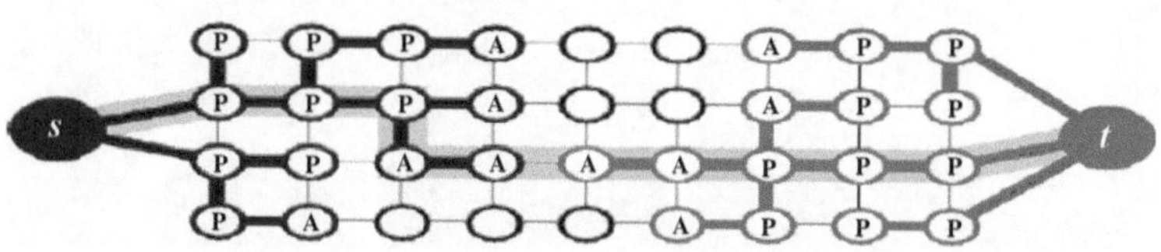

Fig. 4 Path from source to sink

nodes will be converted into passive nodes. Suppose if the nodes which are active come across a adjacent node belonging to the opposite tree, then growth stage ends. Therefore a path is found from source to sink (Fig. 4). Because the push is gutter the largest flow possible, as a result a few of the edges during the path get soaked.

As soon as the stage of the adoption is finishes the algorithm proceeds back to the stage of growth. The algorithm ends when there is no further growth of the search trees S and T. As a result the trees are separated from each other by edges which are saturated. This indicates that a flow is reached to its maximum.

3 Experimental Results

The experimental results using the proposed method are shown in Fig. 5. First column images represent the original input images. The second column contains the enhanced images using the frequency domain method. The third and final column

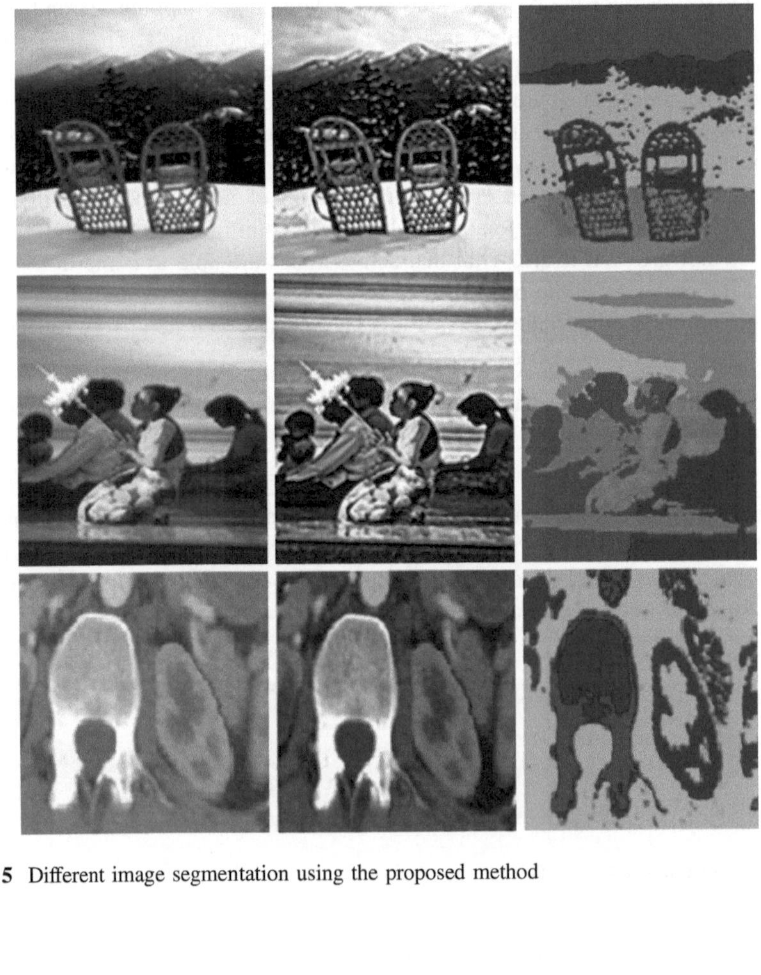

Fig. 5 Different image segmentation using the proposed method

Table 1 Comparison of execution time in seconds of different algorithms with the proposed method

Sl. No.	Method	Performance in second(s)
1	Graph cut	0.6
2	Contour	3
3	Region growing	11
4	K-means	20
5	Pattern Matching	120

represents the segmented color images using graph cuts. We have taken both general and medical images for our experimentation. Over 200 images are tested using the proposed method. The results are very encouraging and useful in both medical and general image analysis such as tumor image analysis, MRI scanned images, endoscopic images, general images such as natural images, building images, flowers and so on and so forth.

There exist many image segmentation algorithms. Among them graph based image segmentation techniques are proved as the efficient and accurate once. Table 1 shows different segmentation approaches compared with graph cut method. We observed that methods such as region growing, k-means clustering, contour are very slow in their segmentation tasks. Whereas graph cut methods are very fast in segmenting the images. We have improved the quality of the segmentation by enhancing the original input images using frequency domain methods. This will improve the image analysis with respect to human perception is get bettered on output images.

4 Conclusion

A color image segmentation using adaptive graph cut method for medical and general has been presented in this paper. The input color image is enhanced using transform method and then processed under graph cut. As all we know that graph based image segmentation techniques yield better results especially for medical images to detect the abnormality caused by diseases in human organs. This technique enhances the detection of abnormal region of the organ caused by various diseases with respect to human perception. Here we tried to improve the graph cut method by giving the enhanced image as input for further improved results. The experimental results we got are tested over 200 images which consist of medical and general images and are found to be very encouraging. The powerful min-cut or max-flow algorithm from combinatorial optimization can be used to minimize certain important energy functions in computer vision. The proposed method can be used in both medical and general image analysis.

References

1. Basavaprasad B, Hegadi RS (2012) Graph theoretical approaches for image segmentation. Aviskar–Solapur Univ Res J 2:7–13
2. Shi J, Malik J (1998) Normalized cuts and image segmentation. In: IEEE Trans PAMI, vol 22, no 8, Aug 2000S
3. Metev M, Veiko VP, Osgood RM Jr (eds) (1998) Laser assisted microtechnology, 2nd ed. Springer, Berlin, Germany
4. Bhambri P, Kaur A (2013) Novel technique for robust image segmentation: new technique of segmentation in digital image processing. Lambert Academic Publishing, Germany
5. Basavaprasad B, Hegadi RS (2014) A survey on traditional and graph theoretical techniques for image segmentation. IJCA Proc Natl Conf Recent Adv Inf Technnol (NCRAIT) (1):38–46
6. Boykov Y, Kolmogorov V (2004) An experimental comparison of min-cut/max-flow algorithms for energy mimization in vision. IEEE Trans Pattern Anal Mach Intell 26(9):1124–1137
7. Umamageswari D, Sivasakthi L, Vani R (2014) Quality analysis and image enhancement using transformation techniques. In: International conference on signal processing, embedded system and communication technologies and their applications for sustainable and renewable energy (ICSECSRE '14), vol 3
8. Boykov Y, Veksler O, Zabih R (2001) Fast approximate energy minimization via graph cuts. IEEE Trans Pattern Anal Mach Intell 23(11):1222–1239

A New Approach for Color Distorted Region Removal in Diabetic Retinopathy Detection

Nilarun Mukherjee and Himadri Sekhar Dutta

Abstract Automatic detection of Diabetic Retinopathy (DR) abnormalities in fundus retinal images can assist in early diagnosis and timely treatment of DR, to avoid further deterioration of vision. Many Fundus Retinal images contain color distorted regions originated due to noise, extremely uneven or poor illumination and improper exposure of fundus camera. These regions are required to be removed to avoid poor results for feature extraction and erroneous DR abnormality detections, as they introduce high amount of false positive detections. In this paper, we have proposed a totally automatic method for segmentation and removal of the color distorted regions in retinal fundus images, using modified Valley Emphasized Automatic thresholding method and morphological operations. The proposed algorithm accurately defines the well illuminated color undistorted retinal region inside the input fundus image, from which both the normal and disease features can be successfully detected. The proposed method has yield an average accuracy of more than 95 % when tested over around 700 fundus images from diaretdb0, diaretdb1, STARE, HRFDB and DRIVE databases.

Keywords Diabetic retinopathy · Retinal fundus image · Automatic thresholding · Color distorted region segmentation

1 Introduction

Diabetic Retinopathy (DR) is one of the most harmful effects of Diabetes, leading to blindness. Diabetic retinopathy (DR) can be defined as the damage of the microvascular system in the retina, due to prolonged Hyperglycemia [1]. Blockages or

N. Mukherjee (✉)
Bengal Institute of Technology, Kolkata, West Bengal, India
e-mail: nilarun.mukherjee@gmail.com

H.S. Dutta
IEEE, Kalyani Government Engineering College, Kalyani, West Bengal, India
e-mail: himadri.dutta@gmail.com

S. Gupta et al. (eds.), *Advancements of Medical Electronics*,
Lecture Notes in Bioengineering, DOI 10.1007/978-81-322-2256-9_9

clots are formed, as blood containing a high level of glucose flows through the small blood vessels in the retina. This in effect raptures the wall of those weak vessels due to high pressure. The leakage of blood on surface of retina leads to blurred vision and can cause complete blindness, known as Diabetic Retinopathy [1]. Study showed that the major systemic risk factors for onset and progression of DR are duration of diabetes, degree of glycemic control and hyper-lipidemia. DR is a vascular disorder affecting the microvasculature of the retina. It is estimated that diabetes mellitus affects 4 % of the world's population, almost half of whom have some degree of DR at any given time [2]. DR occurs both in type 1 and type 2 diabetes mellitus. Earlier epidemiological studies has shown that nearly 100 % of type 1 and 75 % of type 2 diabetes develop DR, after 15 year duration of diabetes [3, 4]. In India, with epidemical increase in type II diabetes mellitus, diabetic retinopathy is fast becoming an important cause of visual disability, as reported by the World Health Organization (WHO) [5]. However, with early diagnosis and timely treatment Diabetic Retinopathy can be well treated.

The regular screening of diabetic retinopathy produces a large number of retinal images, to be examined by the ophthalmologists. The cost and time of manual examination of large number of retinal images becomes quite high. An automated screening system for retinal images has become esteem need for early and in time detection of DR [1]. The automated screening system should be able to differentiate between normal and abnormal retinal images. It should also be able to detect different effects of Diabetic Retinopathy such as Exudates, Micro Aneurysms and Hemorrhages in the retinal images with adequate accuracy. Such a system will greatly facilitate and accelerate the DR detection process and will reduce the workload for the ophthalmologists. Healthy Retina contains Blood Vessels, Optic Disc, Macula and Fovea as main components, as depicted in Fig. 1. An automated system for screening and diagnosis of DR should be able to identify and eliminate all these normal features [6–8] prior to automatically detecting all signs of Diabetic Retinopathy such as Micro-Aneurysms [9–11] and Edema and Hemorrhages [9]

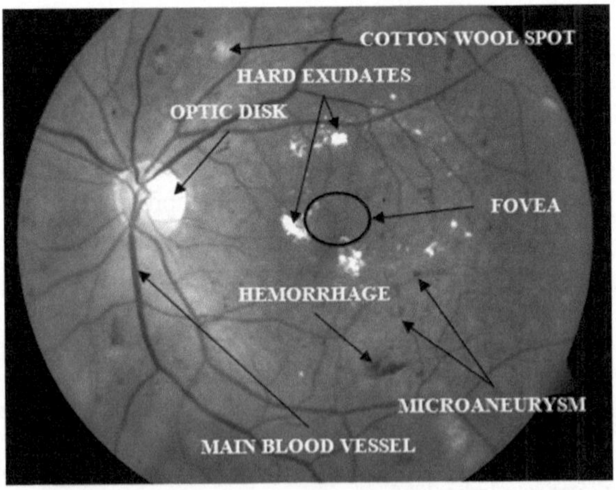

Fig. 1 Illustration of various features on a typical image of retina

and Exudates and Cotton-wool Spots [12, 13]. Accurate detection of DR depends highly on the quality of the retinal fundus images, which practically varies widely due to noise and uneven illumination.

Detection of these abnormalities requires perfect separation of the regions contained in those retinal images, which are outside the retina and belong to the background. In many cases, due to noise, uneven and poor illumination and degradation of illumination away from the center and improper exposure of the fundus camera, certain regions inside the retina become totally unrecoverable or unusable due to color distortion. These color distorted retinal regions are required to be removed during the preprocessing steps, to extract and define the actual Region of Interest (ROI) before applying any DR detection algorithm. Otherwise it may lead to poor results for feature extraction and erroneous abnormality detections. These preprocessing steps are applied to enhance the quality of the retinal images to make them suitable for reliable detection of DR abnormalities by applying any DR detection algorithms.

In this paper, we have proposed a new fully automated algorithm for color distorted unrecoverable retinal region segmentation for retinal fundus images. The paper has been organized as follows: In Sect. 2 we have discussed all the related works. Then in Sect. 3, the proposed method has been depicted along with actual output images corresponding to each processing step, from its software implementation. In Sect. 4, we have provided both subjective and analytical accuracy and performance analysis of the proposed method with supporting experimental results and sensitivity-specificity analysis.

2 Related Work

Design and development of automated and accurate Diabetic Retinopathy detection systems, has gained significant research interests in recent times. Many contributions has been made till date, on preprocessing and background segmentation mask creation for retinal images. Which are necessary for accurate detection of DR. The aim of preprocessing is to increase the quality of an image by reducing the amount of noise appearing in the image and highlighting features that are used in image segmentation. Chaudhuri et al. has used a fixed global threshold on the I channel of the HSI format of the retinal images [8]. Two typical techniques used in preprocessing are filtering and contrast enhancing. Lee et al., Goldbaum et al. and Osareh et al. have applied standard contrast stretching techniques for segmentation and noise reduction [9, 14, 15]. Usher et al., Sinthanayothin et al. and Firdausy et al. have used local contrast enhancement method for equalizing uneven illumination in the intensity channel of retinal images [16–18]. A large mean filter and a large median filter are collectively used on the intensity channel values to detect the dark regions from retinal images. Thresholding is also an important and widely used technique in image segmentation [19], because thresholding is effective and simple to implement. In thresholding, pixels having gray level intensity values within a defined intensity range are selected as belonging to the foreground objects, whereas pixels having gray levels outside that range are rejected to be the background [19]. Hoover et al. and Goldbaum et al. have used thresholding on the Mahalanabis Distance over a neighborhood for each pixel, for background estimation [20, 21]. Jamal et al. has used thresholding on the Standard Deviation over a neighborhood for each pixel for background estimation and removes noise using HSI color space [22]. The threshold is chosen in an empirical basis. Kuivalainen et al. has thresholded the Intensity (I) channel in the HSI converted retinal fundus images using a sufficiently low I value, to form the background segmentation mask [23]. The I channel threshold is experimentally or empirically chosen. From the training image set, Kuivalainen et al. has found that the regions of distorted color due to inadequate illumination, have high hue values (H) and relatively low intensity values (I) in the HSI color system. Thus, regions having distorted color was found by, first dividing the hue channel by the intensity channel and then thresholding the result with a preset threshold, to form the distorted region segmentation mask [23]. The preset threshold is also empirically chosen.

Although, there have been a lot of work in background and color distorted region segmentation in retinal fundus images, but all of them have used some empirically selected threshold values to create the masks, which requires manual intervention. This is turn, restricts the entire process of DR abnormality detection from becoming totally automatic. In this paper, we have proposed a totally automated and dynamic threshold selection method to create the color distorted region segmentation mask for any given retinal fundus image.

3 Proposed Method

In this paper, an intuitive and fully automatic technique for detection and removal of the color distorted unrecoverable retinal regions has being proposed, which takes advantage of the bimodal nature of the red channel histograms of the input retinal images. It has been found through rigorous testing on the retinal images from STARE [20, 21], DRIVE [24], diaretdb0 [25], diaretdb1 [26] and HRFDB [27] databases, that the red channel histograms of most of the retinal fundus images exhibits clear bimodal nature, having a clear separation i.e. valley region between the background and object regions. Moreover, it is established that the red channel does not contain much information regarding the retinal features and abnormalities. It only contains the illumination difference information of the retinal region and the background, whereas green channel or the intensity (I) channel for HSI converted retinal images, contains most of the retinal component information. This opens the opportunity to apply a modified version of Valley Emphasis Method [28] for Automatic Threshold Selection, to segment the color distorted object regions. Therefore, the red channel threshold determined by the modified Valley Emphasized Method [28] is used to tune the I channel threshold, determined by the same method, to determine the optimized threshold level in the red channel for color distorted region segmentation.

In Valley Emphasis Method [28] for Automatic Threshold Selection, an image is represented by a 2D gray level intensity function f(x, y). The value of f(x, y) is the gray level intensity value, ranging from 0 to L − 1, where L is the number of distinct gray levels. Let the number of pixels with gray-level i be n_i and n be the total number of pixels in a given image, the probability of occurrence of gray-level i is defined as:

$$p_i = \frac{n_i}{n} \tag{1}$$

The average gray-level of the entire image is computed as:

$$\mu_T = \sum_{i=0}^{L-1} ip_i \tag{2}$$

In the case of single thresholding, the pixels of an image are divided into two classes $C_1 = \{0, 1,\ldots, t\}$ and $C_2 = \{t + 1, t + 2,\ldots, L - 1\}$, where t is the threshold value. C_1 and C_2 normally correspond to the foreground (objects of interest) and the background. Probabilities of the two classes are:

$$\lambda_1 = \sum_{i=0}^{t} p_i \text{ and } \lambda_2 = \sum_{i=t+1}^{L-1} p_i \tag{3}$$

The mean gray level values of the two classes are computed as:

$$\mu_1(t) = \sum_{i=0}^{t} \frac{ip_i}{\lambda_1(t)} \text{ and } \mu_2(t) = \sum_{i=t+1}^{L-1} \frac{ip_i}{\lambda_2(t)} \tag{4}$$

In the proposed algorithm, a modified version of the Valley Emphasis Method [28] is used on the Red and Intensity Channel of the fundus retinal images. The average intensity level of the original red and intensity channel image, have been incorporated in the formulation for the Valley Emphasis Method [28]. The modified Valley Emphasis threshold selection formula used in the proposed method is given below. The optimal threshold τ can be determined by maximizing the between-class variance; that is:

$$\begin{aligned}\tau = \operatorname*{ArgMax}_{0\le t<L}\Big\{&(1 - p_t)\Big(\lambda_1(t)(\mu_1(t) - \mu_T)^2\Big)\\ &+ \lambda_2(t)(\mu_2(t) - \mu_T)^2\Big)\Big\}\end{aligned} \tag{5}$$

The key to this formulation is the application of a weight, $(1 - p_t)$. The smaller the p_t value (low probability of occurrence), the larger the weight will be. This weight ensures that the result threshold will always be a value that resides at the valley or bottom rim of the gray-level distribution. The objective of automatic thresholding is to find the valley in the histogram that separates the foreground from the background. For single thresholding, such threshold value exists at the valley of the two peaks (bimodal), or at the bottom rim of a single peak (unimodal). The modified Valley Emphasis Method exploits this observation to select a threshold value that has a small probability of occurrence and also maximize the between group variance.

Morphological operations such as Erosion, Dilation, Opening and Closing are used to remove the boundary by inclusion of it inside the masked background region and to minimize and remove small and medium sized white islands and black holes both from the object and background regions. Morphological Erosion, Dilation, Opening and Closing operations [19] in binary image is defined as:

Morphological Erosion: $A \ominus B = \{x \mid (B)_x \subseteq A\}$
Morphological Dilation: $A \oplus B = \{x \mid (\widehat{B})_x \cap A \not\subseteq \emptyset\}$
Morphological Opening: $A \bigcirc B = (A \ominus B) \oplus B$
Morphological Closing: $A \bullet B = (A \oplus B) \ominus B$

where A is the input image and B is the structuring element. Then connected component extraction is used on the resultant image to retain the largest connected object, removing any white patches left inside the background. Then a second connected component extraction is used on the negative of the resultant image to retain the largest connected object, removing any black holes left inside the object. Final color distorted unrecoverable retinal region segmentation mask is obtained by smoothing the boundary of the negative of the resultant image.

The Proposed Method:

Step 1: The Red channel from the original RGB fundus image and the Intensity (I) channel from the HSI [19] converted original fundus image are extracted.

Step 2: Noise Removal: A low pass filtering is done in spatial domain on the resultant Red and I channel images using 7 × 7 Median filter [19], to remove any sand paper noise present in the image.

Step 3: Threshold Selection: The modified version of the Valley Emphasis Method depicted in Eq. 5 is applied on both of the resultant images to get the Red and I channel threshold intensity levels. The final threshold is obtained by averaging the Red and I channel threshold levels, which is used to threshold the Red channel image. Figure 2 shows, the red channel (Black Bar) and I channel (Magenta Bar) thresholds determined by the modified Valley Emphasized method, drawn over the Red channel histogram of a sample fundus image from the image databases. The resultant binary masks contain the well illuminated retinal boundary inside the object region. Figure 3 shows, sample original color fundus images from diarectdb0 database in Column (a) and Corresponding Thresholded Red Channel Images in Column (b).

Step 4: It has been observed through thorough examination that, the maximum expansion of the retinal border approximately ranges to 15–17 pixels. The retinal

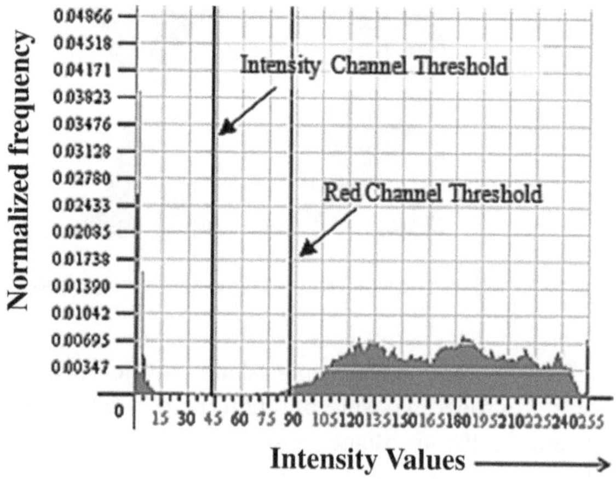

Fig. 2 The red channel (*Black bar*) and I channel (*Magenta bar*) thresholds determined by the modified valley emphasized method drawn over the red channel histogram of a sample fundus image

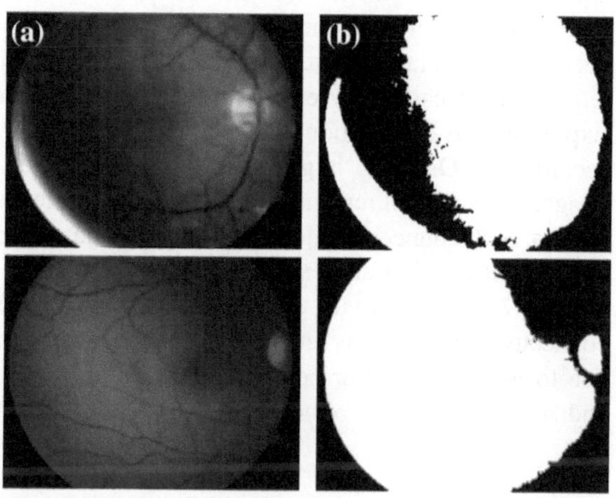

Fig. 3 Threshold selections for fundus images with bimodal distribution in red channel: column **a** original color fundus images from database; column **b** thresholded images

boundary inside the object region is eliminated and included inside the masked background region using morphological Erosion operation [19] with a disc shaped structuring element of size 17 × 17. This also helps to isolate or separate the object regions corresponding to the high intensity color distorted regions inside the retina, from the object regions corresponding to the color undistorted well-illuminated regions inside the retina, in certain thresholded output images from step 3.

Step 5: The resultant mask output may also contain white islands in the background regions originated due to noise, improper exposure of fundus camera and uneven illumination and black holes in the object region originated due to noise, uneven illumination and dark abnormalities such as lesions and hemorrhages inside retina. Small and medium sized white islands and black holes are removed using morphological Opening and Closing operations [19], with a disc shaped structuring element of size 15 × 15, as shown in Fig. 4.

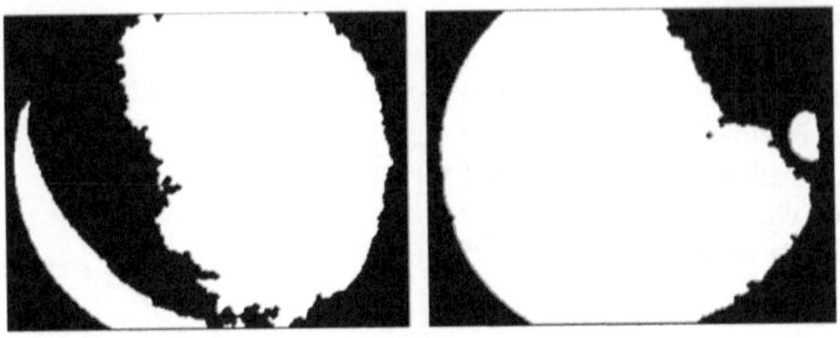

Fig. 4 Result of morphological opening—closing operation

Step 6: Large sized white islands in the background regions originated due to noise, improper exposure of fundus camera and uneven illumination, which are still present in the resultant mask image are removed using connected component labeling and extraction algorithm [19].

It is evident, that the largest connected component in the thresholded, opened and closed mask image, will be the actual ROI i.e. the most prominent, well illuminated, color undistorted useful region of the retina. All other smaller connected components present, will either be noise or disjoint color distorted part of the retinal region caused by extreme uneven illumination or/and abnormal exposure. Exploiting this observation, all the connected components are extracted and labeled accordingly. Only the largest connected component is preserved in the resultant image, removing all remaining white islands in the background region, as shown in Fig. 5a, b. 8–connectivity [19] among pixels have been considered in the connected component labeling and extraction algorithm.

Step 7: After removing the white islands in the background region, a single large connected component is retained. It may contain large sized black holes, originated due to noise, uneven or poor illumination and dark abnormalities such as lesions and hemorrhages. To remove them, the resultant image from step 6 is negated.

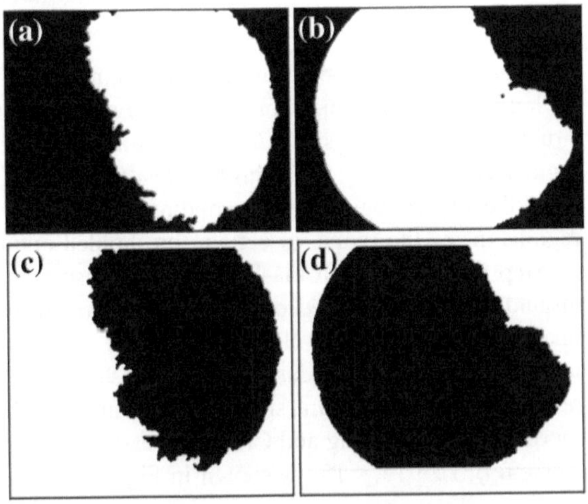

Fig. 5 Result of white island removal is depicted in (**a** and **b**) and result of black hole removal is depicted in (**c** and **d**)

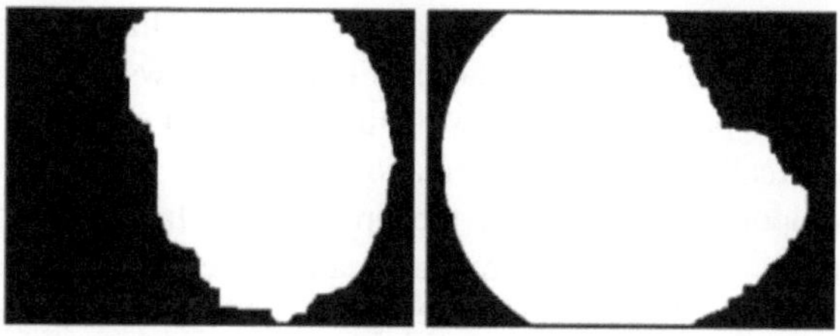

Fig. 6 Final color distorted unrecoverable retinal region segmentation masks

In the negative image, original background becomes the largest spotless connected component and the original object region becomes the background. The large sized black holes inside the original object region in the previous image become smaller connected components. Exploiting this observation, all the connected components in the negative image are extracted and labeled accordingly, using connected component labeling and extraction algorithm [19]. Only the largest connected component is preserved in the resultant image, thus removing all the remaining large sized black holes in the original object region, as shown in Fig. 5c, d.

Step 8: In the resultant mask image, every black pixel having white pixels either at both left and right in the same row or at both top and bottom in the same column is replaced by a white pixel, to get the final boundary smoothed background segmentation mask, as shown in Fig. 6. These final segmentation masks are intersected with the original images to get the color distorted region segmented images.

4 Experimental Results

In medical image processing, it is crucial to verify the validity and evaluate any newly proposed algorithm contributing to the automated diagnosis of a disease. We have used five standard retinal image databases i.e. STARE [20, 21], DRIVE [24], Diaretdb0 [25], Diaretdb1 [26] and HRFDB [27] as depicted in Table 1, to extensively verify and validate the proposed method for color distorted unrecoverable retinal region segmentation. These databases contain both normal and DR affected retinal images with different qualities in terms of noise and illumination.

Table 1 Retinal fundus image databases

Retinal image database	No. of images	Resolution
STARE [20, 21]	402	700 × 605
DRIVE [24]	40	565 × 585
DIARETDB0 [25]	130	1,500 × 1,152
DIARETDB1 [26]	89	1,500 × 1,152
HRFDB [27]	44	3,504 × 2,336

Figure 7, depicts the subjective validity of proposed method, it well represent the diverse characteristics of the test images. Figure 7 shows the manually created segmentation masks provided along with the Diaretdb0 database and the corresponding automatically created segmentation masks by the proposed algorithm. These results support the validity of the proposed technique and show that the proposed automatic segmentation mask creation technique gives considerably acceptable results for both well illuminated good quality fundus images and poorly illuminated, noisy and color distorted fundus images. The quantitative accuracy analysis of the proposed background and color distorted unrecoverable retinal region segmentation algorithm is performed on the images from DRIVE, Diaretdb0 and HRFDB databases, for which manually labeled masks are available. The manually labeled segmentation masks provided along with these image databases, serve as ground truths and are used to calculate the accuracy of the proposed algorithm.

For each of the fundus image in the set, the following metrics are calculated by pixel by pixel comparison between the mask created by the proposed method and the corresponding manually created mask provided with the respective image databases:

$$\text{Sensitivity} = \text{TPR (True Positive Rate)} = \frac{TP}{(TP+FN)}$$

$$\text{Specificity} = \text{TNR (True Negative Rate)} = \frac{TN}{(TN+FP)}$$

$$\text{FPR (False Positive Rate)} = 1 - \text{Specificity} = \frac{FP}{(FP+TN)}$$

$$\text{Accuracy} = \text{Sensitivity} \times \frac{P}{T} + \text{Specificity} \times \frac{N}{T} = \frac{TP+TN}{T}$$

Here, TP = The total number of those Pixels, which are detected as Object Pixels, by the proposed automatic mask creation method, which are also Object

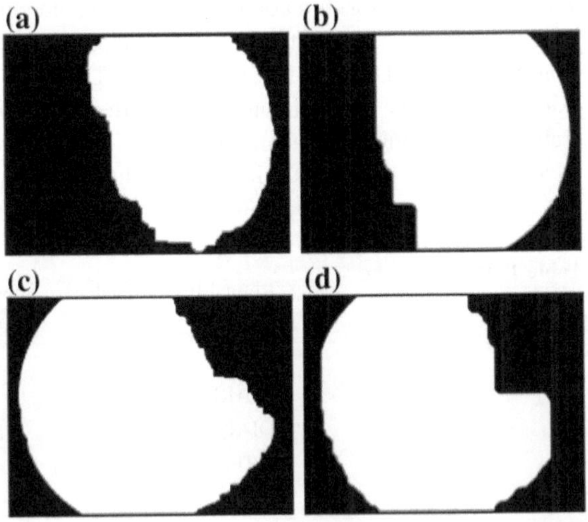

Fig. 7 Subjective validity of the proposed algorithm: **a** and **c** depict the final color distorted region segmentation masks created by the proposed algorithm and **b** and **d** depict the manually created masks

Table 2 Accuracy of the proposed automatic color distorted region segmentation method

Image database	No. of images	Accuracy
HRFDB [27]	44	99.93
DRIVE [24]	40	99.49
DIARETDB0 [25]	130	94.98

Pixels in the manually created masks. TN = The total number of those Pixels, which are detected as Background Pixels, by the proposed automatic mask creation method, which are also Background Pixels in the manually created masks. FN = The total number of those Pixels, which are detected as Background Pixels, by the proposed automatic mask creation method, but which are Object Pixels in the manually created masks. FP = The total number of those Pixels, which are detected as Object Pixels, by the proposed automatic mask creation method, but which are Background Pixels in the manually created masks. T = Total number of Pixels in the Image. P = Total number of Object Pixels in the manually created mask. N = Total number of Background Pixels in the manually created mask.

The Sensitivity, Specificity and thereafter Accuracy is calculated for each individual automatically created mask by the proposed method. The overall accuracy of the proposed method for the three image databases i.e. DRIVE, Diaretdb0 and HRFDB databases are calculated separately, by taking the average accuracy of all the images belonging to a particular database, which is depicted in Table 2.

It is evident from Table 2, that our proposed algorithm has worked quite efficiently and correctly for the fundus images in all the three databases. The proposed technique has failed for only one image in Diarectdb0 database and for some Diarectdb0 fundus images it has resulted larger object area than the corresponding manual masks. Although, it has been found that the automatic masks created by the proposed algorithm have successfully rejected the really color distorted or poorly illuminated portions inside the retinal regions for those particular Diarectdb0 fundus images. However, the automated masks created by the proposed technique, have considered some more regions inside the retina as object region than their corresponding manual masks. These regions are found to contain certain important DR abnormality information with adequate illumination, which have been unnecessarily rejected in the corresponding manual masks. A Java based standalone application has been built to implement the proposed Color Distorted Unrecoverable Retinal Region segmentation technique. All the outputs and histograms shown in this paper are captured from the running instance of the application.

5 Conclusion

In this paper, a fully automated technique for segmentation of the color distorted regions in retinal fundus images have been proposed. These color distorted retinal regions originate due to noise and extremely uneven and poor illumination and improper exposure of the fundus camera. These regions are required to be removed to avoid poor results for feature extraction and erroneous DR abnormality detections, as they introduce high amount of false positive detections. Although, there have been a lot of contributions in background and color distorted region segmentation in retinal fundus images, but all of them have used some empirically selected threshold values to create the masks, which requires manual intervention. This is turn, restricted the entire process of DR abnormality detection from becoming totally automatic. In this paper, we have overcame that limitation by developing a totally automated and dynamic threshold selection method to create the color distorted region segmentation mask for any given retinal fundus image. The proposed algorithm accurately defines the well illuminated and color undistorted retinal region inside the input fundus image, from which both the normal and disease features can be successfully detected. The proposed method yields satisfactory results to segment color distorted retinal regions, when tested over around 700 images from diaretdb0, diaretdb1, STARE, HRFDB and DRIVE retinal image databases, with an average accuracy of more than 95 %. Our technique may further be combined with some automated learning methods to achieve better results, for the few fundus images, for which the proposed algorithm has failed to produce accurate segmentation masks.

References

1. Sussman EJ, Tsiaras WG, Soper KA (1982) Diagnosis of diabetic eye disease. JAMA Ophthalmol 247(23):3231–3234
2. Rema M, Pradeepa R (2007) Diabetic retinopathy: an Indian perspective. Indian J Med Res 125:297–310
3. Klein R, Klein BE, Moss SE, Davis MD, DeMets DL (1984) The Wisconsin epidemiologic study of diabetic retinopathy II. Prevalence and risk of diabetic retinopathy when age at diagnosis is less than 30 years. Arch Ophthalmol 102:520–526
4. Klein R, Klein BE, Moss SE, Davis MD, DeMets DL (1984) The Wisconsin epidemiologic study of diabetic retinopathy III. Prevalence and risk of diabetic retinopathy when age at diagnosis is 30 or more years. Arch Ophthalmol 102:527–532
5. Wild S, Roglic G, Green A, Sicree R, King H (2004) Global prevalence of diabetes, estimates for the year 2000 and projections for 2030. Diab Care 27:1047–1053
6. Sinthanayothin C, Boyce JF, Cook HL, Williamson TH (1999) Automated localization of the optic disc, fovea and retinal blood vessels from digital color fundus images. Br J Ophthalmol 83(8):231–238
7. Foracchia M, Grisan E, Ruggeri A (2004) Detection of optic disc in retinal images by means of a geometrical model of vessel structure. IEEE Trans Med Imaging 23(10):1189–1195
8. Chaudhuri S, Chatterjee S, Katz N, Nelson M, Goldbaum M (1989) Detection of blood vessels in retinal images using two dimensional matched filters. IEEE Trans Med Imaging 8(3):263–269
9. Lee SC, Lee ET, Kingsley RM, Wang Y, Russell D, Klein R, Warner A (2001) Comparison of diagnosis of early retinal lesions of diabetic retinopathy between a computer system and human experts. Graefe's Arch Clin Exp Ophthalmol 119(4):509–515
10. Spencer T, Phillips RP, Sharp PF, Forrester JV (1991) Automated detection and quantification of micro-aneurysms in fluoresce in angiograms. Graefe's Arch Clin Exp Ophthalmol 230 (1):36–41

11. Frame AJ, Undill PE, Cree MJ, Olson JA, McHardy KC, Sharp PF, Forrester JF (1998) A comparison of computer based classification methods applied to the detection of micro aneurysms in ophthalmic fluoresce in angiograms. Comput Biol Med 28(3):225–238
12. Osareh A, Mirmehdi M, Thomas B, Markham R (2001) Automatic recognition of exudative maculopathy using fuzzy c-means clustering and neural networks. In: Proceedings of conference on medical image understanding analysis, pp 49–52
13. Phillips R, Forrester J, Sharp P (1993) Automated detection and quantification of retinal exudates. Graefe's Arch Clin Exp Ophthalmol 231(2):90–94
14. Goldbaum MH, Katz NP, Chaudhuri S, Nelson M, Kube P (1990) Digital image processing for ocular fundus images. Ophthalmol Clin N Am 3(3):447–466
15. Osareh A, Mirmehdi M, Thomas B, Markham R, Classification and localization of diabetic-related eye disease. In: Proceedings of 7th european conference on computer vision, vol 2353. Springer LNCS, Copenhagen, Denmark, pp 502–516
16. Usher D, Dumskyj M, Himaga M, Williamson TH, Nussey S, Boyce J (2003) Automated detection of diabetic retinopathy in digital retinal images: a tool for diabetic retinopathy screening, diabetes UK. Diab Med 21(1):84–90
17. Sinthanayothin C, Kongbunkiat V, Ruenchanachain SP, Singlavanija A (2003) Automated screening system for diabetic retinopathy. In: Proceedings of the 3rd international symposium on image and signal processing and analysis, pp 915–920
18. Firdausy K, Sutikno T, Prasetyo E (2007) Image enhancement using contrast stretching on RGB and IHS digital image. TELKOMNIKA 5(1):45–50
19. Gonzalez RC, Woods RE (2002) Digital image processing, 2nd edn. Prentice Hall, New Jersey
20. Hoover A, Kouznetsova V, Goldbaum M (2000) Locating blood vessels in retinal images by piece-wise threshold probing of a matched filter response. IEEE Trans Med Imaging 19 (3):203–210
21. Hoover A, Goldbaum M (2003) locating the optic nerve in a retinal image using the fuzzy convergence of the blood vessels. Trans Med Imaging 22(8):951–958
22. Jamal I, Akram MU, Tariq A (2012) Retinal image preprocessing: background and noise segmentation. TELKOMNIKA 10(3):537–544
23. Kuivalainen M (2005) Retinal image analysis using machine vision, Master's Thesis, 6 June 2005, pp 48–54
24. Staal JJ, Abramoff MD, Niemeijer M, Viergever MA, Ginneken BV (2004) Ridge based vessel segmentation in color images of the retina. IEEE Trans Med Imaging 23:501–509
25. Kauppi T, Kamarainen V, Lensu JK, Sorri L, Uusitalo I, Kälviäinen H, Pietilä J (2006) DIARETDB0, evaluation database and methodology for diabetic retinopathy algorithms, Technical Report
26. Kauppi T, Kamarainen V, Lensu JK, Sorri L, Raninen A, Voutilainen R, Uusitalo I, Kälviäinen H, Pietilä HJ (2007) DIARETDB1, diabetic retinopathy database and evaluation protocol, Technical Report
27. Köhler T, Budai A, Kraus M, Odstrcilik J, Michelson G, Hornegger J (2013) Automatic no-reference quality assessment for retinal fundus images using vessel segmentation. In: 26th IEEE international symposium on computer-based medical systems, Porto
28. Hui-Fuang N (2006) Automatic thresholding for defect detection. Pattern Recogn Lett 27 (15):1644–1649

Part II
Biomedical Instrumentation and Measurements

A New Heat Treatment Topology for Reheating of Blood Tissues After Open Heart Surgery

Palash Pal, Pradip Kumar Sadhu, Nitai Pal and Prabir Bhowmik

Abstract This paper presents the human blood reheating technique for the blood tissues after surgery as 37–51 °C temperature is required during surgery with high frequency induction heating system. The surgeon opens the chest by dividing the breastbone (sternum) and connects with the heart-lung machine to operate on the heart. This machine allows the surgeon to operate directly on the heart by performing the functions of the heart and lungs. The length of the operation will depend on the type of surgery that is required for patient. Most surgeries take at least 4–5 h. The preparation for surgery, which requires approximately 45–60 min included in this time. After this operation the patients has required high temperature blood for the continuation of blood flow to the heart as the human body temperature is decreases after operation. Here high frequency converter technique can give better topology for reheating the blood after open heart surgery by taking least time than conventional system.

Keywords Open heart surgery · Blood reheating · Modified half bridge inverter · Induction heating · MATLAB

P. Pal (✉) · P.K. Sadhu · N. Pal
Department of Electrical Engineering, Indian School of Mines
(Under MHRD Government of India), Dhanbad, Jharkhand, India
e-mail: mail2palash@gmail.com

P.K. Sadhu
e-mail: pradip_sadhu@yahoo.co.in

N. Pal
e-mail: nitai_pal@rediffmal.com

P. Bhowmik
AMRI Mukundapur, Kolkata 700099, India
e-mail: prabir_bhowmik@yahoo.com

S. Gupta et al. (eds.), *Advancements of Medical Electronics*,
Lecture Notes in Bioengineering, DOI 10.1007/978-81-322-2256-9_10

1 Introduction

In now a day's for clean heat production high frequency induction heating are more efficiently applicable for increasing quality of industrial equipments, domestic equipments as well as medical purposes. The general purpose of a heat treatment is to enhance human blood flow before transfused of blood to the human body [1, 2]. Basically superficial and deep heat treatment processes are used in medical system. Superficial heat treatments apply heat to the outside of the body. Deep heat treatments direct heat toward specific inner tissues through ultrasound technology and by electric current. High frequency heat treatment is beneficial for reheating of blood before transfusion to human body [3, 4].

Now the fact is that at the time of open heart surgery heart-lung machine is used [5, 6]. This machine does the work for the heart and lungs to oxygenate and circulate the blood through the body while allowing the surgical team to perform the detailed operation on a still, non-beating heart. In that time the body temperature is too much decreased which, is near about 10–15 °C [7–9]. If the reheated blood is inject to the human body after open heart surgery, then the heart beat will be recovered and corresponding all organs of human body will be well functioning [10–14].

There are different ways to convey heat such as conduction is the transfer of heat between two objects in direct contact with each other, conversion is the transition of one form of energy to heat, radiation involves the transmission and absorption of electromagnetic waves to produce a heating effect, convection occurs when a liquid or gas moves past a body part creating heat are used for heat treatment purposes [15].

Prior to the development of induction heating, microwave provided the prime means of heating human blood [16]. Induction heating offers a number of advantages over that heating such as quick heating, uniform heat distribution, Smooth and easy temperature control, good compactness and high reliability, high energy density. Moreover the high frequency induction heating provides other advantages such as easy of automation and control, requirement of less maintenance, safe and clean working conditions [17].

2 Methodology

High frequency induction heating is incorporated for human blood reheating. Basically, Induction heating involves applying an ac electric signal to a coil placed near specific place in the heating loop and the metallic object will be heated [18]. The alternating current creates an alternating magnetic flux within the metal to be heated in the loop. Eddy emf is induced in the metal by the electromagnetic flux and heats up the material. Fundamental theory of induction heating is similar as transformer operation, where primary coil is treated as heating coils and the current induced in secondary is directly proportional to primary current according to turn ratio [17, 19].

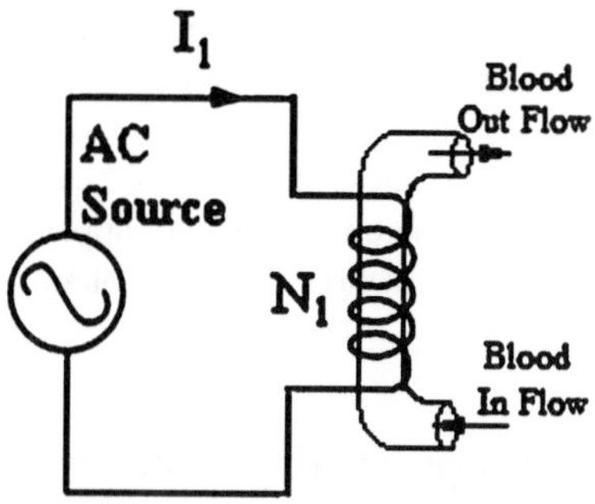

Fig. 1 Diagram of blood heating through non-metallic pipe line

In this proposed induction heating topology blood is taking as a secondary element of blood itself and secondary current will be flow through the human blood, there by eddy current will be generated and blood will be heated as with required temperature. Figure 1 shows the diagram of blood heating through non-metallic pipe line.

Here the human blood is excited by the high frequency alternating current. The simulation shows that the heating area can be effectively controlled by using the cylindrical shield with adjustable space and thereby the temperature will be controlled [20]. Figure 2 depicts that the proposed topology of blood reheating in the non-metallic pipe line. If the small heating area is needed, it may require longer treatment time. However, the efficiency of heat can be increased by varying the radius size of cylinder there by more flux will be passes and more eddy emf is induced. As a result the eddy current flows through blood cell and heat will be generated within the blood cell due to presence of iron particle in the blood [21].

Generally the rate of change in total body heat (TBH) is calculated as follows,

The rate of change in total body heat = (total body heat at the end of reheating) − (total body heat before reheating)/time taken for reheating.

In this topology the temperature can be controlled before transfusion of blood to the human body which is required for operating all organs [22, 23]. Here the inductive coil is a primary working coil of induction heating system which is

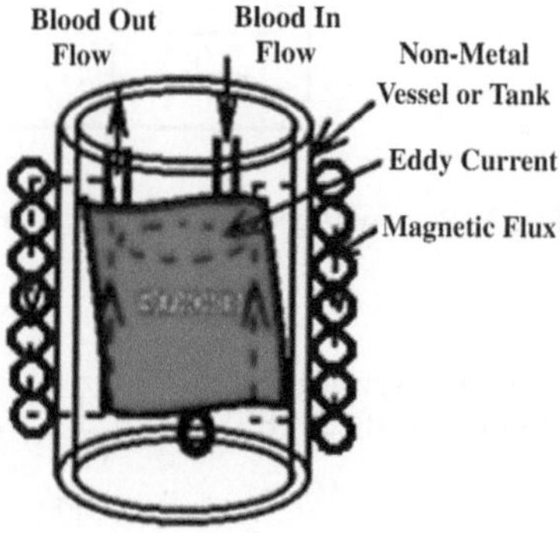

Fig. 2 Proposed blood reheating topology through the non-metallic pipe line

energized by modified half bridge high frequency inverter and the human blood will be working as secondary element of this high frequency inverter. The heating area can be effectively controlled by using the cylindrical shield with adjustable space. However, the efficiency of heat can be increased by varying the radius size of cylinder thereby more flux appears and more eddy emf is induced. As a result the eddy current flows through blood cell and blood is reheated [19].

3 Proposed High Frequency Modified Half Bridge Inverter for Blood Reheating

In the circuit operation has been discussed in detail. Here human blood is considered secondary coil of heating element which can be passes through the vessel or placed in the vessel thereby it can be reheated with this proposed inverter [19, 24]. The exact circuit diagram of the Modified Half Bridge inverter is shown in Fig. 3.

Modified half bridge circuit is normally used for higher power output. Four solid state switches are used and two switches are triggered simultaneously. Here MOSFETs (BF1107) are used as solid state switches because it can be exist at high frequency applications. Anti-parallel diodes D1 and D2 are connected with the switches S1 and S2 respectively that allows the current to flow when the main switch is turned OFF. According to Table 1, when there is no signal at S_1 and S_2, capacitors C_1 and C_2 are charged to a voltage of $V_i/2$ each. The Gate pulse appears at the gate G1 to turn S1 ON. Capacitor C_1 discharges through the path NOPTN. At the same time capacitor C_2 charges through the path MNOPTSYM. The discharging current of C_1 and the charging current of C_2 simultaneously flow from P to T. In the next slit of the gate pulse, S_1 and S_2 remain OFF and the capacitors charge to a voltage

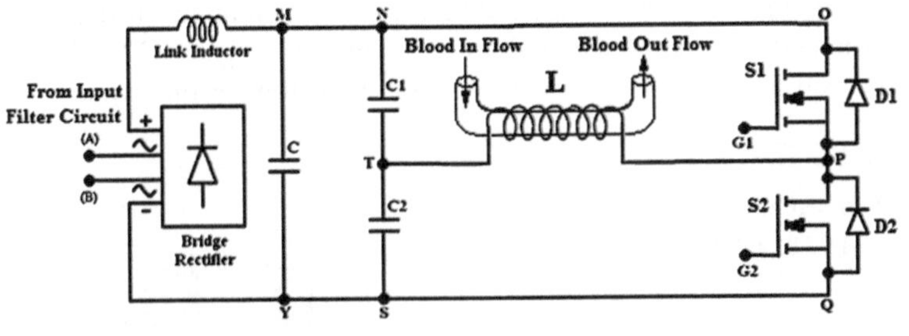

Fig. 3 Proposed modified half-bridge inverter circuit diagram

Table 1 Switching ON–OFF chart of MOSFETs (BF1107)

S1	S2	V_{out}
ON	OFF	$+V_i/2$
OFF	ON	$-V_i/2$

$V_i/2$ each again. The Gate pulse appears at the gate G_2, so turning on S_2. The capacitor C_2 discharges through the path TPQST and the charging path for capacitor C_1 is MNTPQSYM. The discharging current of C_2 and the charging current of C_1 simultaneously flow from T to P. The both switches must operate alternatively otherwise there may be a chance of short circuiting. In case of resistive load, the current waveform follows the voltage waveform but not in case of reactive load. The feedback diode operates for the reactive load when the voltage and current are of opposite polarities.

In new topology, the modified half bridge inverter is operating at high frequency (above 30 MHz) and feeding the blood for reheating. The high frequency alternating current is created by switching two MOSFETs sequentially by an appropriate logic circuit which keeps track of the frequency. The frequency can be varied by varying the pulse rate of the logic train. The load is represented as a series combination of resistance and inductance. Both these parameters vary along with temperature rise. An inductance is placed in series with the rectifier output to smooth out the ripples as far as possible so as to realize a current source.

Here the proposed high frequency induction heating topology can give clean heated blood without damaging the blood cell components [17]. This is required after surgery as human body temperature is decreased after open heart surgery as 37–51 °C temperature is required during surgery [25]. As the blood temperature is decreased about 10–15 °C after open heart surgery so, this proposed scheme will be too much for blood reheating which is required for human body to functioning all organs after open heart surgery.

4 Simulation Results and Discussion

Here the proposed modified half-bridge high frequency inverter for blood reheating after open heart surgery is simulated with MATLAB for getting output voltage and current waveforms. Thereby the output harmonic current will be get and blood will be reheated due to high harmonic current.

Figure 4 shows the wave-form of the output voltage for the proposed modified half-bridge high frequency inverter. The rms value of output voltage is 309.38 V across T and P point of Fig. 3. Figure 5 depicts that the output current wave form for the same. The rms value of output current is 20.31 ampere through load working coil, which is taken in the platform of PSIM.

The Fig. 6 depicts that the harmonic content of output that is for the load current. The 3rd, 5th, 7th and 9th harmonics have a magnitude of 151, 119.7, 40 and 19.8 % respectively of the fundamental component.

It is observed from the wave-shapes that the harmonics of the output current is very high. In this proposed high frequency inverter high eddy current will be generated at the output due to generation of high harmonics at the output. And the blood temperature will be reached to 37 °C or above as per requirement. It is required for blood reheating with a short period of time without damaging the blood

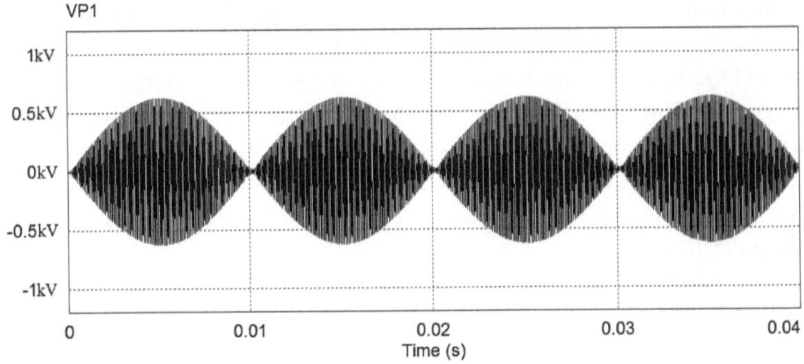

Fig. 4 Output voltage waveform

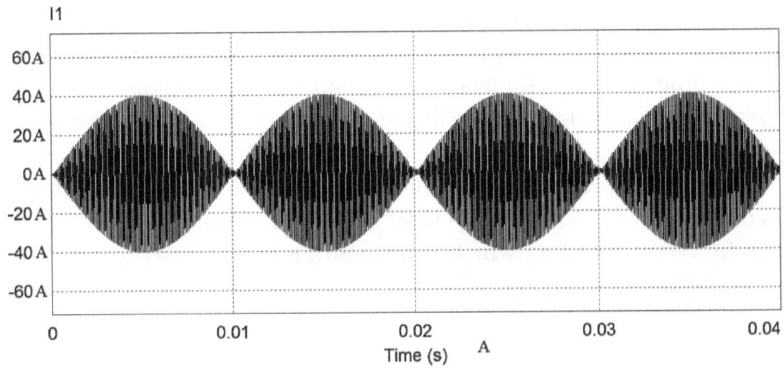

Fig. 5 Output current waveform

Fig. 6 Output harmonic content

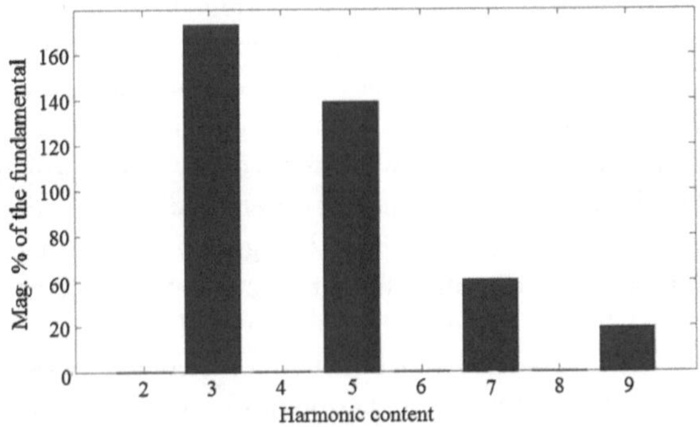

cell components after open heart surgery. This temperature will be controlled by adjusting space of cylindrical vessel. It is clear that the temperature field follows the heat-source distribution quite well. That is, near the projection the heat source is strong, which leads to high temperatures and the blood manages to keep the tissue at normal body temperature after surgery. The eddy currents in a conductive cylinder produce heat. Here, the ohmic losses and temperature distribute in the vessel, the heat transfer and electric field simulations must be carried out simultaneously. Finally the results are in good conformity with this proposed inverter scheme.

5 Conclusion

Hence this proposed high frequency induction heating topology will be more suitable for blood reheating after open heart surgery. As these heat treatments have the potential of human blood reaming without damaging blood composition from excessive temperatures without creating any hazards due to easy control can be possible where as other system can damage the blood cells. Again during heat treatment results in easy, pollution free and clean heat production for the present of power electronics converter. However proposed modified half bridge inverter will give new setup in medical sciences for blood reheating before blood transfusion to human body after open heart surgery.

Acknowledgments Authors are thankful to the UNIVERSITY GRANTS COMMISSION, Bahadurshah Zafar Marg, New Delhi, India for granting financial support under Major Research Project entitled "Simulation of high-frequency mirror inverter for energy efficient induction heated cooking oven" and also grateful to the Under Secretary and Joint Secretary of UGC, India for their active co-operation.

References

1. Sharkey A, Gulden RH, Lipton JM, Giesecke AH (1993) Effect of radiant heat on the metabolic cost of postoperative shivering. Br J Anaesth 70:449–450
2. Sessler DI, Moayeri A (1990) Skin-surface warming: heat flux and central temperature. Anesthesiology 73:218–224
3. Shander A, Hofmann A, Gombotz H, Theusinger OM, Spahn DR (2007) Estimating the cost of blood: past, present, and future directions. Best Pract Res Clin Anaesthesiol 21:271–289
4. Burns JM, Yang X, Forouzan O, Sosa JM, Shevkoplyas SS (2012) Artificial micro vascular network: a new tool for measuring rheologic properties of stored red blood cells. Transfusion 52(5):1010–1023
5. Van Beekvelt MC, Colier WN, Wevers RA, van Engelen BG (2001) Performance of near-infrared spectroscopy in measuring local O2 consumption and blood flow in skeletal muscle. J Appl Physiol 90(2):511–519. PMID 11160049
6. Sinha D, Sadhu PK, Pal N (2012) Design of an induction heating unit used in Hyperthermia treatment- advances in therapeutic engineering, CRC Press, Taylor and Francis Group, ISBN:978-1-4398-7173-7, pp 215–266 (Chapter 11)
7. Gersh BJ, Sliwa K, Mayosi BM, Yusuf S (2010) Novel therapeutic concepts: the epidemic of cardiovascular disease in the developing world: global implications. Eur Heart J 31(6):642–648
8. Logmani L, Jariani AL, Borhani F (2006) Effect of preoperative instruction on postoperative depression in patients undergoing open heart surgery. Daneshvar pezeshki 14(67):33–42. [Persion]

9. McAlister FA, Man J, Bistritz L, Amad H, Tandon P (2003) Diabetes and coronary artery bypass surgery: an examination of perioperative glycemic control and outcomes. Diabetes Care 26:1518–1524
10. Padmanaban P, Toora B (2011) Hemoglobin: emerging marker in stable coronary artery disease. Chron Young Sci 2(2):109. doi:10.4103/2229-5186.82971
11. Handin RI, Lux SE, Stossel B, Thomas PB (2003) Principles and practice of hematology. Lippincott Williams and Wilkins. ISBN:0781719933
12. Minic Z, Hervé G (2004) Biochemical and enzymological aspects of the symbiosis between the deep-sea tubeworm Riftia pachyptila and its bacterial endosymbiont. Eur J Biochem 271 (15):3093–3102. doi:10.1111/j.1432-1033.2004.04248.x. PMID 15265029
13. Newton DA, Rao KM, Dluhy RA, Baatz JE (2006) Hemoglobin is expressed by alveolar epithelial cells. J Biol Chem 281(9):5668–5676. doi:10.1074/jbc.M509314200. PMID 16407281
14. Marik PE, Corwin HL (2008) Efficacy of red blood cell transfusion in the critically ill: a systematic review of the literature. Crit Care Med 36:2667–2674
15. Charles P, Elliot P (1995) Handbook of biological effects of electromagnetic fields. CRC Press, New York
16. Sang V (2010) The use of the mechanical fragility test in evaluating sublethal RBC injury during storage 99(4):325–331
17. Sadhu PK, Mukherjee SK, Chakrabarti RN, Chowdhury SP, Karan BM, Gupta RK, Reddy CVSC (2002) High efficient contamination free clean heat production. Indian J Eng Mater Sci 9:172–176
18. Sadhu PK, Pal N, Bhattacharya A (2013) Design of working coil using Litz Wire for industrial induction heater. Lap Lambert Academic Publishing, ISBN:978-3-659 -35853-1, pp 1–65
19. Inayathullaah MA, Anita R et al (2010) Single phase high frequency Ac converter for induction heating application. Int J Eng Sci Technol 2(12):7191–7197
20. Kotsuka Y, Hankui E, Shigematsu E (1996) Development of ferrite core applicator system for deep-induction Hyperthermia. IEEE Trans Microw Theor Tech 44(10):1803–1810
21. Chen H, Ikeda-Saito M, Shaik S (2008) Nature of the Fe-O_2 bonding in oxy-myoglobin: effect of the protein. J Am Chem Soc 130(44):14778–14790. doi:10.1021/ja805434m. PMID 18847206
22. Hohn L, Schweizer A, Kalangos A et al (1998) Benefits of intraoperative skin surface warming in cardiac surgical patients. Br J Anaesth 80:318–323
23. Kotsuka Y et al (2000) Development of inductive regional heating system for breast Hyperthermia. IEEE Trans Microw Theor Tech 48(2):1807, 1813
24. Burdio JM, Fernando M, Garcia JR, Barragan LA, Abelardo M (2005) A two-output series-resonant inverter for induction-heating cooking appliances. IEEE Trans Power Electron 20 (4):815–822
25. Rajek A, Lenhardt R, Sessler DI et al (1998) Tissue heat content and distribution during and after cardiopulmonary bypass at 31 °C and 27 °C. Anesthesiology 88:1511–1518

Real Time Monitoring of Arterial Pulse Waveform Parameters Using Low Cost, Non-invasive Force Transducer

S. Aditya and V. Harish

Abstract Cardiovascular disease is currently the biggest single cause of mortality in the developed world (Alty, IEEE Trans Biomed Eng 54:2268–2275, 2007) [1], (Clerk Maxwell, A treatise on electricity and magnetism, Clarendon, 1892) [2]. Hence, the early detection of its onset is vital for effective prevention. Aortic stiffness as measured by aortic Pulse wave velocity (PWV) has been shown to be an important predictor of Cardiovascular disease. However, the measurement of the same is complex and time consuming (Alty, IEEE Trans Biomed Eng 54:2268–2275, 2007) [1], (Clerk Maxwell, A treatise on electricity and magnetism, Clarendon, 1892) [2]. This paper presents a simple, low-cost, speedy and non-invasive method using Force Sensing Resistors (FSR) strategically placed over the Carotid and Radial arteries to evaluate various arterial wave pulse parameters like heart-rate, Stiffness Index (SI), Reflectivity Index (RI) and Pulse Wave velocity. The pulse rate and shape is used as an estimate of heart-rate. This is used for diagnosis of arrhythmias, tachycardia and bradycardia. The proposed method could be employed as a cheap and effective Cardiovascular disease screening technique, to be later integrated with small wrist watch-like monitors for suitable commercial purposes.

Keywords Cardiovascular disease (CVD) · Pulse wave velocity (PWV) · Stiffness index (SI) · Reflectivity index (RI) · Arterial wave pulse · Force sensing resistor (FSR)

S. Aditya (✉)
Department of Electronics, Electrical and Instrumentation, BITS Pilani, Pilani, Goa, India
e-mail: f2011350@goa.bits-pilani.ac.in

V. Harish
Department of Electronics and Instrumentation, Madras Institute of Technology, Chennai, India
e-mail: harishvenka@gmail.com

S. Gupta et al. (eds.), *Advancements of Medical Electronics*,
Lecture Notes in Bioengineering, DOI 10.1007/978-81-322-2256-9_11

Radial Artery	In human anatomy, the radial artery is the main artery of the lateral aspect of the forearm
Carotid Artery	In human anatomy, the left and right carotid arteries the head and neck with oxygenated blood
Arteriosclerosis	The thickening, hardening and loss of elasticity of the walls of arteries
Atherosclerosis	A specific form of arteriosclerosis in which an artery wall thickens as a result of invasion and accumulation of white blood cells
Hemodynamic	Relating to the flow of blood within the organs and tissues of the body
Systolic Hypertension	Refers to elevated systolic blood pressure

1 Introduction

Cardiovascular Disease (CVD) is the leading cause of mortality in the developed world. An estimated 17 million people die every year from CVD (mainly from myocardial infarction and stroke; source: World Health Organization). Stiffness Index and Pulse Wave velocity are an important parameters used in the assessment of risk of CVD. Arterial stiffening leads to systolic hypertension and increased load on the heart. Arterial stiffness is not only a marker of the effects of atherosclerosis/arteriosclerosis on the arterial wall, but in itself lead to adverse hemodynamic effects that increase CVD risk [1, 2]. Studies performed in [3, 4] indicated large artery stiffness, as measured by PWV has proven to be a powerful predictor of CVD events [4]. Pulse wave velocity was also recognized by the European Society of Hypertension as integral to the diagnosis and treatment of hypertension [5]. Thus it is important to estimate these biological parameters for early diagnosis and prevention of CVD.

Through the proposed method the Arterial Pulse Waveform and critical parameters related to it such as heart rate, SI and PWV can be extracted. The current method of measuring the arterial pulse waveform parameters is through Photoplethysmography and analysis of the obtained Digital Volume Pulse (DVP) [1, 2]. Other methods of measuring PWV use invasive catheters [3] or mechanical tonometers [6] to measure the transit time between pressure waves at two different points. Some of these sensors are not only invasive but also expensive. The proposed method would greatly simplify measurement of the arterial wave pulse and its parameters using a low-cost and non-invasive transducer.

10 Consenting test subjects participated in the study. 2 Healthy males in the age group of 20–30 and two males in the age group of 40–50 one healthy and the other a patient and two males in the age group of 70–80 one healthy and the other a patient. 4 females participated in the study 2 healthy in the age group of 40–50 and 2 in the age group of 70–80, one healthy and the other a patient.

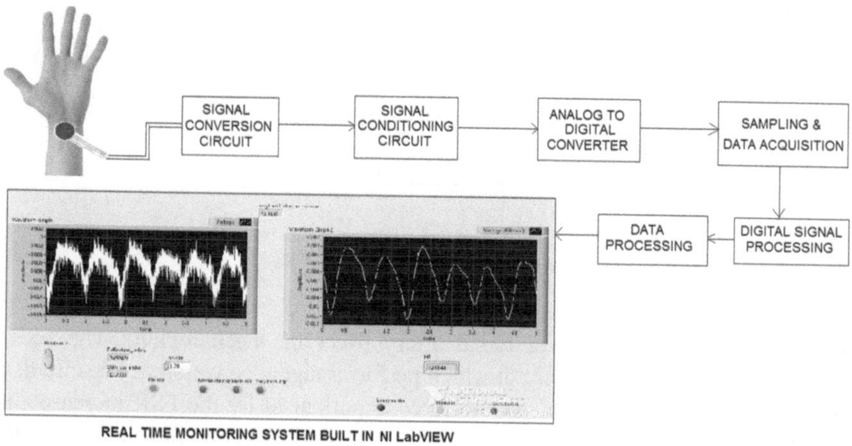

Fig. 1 Block diagram of proposed system

Firstly, from the arterial wave pulse, Heart Rate of a patient can be estimated through the proposed method. The count of number of pulses with a given time interval gives heart rate information. Diagnosis of Tachycardia and Bradycardia can also be performed using heart-rate information. Through the shape of the wave pulse, Stiffness Index and Reflectivity index are estimated. Pulse Wave Velocity is measured using 2 FSRs placed appropriately over the Carotid artery. The integrity of the system is verified through comparison with PPG Analysis of the same test subject. The variation of PWV and SI estimated using the proposed method with age has also been measured. The working of the system is as described in the diagram below (Fig. 1).

2 Methodology

2.1 *Signal Conversion and Conditioning*

2.1.1 Signal Sensing

During these measurements the patient is requested to remain still to avoid errors due to motion artifact although a compensation has been provided for these in the proposed algorithm later on. Force Sensing Resistors, are robust polymer thick film devices that exhibit a decrease in resistance with increase in force applied to the surface of the sensor. A standard Interlink Electronics FSR 402 sensor, a round sensor 18.28 mm in diameter is used to sense the bio signal. The FSR is placed over

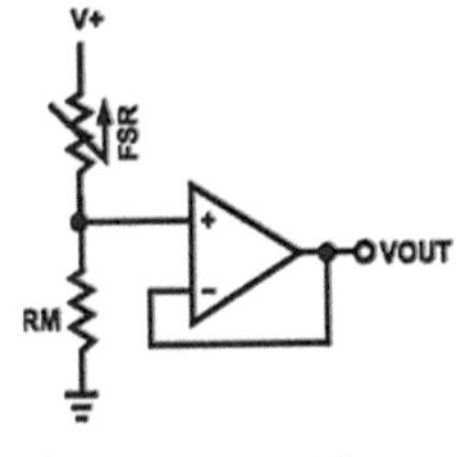

Fig. 2 Signal conversion circuit with $V_+ = 10$ V, RM = 120 KΩ

the carotid (over the wrist) or radial artery (over the neck region). The FSR's terminals are connected to a circuit which performs signal conversion. The circuit in Fig. 2 Performs this function. As the force experienced by the FSR increases, its resistance decreases consequently the voltage across R_M increases and since the amplifier is connected in a buffer configuration, its output voltage increases. Voltage V_+ was chosen to be 10 V and split power supply of ±15 V is provided to the amplifier circuit.

2.1.2 Signal Conditioning

The typical human heart beats at about a minimum 60 up to a maximum of 300 beats/min which corresponds to a frequency range of 1–5 Hz. The output voltage signal from the circuit in Fig. 2 is required to be conditioned to extract the signal of desired frequency. This is done using a Simple filter circuit as shown in Fig. 3. This signal is filtered using a band pass filter of 1–5 Hz. Although this filtering should give us the signal in the desired frequency, we also use a Notch Filter as a precaution to eliminate power line noise. Interference from power lines (50 Hz) is the largest source of extraneous noise in bio-electric signals. Since the device is to be used in environments such as homes and hospitals where power line interference is unavoidable, there is further a need to eliminate distortions due it. Thus the signal is also passed through a notch filter of Bandwidth 40–60 Hz to specially eliminate power line 50 Hz noise. The frequency response of the filter

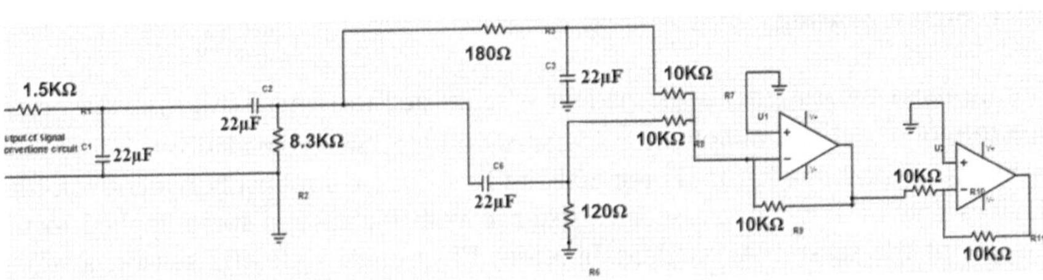

Fig. 3 Signal conditioning circuit with band pass: LPF R1 = 1.5 KΩ C1 = 22 μF and HPF R2 = 8.3 KΩ C2 = 22 μF and Notch: LPF R3 = 180 Ω C3 = 22 μF and HPF R4 = 120 Ω C4 = 22 μF and R7 = R8 = R9 = R10 = R11 = 10 KΩ

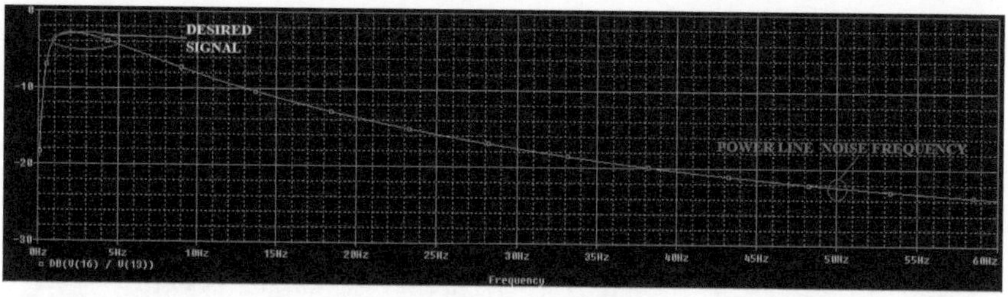

Fig. 4 Frequency response of filter as simulated using Pspice

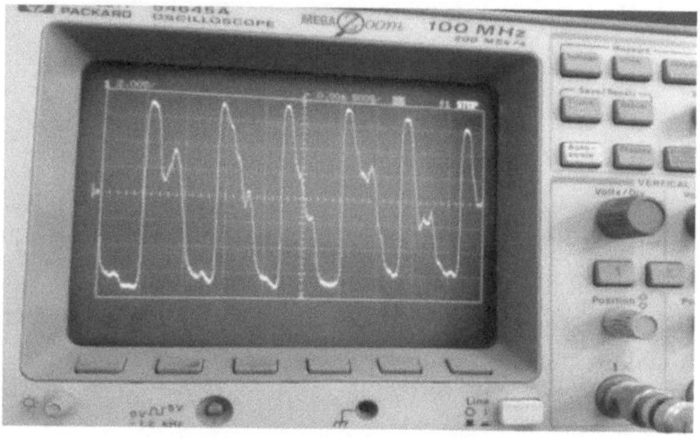

Fig. 5 Smoothened waveform at the output of the circuit as seen on oscilloscope

circuit is as shown in Fig. 4. Y axis is magnitude response in dB and X axis frequencies from 0 to 60 Hz. Simulations were carried out in Pspice.

From the frequency response plot of the signal it in the desired frequency range (1–5 Hz) is attenuated by only 2.3 dB as indicated in Fig. 4 (yellow lines) while power line noise frequency (50 Hz) is attenuated by about 23 db (Red Line) hence being able to extract the signal of desired frequency. The output of the circuit is observed using an oscilloscope and after smoothing a waveform is obtained as shown in Fig. 5.

2.2 Sampling and Data Acquisition

The analog output of this circuit is required to be converted to digital and the voltage values are required to be sampled to process the waveform data. This is done using the NI- My DAQ Data acquisition system. The output of the signal conditioning circuit Fig. 3. Is connected to analog input channel 1 of the My DAQ and its USB interface with a laptop allows data to be acquired. The DAQ assistant in LabVIEW is then configured to acquire the data. A correctly acquired heart beat signal can have a maximum frequency of 5 Hz and hence by Nyquist Criterion,

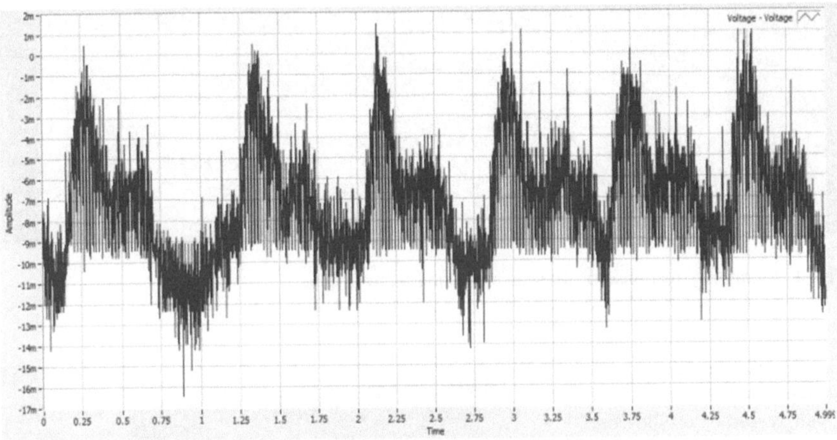

Fig. 6 Noisy waveform as acquired using DAQ assistant

sampling should be done at at least 10 Hz. A large sampling rate of 1 K sample/s is chosen to acquire large number of samples and for a more accurate digital reconstruction of the waveform. The reconstructed waveform acquired through DAQ assistant can be seen in Fig. 6.

2.3 Assessment of Heart-Rate

For calculation of parameters, the noisy waveform obtained as shown in Fig. 6 is difficult to work with. Thus Digital Signal Processing is required. First 6 Hz low pass filtering of this data is done using signal processing tool in LabVIEW. This filtered output is given as an input to the Math-Script Node in LabVIEW. The MATLAB code processes the filtered waveform at the input and extracts the necessary information. Now that a smoother waveform was obtained, it is required to count the number of pulses. A simple method to do this would be counting the number of peaks within a given time interval. The following counting method is followed. The derivative of the digitally reconstructed signal is taken to find out the maxima points. Now since the arterial wave pulse by nature is a double-peaking pulse, i.e. it has a end of systolic peak and end of systole peak. Let the number of maxima be N. Thus the number of pulses can be found as

$$\text{No of Pulses} = N/2$$

where N is the number of maxima points.

The heart rate in BPM would be = N/2 * 60/5 = 6 N (bpm)

In the waveform in Fig. 7, the number of maximas = 11. Thus heart rate = 66 bpm which is the correct heart rate. In some cases however due to artifacts

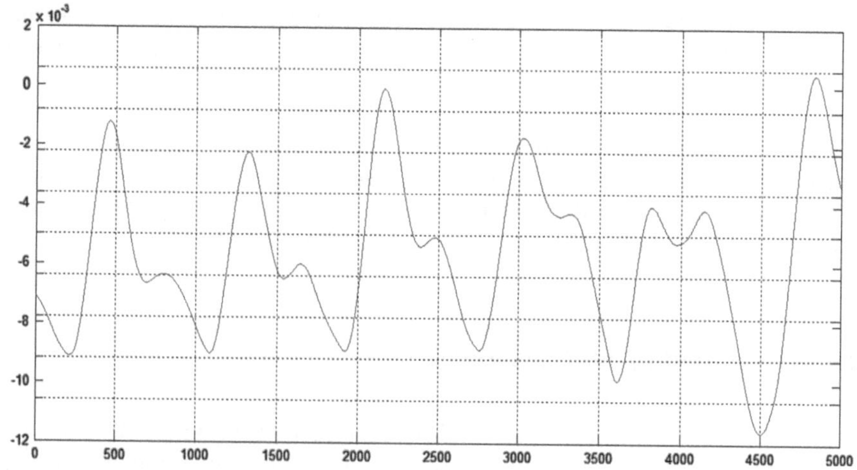

Fig. 7 Arterial pulse waveform as seen MATLAB

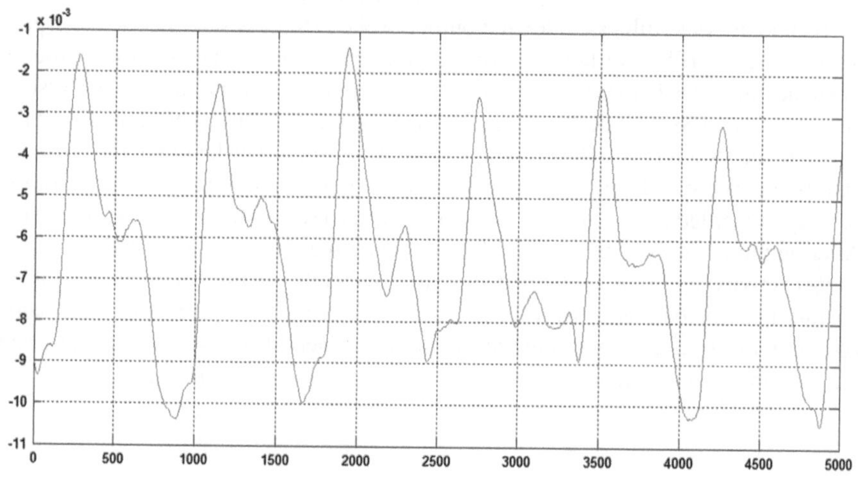

Fig. 8 Erroneous waveform obtained as seen in MATLAB

waveforms like those in Fig. 8 are obtained. In Fig. 8 there are more than 2 maxima per pulse. This will lead to incorrect estimation of heart rate. The same code when run for the waveform in Fig. 8 counts 16 maxima, i.e. $6 \times 16 = 96$ bpm. However the actual heart rate is 72 bpm.

A compensation for such artifacts is provided by checking closeness of maxima points in time. If two maxima points are closer than 100 ms apart in time, then there is highly likely to be an error and the code provides a compensation for the same, by ignoring such maxima while counting. Hence miscounting of maxima is avoided using this method.

Artifacts due to Power Line interference have been removed by the notch filtering performed earlier. Even after compensation some errors due to patient movements and irregular breathing may still exist as shown in Fig. 8. High frequency artifacts caused by sudden movements or spike in circuit voltage won't cause a problem as the signal is already pre-filtered. Detection and removal of artifacts requires further analysis. Although artifact detection and compensations is not covered in the paper, this can also be performed through extensive data collection, signal processing, feature extraction and learning machines such as SVMs or ANNs.

2.4 Assessment of Stiffness Index

Stiffness is defined as the ratio of the height of a subject (h) to the time difference (ΔT) or Peak to Peak time (PPT) between systolic peak and end of systole peak [1, 2]. This information is usually obtained from the Digital Volume Pulse (DVP) measured using Photoplethysmography. However a simpler and low-cost method to obtain the same information is from the waveform obtained using an FSR.

To find arterial stiffness index the output of the DAQ assistant is 6 Hz low-pass filtered in LabVIEW. The data points corresponding to maxima and minima are found. Then the difference between consecutive maxima points corresponding to the Systole peak and End of Systole peak is found. This gives us ΔT or PPT. This done for all the pulses in the 5 s interval and the average value is recorded. The height of the peaks 'a' and 'b' are calculated with respect to the closest previous minima. The height of the subject is recorded previous to the experiment using a simple height- measuring scale. After inputting the subject's height, the arterial stiffness can be calculated.

Reflectivity index is as defined the ratio of height of End of Systole peak to Systole Peak or b/a, as is seen from Fig. 9. To obtain this information, the same code previously used is modified to obtain the values of the waveform peaks and the bottom. Then the ratio of the difference is found and calculated.

2.5 Assessment of Pulse Wave Velocity

Pulse Wave velocity is also another important indicator of arterial stiffness. For calculation of Pulse wave velocity, the arterial pulse waveform is measured at two points about 5 cm apart over the carotid artery simultaneously. Two FSR's were used and analog and digitally conditioned in the same manner as done previously. The outputs of the circuit were given to channel 0 and channel 1 of the Ni-My DAQ. The data was acquired at the same sampling rate of 1 K sample/s and 5 s. The signals are low pass filtered at 6 Hz in LabVIEW and then fed to a math script node. The Algorithm devised basically extracts the data samples, corresponding to the systolic maxima (upper-peak) of the waveform. The time value of data sample corresponding to the first peak that occurs of each waveform is stored. Let us

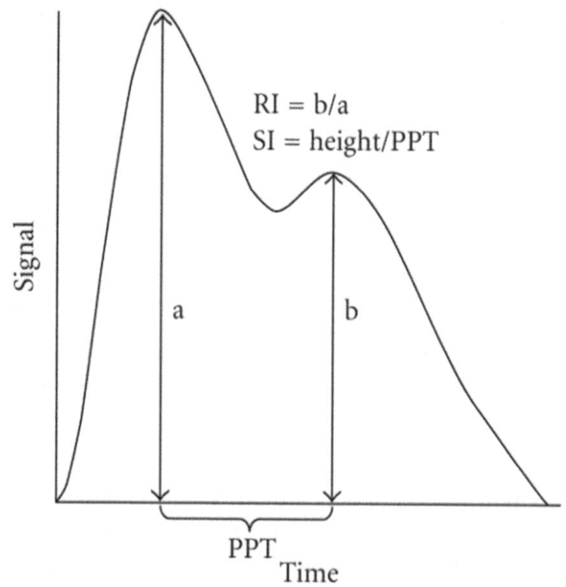

Fig. 9 Estimation of SI and RI from arterial pulse wave

assume that measurements start from time t = 0 and the first maximas occur at time t1 and t2 respectively. Thus |t1–t2| would give us the time difference. Now that the time difference is known and the distance between two points is known,

$$\text{PWV} = \text{Distance}\,(5\ \text{cm})/\text{Time Difference}.$$

Although the above measurements could be performed by placing the sensor over two points over the radial artery, in case of the radial artery, the sensitivity was found to be far less than when placed over the carotid artery.

3 Results and Statistical Analysis

The table below presents the average value of different biological parameters- heart rate, Stiffness Index and Reflectivity Index obtained for different test subjects in 10 estimation trials.

From the data in Table 1, a graph of Stiffness Index versus Age is plotted. This can be seen in Fig. 10.

The best fit line for the plot gives linear equation SI = 0.08805 * Age + 4.2997. Here we observe a positive correlation between Stiffness Index and age, i.e. as the age of the test subject increases, the stiffness index increases. This is in close agreement to the results presented in [7]. Mean Arterial Pressure (MAP) is another variable which affects the Stiffness Index, measuring Blood pressure is not in the scope of the proposed hence is neglected while performing regression analysis.

Table 1 Estimated value of Biological Parameters for test subjects using the proposed method

Test subject	Gender	Age	Heart rate (bpm)	Stiffness index (SI) (m/s)	Reflectivity index (RI)	Pulse wave velocity (PWV) (m/s)
I	Male	21	74.2	5.93	0.47	7.41
II	Male	22	68.3	6.22	0.53	7.50
III	Male	45	80.5	7.97	0.38	9.32
IV	Male	48	92.4	8.69	0.54	9.65
V	Male	67	84.7	10.15	0.46	11.30
VI	Male	65	71.8	9.83	0.56	11.43
VII	Female	42	76.6	8.32	0.72	9.28
VIII	Female	45	83.1	8.54	0.45	9.36
IX	Female	63	80.8	9.84	0.84	10.76
X	Female	68	70.7	10.28	0.67	11.25

Mean of the Stiffness Index = 8.57 m/s and Standard Deviation (SD) = 1.547 m/s. The mean heart rate for the 10 test subjects was found to be 78.3 bpm with SD = 7.4058 bpm. Range of heart rate = 24.1 bpm.

Table 1 also presents Pulse Wave velocity of different test subjects. Regression analysis of Pulse Wave velocity versus age obtained using the proposed method, yields best fine line y = 0.0843 * Age + 0.527. This is also in agreement with results obtained in [7]. The mean PWV was found to be 9.726 m/s and SD = 1.475 m/s. A better sensitivity was observed when the FSR was placed over the carotid artery than the radial artery during Pulse Wave Velocity Measurement.

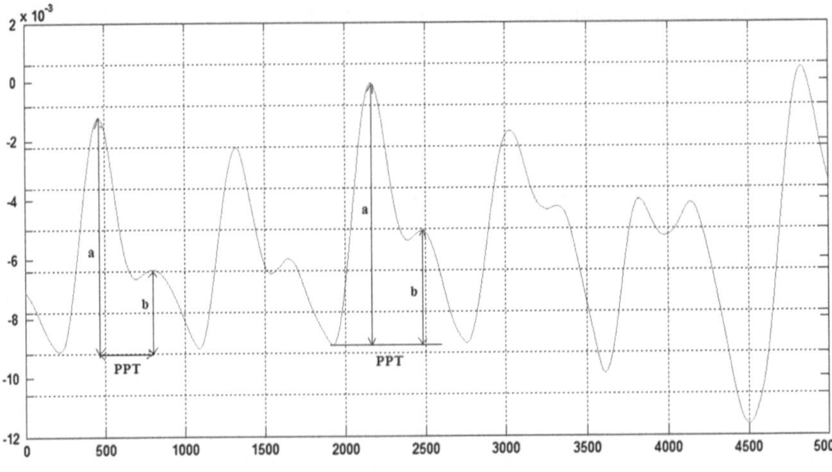

Fig. 10 A 6 Hz low pass filtered waveform used for estimation of Reflectivity index (RI) and Stiffness in dex (SI)

Table 2 Comparison of Biological Parameters obtained from proposed method and Photoplethsymography

Biological parameters from different sensors		
Parameter	FSR sensor	PPG signal
Heart rate	78.3 BPM	78.3 BPM
Peak to peak time	376 ms	364 ms
Stiffness index	8.57 m/s	8.85 m/s
Reflectivity index	0.562	0.546

4 Photoplethysmography Analysis

To validate the results obtained using the existing sensor, Photoplethysmogram measurements were performed simultaneously on the same test subject. The standard PPG probe was placed on the finger tip of the test subject and the arterial pulse waveform was obtained from PPG signal acquired in LabVIEW and imported. This data is processed using the same algorithm as before after making some changes with respect to sampling rate.

A comparison of the average value parameters obtained for the 10 test subjects from PPG analysis of all test subjects and the proposed method is presented in the Table 2.

Analysis shows that there is a 3.3 % error in values of Peak to Peak Time and Stiffness Index, and 2.84 % error in values of reflectivity index obtained using FSR sensor and from PPG signal processing. Hence the parameters are in agreement and can be estimated with reasonably high accuracy.

5 Discussion: Hardware and Software Integration

After processing the signal and finding techniques to accurately extract information about these biological parameters, a Real time monitoring System was built in LabVIEW which monitors these parameters every 5 s. Further based on the value of these parameters, and a diagnosis of medical conditions such as arrhythmias, Tachycardia and Bradycardia and risk of cardiovascular disease based on Pulse Wave Velocity and Stiffness Index is done. Hear rate in the range of 60–100 bpm are classified as normal. Heart rates below 60 bpm are classified as bradycardia and those over 100 bpm are classified as tachycardia. Pulse wave velocities above 10 m/s are classified as high risk of CVD. The front-panel View of the integrated system is shown in Fig. 11. The final system comprises of a simple User friendly GUI in LabVIEW where the user simply has to enter his/her enter age and height based on which, his/her heart rate, pulse wave velocity, stiffness index and reflectivity will be displayed every 5 s. The transducer is also fast in response and negligible delays are observed during measurements An averaging feature is also introduced in the system which calculates, the average heart rate per minute. The user can know about his/her risk of CVD or Arrhythmias. Alarms are also introduced in case of abnormally high heart rates or PWVs (Figs. 12 and 13).

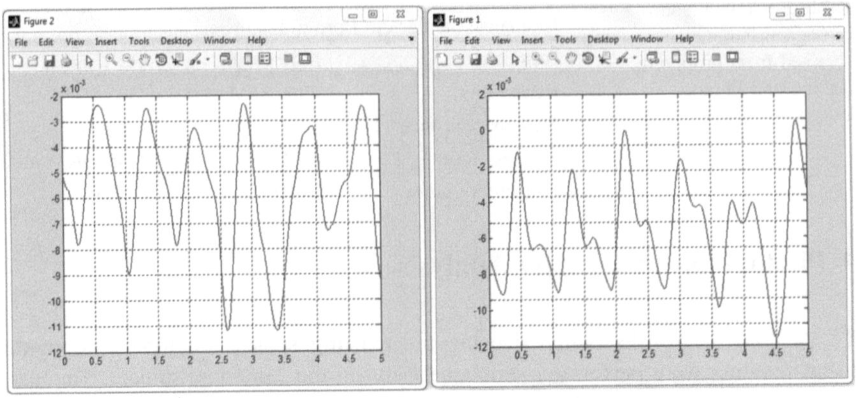

Fig. 11 Arterial pulse waveforms obtained 5 cm apart above the carotid artery

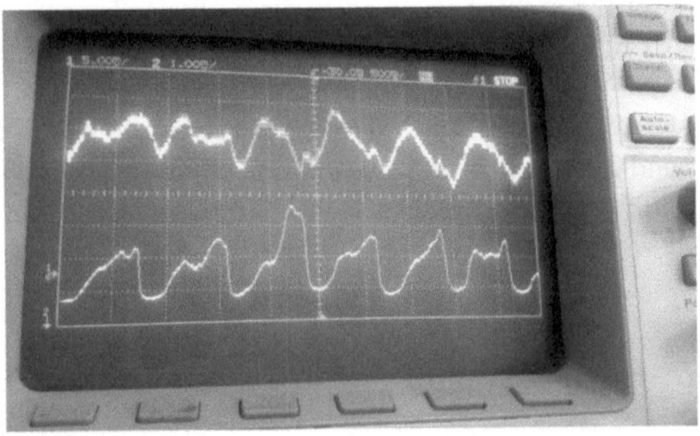

Fig. 12 Arterial pulse waveforms obtained 5 cm apart above the carotid artery

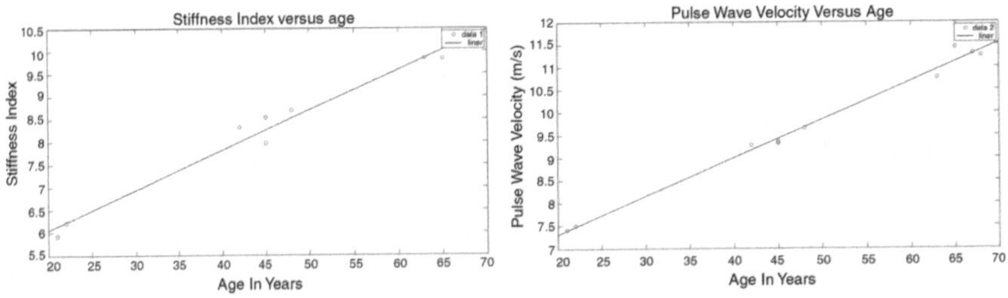

Fig. 13 Variation of Stiffness Index (*left*) and Pulse Wave Velocity (*right*) obtained using the proposed method with age

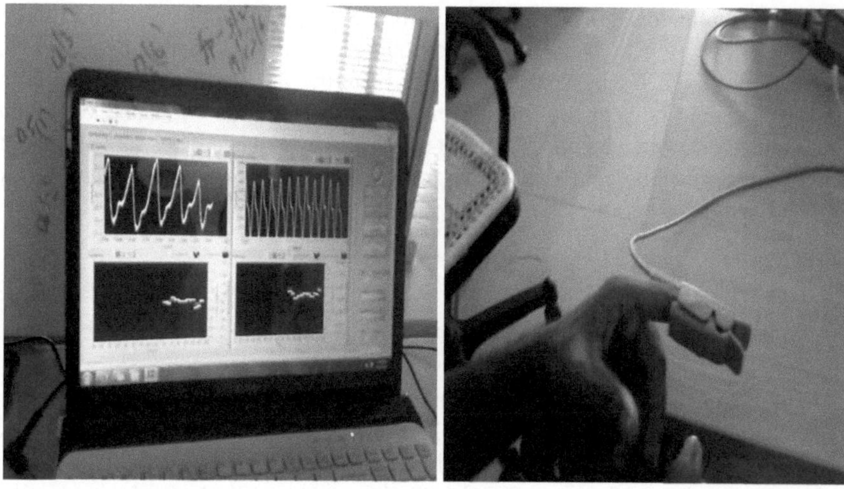

Fig. 14 Acquisition of PPG signal

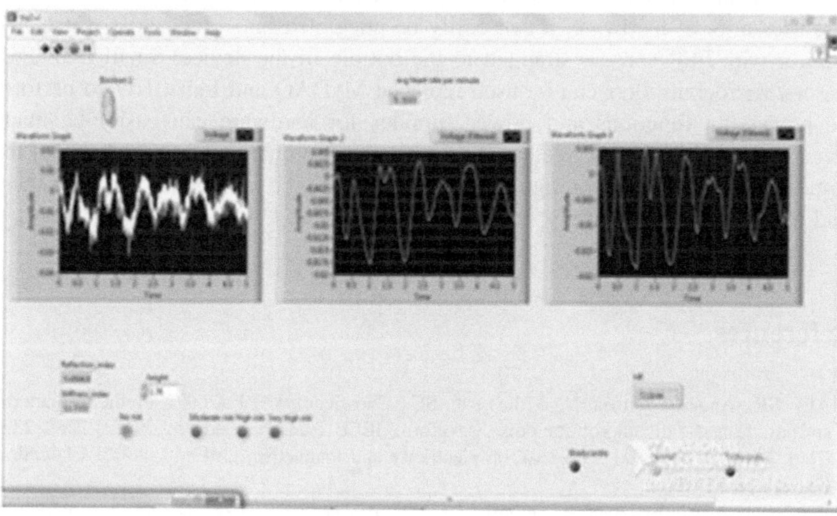

Fig. 15 Front panel view of integrated system

6 Calibration

Since the proposed system is interested in measurement of parameters (hear rate, SI, PWV) that are sensitive to the time variation of the signal as opposed to amplitude, thus a calibration of amplitudes is not required. Even RI remain unaffected as it is simply an amplitude ratio. However an Omp amp power supply of ±15 V and a 10 V for V_+ supply to the signal conditioning circuit will yield proper results. Power supply voltage can be increased, for bettering the sensitivity of the circuit however it should be kept within safe limits as IC 741 cannot withstand very high input voltages. Also the height of the test subject measured accurately at the time of measurement must be given as input for correct estimation of Stiffness Index (Figs. 14 and 15).

7 Conclusion

Accurate estimation of heart-rate, Stiffness Index and Pulse Wave velocity was done using a low-cost, speedy and non invasive FSR. Variation of Stiffness Index and Pulse Wave Velocity with age, estimated using the proposed method is also performed and the results were found to be in agreement to those obtained using previous methods. Offline monitoring of these parameters can be performed using TDMS logging option available in the DAQ assistant and the same algorithm can be implemented again to extract these parameters. Although the system proposed here uses a lot of hardware, it can easily be integrated into small wrist watch like monitor with FSR sensors strapped to the bottom of the device. Small DSP Processors/Microcontrollers can be used replaced MyDAQ and LabVIEW to perform the processing functions and power supplies for hardware can easily be made portable through batteries. We also hope that the affordability and simplicity of the proposed system will encourage more people to use the instrument and help them lead a safe and healthy life.

References

1. Alty SR, Angarita-Jaimes N, Millasseau SC, Chowienczyk PJ (2007) Predicting arterial stiffness from the digital volume pulse waveform. IEEE Trans Biomed Eng 54(12):2268–2275
2. Clerk Maxwell J (1892) A treatise on electricity and magnetism, 3rd edn, vol 2. Clarendon, Oxford, pp 68–73
3. Blacher J, Asmar R, Djane S, Gerard M (1999) Aortic pulse wave velocity as a marker of cardiovascular risk in hypertensive patients. Hypertension 33:1111–1117
4. Boutouyrie P et al (2002) Aortic stiffness is an independent predictor of primary coronary events in hypertensive patients: a longitudinal study. Hypertension 39:10–15
5. Mancia G, de Backer G, Dominiczak A et al (2007) Guidelines for the management of arterial Hypertension: the task force for the management of arterial Hypertension of the European Society of Hypertension (ESH) and of the European Society of Cardiology (ESC). J Hypertens 25(6):1105–1187

6. Salvia P, Liob G, Labatc C, Riccid E, Panniere B, Benetosc A (2004) Validation of a new non-invasive portable tonometer for determining arterial pressure wave and pulse wave velocity: the Pulse Pen device. J Hypertens 22:2285–2293
7. Millasseau SC, Kelly RP, Ritter JM, Chowienczyk PJ (2002) Determination of age-related increases in large artery stiffness by digital pulse contour analysis. Department of Clinical Pharmacology, St. Thomas' Hospital, Centre for Cardiovascular Biology and Medicine, King's College London SE1 7EH, UK, Clinical Science

Selection of Relevant Features from Cognitive EEG Signals Using ReliefF and MRMR Algorithm

Ankita Mazumder, Poulami Ghosh, Anwesha Khasnobish, Saugat Bhattacharyya and D.N. Tibarewala

Abstract Cognition may be defined as a set of mental activities or processes which deals with knowledge, attention, memory and working memory, reasoning and computation, and judgement and evaluation. In this paper, we aim to study two distinctive cognitive processes dealing with evaluation of two similar stimuli and reasoning and computation of some mathematical problem. Here, we have used Wavelet Transforms and Distance Likelihood Ratio Test for feature extraction and classification respectively. We have also used two feature selection algorithm: ReliefF and Minimum Redundancy Maximum Relevance to select only the best relevant features for classification. The results show a 15 % improvement on accuracy when feature selection algorithms are used in the process. The results also suggests that the brain activation is dominant at the frontal, parietal and temporal region.

Keywords Cognition · Electroencephalography · ReliefF · Minimum redundancy maximum relevance · Feature selection

A. Mazumder (✉) · P. Ghosh · A. Khasnobish · S. Bhattacharyya · D.N. Tibarewala
School of Bioscience and Engineering, Jadavpur University, Kolkata, India
e-mail: mazumderankita@gmail.com

P. Ghosh
e-mail: pologhosh58@gmail.com

A. Khasnobish
e-mail: anweshkhasno@gmail.com

S. Bhattacharyya
e-mail: saugatbhattacharyya@live.com

D.N. Tibarewala
e-mail: biomed.ju@gmail.com

S. Gupta et al. (eds.), *Advancements of Medical Electronics*,
Lecture Notes in Bioengineering, DOI 10.1007/978-81-322-2256-9_12

1 Introduction

The entire human brain is divided into number of regions and each of these regions has separate sets of functions. Cognition is one of these functions which is basically related to tasks like decision making, memorization, perception, consciousness and likewise [1]. This study is mainly focused on the cognitive capabilities, or to be more precise, on the problem-solving capabilities of the human brain. It has been confirmed in a number of previous studies that the bio-signals generated while performing the cognitive tasks fall within the alpha (α) and beta (β) frequency bands [2], which originates from the parietal and temporal regions of the brain. It is also be concluded from literature that the frontal lobe takes a major part in the process of cognition [3]. So, it can be said that the activation will take place mostly in these three regions while performing any cognitive task.

The ability of the human brain can get deteriorated due to the advent of number of diseases such as Parkinson's disease, Alzheimer's disease, stroke, multiple sclerosis, Lupus, severe brain injury and many more [4, 5]. Through advances in neuroscience and computing, it is now possible to provide such patients with rehabilitative treatments. A number of different methods for cognitive rehabilitation have been devised over the years in form of Brain Computer Interface (BCI) technologies. The main aim of the BCI system is to decode the signals acquired from human brain and then decode those into signals which can be used for controlling a device (for example a rehabilitative aid and likewise). The principal components of the BCI technologies are feature extraction [6], selection of the features [7] of interest and finally classification [8] of those signals.

The outputs of the BCI system can be optimized by selecting a proper combination of these algorithms. The feature extraction and classification stages are mandatory for any BCI application. The feature selection stage is optional but it is important for obtaining precise and error free outputs for a number of applications. Sometimes, it is noticed that the performance of the BCI system is affected because of the high dimensionality of the feature vector. Because of the presence of large number of irrelevant features (features which are not discriminable among classes), the effect of relevant features are negated which reduces the performance of the BCI [9].

In this paper we aim to study the performance of ReliefF [10], and Minimum Redundancy Maximum Relevance (MRMR) [11] for feature selection of the best relevant features in a cognitive task experiment and its effect on the performance of the classifier. The cognitive task experiment comprises two separate assignments. First, the subject performs a task where he/she has to spot the difference between two similar pictures. This corresponds to the evaluation related cognitive process. The second task presents a mathematical puzzle to the subject, who must solve it in a given time period. This corresponds to the reasoning and computation cognition.

The rest of the paper is organized as follows: Sect. 2 provides a description on the experimental methods employed in this study. It also gives a brief description on the feature extractor, feature selector and classifier algorithm. Section 3 presents the results produced by this experiment and the concluding remarks are mentioned in Sect. 4.

2 Materials and Methods

In this experiment, the subjects are instructed to spot the difference between two sets of 'look alike' pictures, and solve some mathematical tasks (as shown in Fig. 1). Through these tasks, we aim to understand the underlying processes taking place in the brain when the subjects perform these tasks. 7 healthy subjects (4 female and 3 male) in the age group of 22–28 years have participated in this experiment. The EEG signals from the subjects were recorded using a 19 channel EEG amplifier (NeuroWin, Make-NASAN). Based on the nature of the experiment, we have selected the electrodes: F3, F4, Fz, P3, P4, T3, and T4 for our study as because these electrode locations coincide with those regions of the brain which are responsible for such cognitive tasks, which are the parietal, frontal and temporal lobe. After that these data are filtered using a band-pass filter. The next step is the extraction of features which is performed by Wavelet Transform [12]. Features pertaining to wavelet transforms yields high dimensional features which is suitable for this paper. The next step is the feature selection step which uses ReliefF and Minimum Redundancy Maximum Relevance (MRMR) to select the best N features from the original feature set. After that the selected features are classified using Distance Likelihood Ratio Test (DLRT) [13]. The same steps are performed again but on the second trial the feature selection step is excluded. The results suggest that the performance of the classifier is better while incorporating the feature selection step than without it.

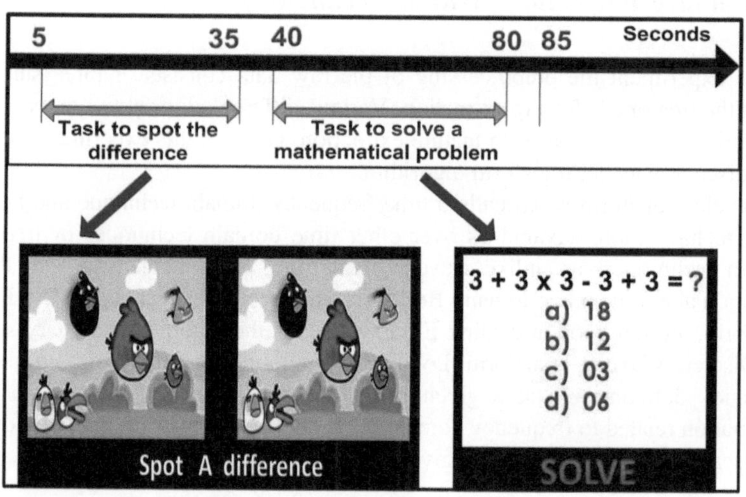

Fig. 1 A example of a portion of the timing sequence of the visual stimuli

2.1 Design of the Visual Stimuli

The comprehensive format for the visual cue is given below: For the first 5 s of a session, the subject is made to relax during which the baseline EEG of the subject is recorded. Then for 30 s a set of two 'look alike' pictures which has only one single difference, are shown to the subject and he is required to identify that particular difference. For the next 40 s a mathematical puzzle appears in front of the subject and he is asked to solve that. Then again a set of picture appears for 30 s and so on. In between each of the subsequent task slides there is a blank slide of 5 s during which the subject is asked to answer. This collective set of 80 s (30 + 5 + 40 + 5) is repeated for 5 times in the experiment with different pictures and mathematical puzzles in each set. A visual cue for a single set is shown in Fig. 1.

2.2 Filtering the EEG Signal

It is known from standard literature that cognitive signals are dominant in the α (8–12 Hz) and β (16–30 Hz) band. Thus, for this study, we have designed an IIR elliptical filter of bandwidth 8–30 Hz to filter the EEG signals acquired from the amplifier. The selection of an elliptical filter was made on the basis that it possesses good frequency domain characteristics with sharp roll off and also has good attenuation of the pass- and stop-band ripples.

2.3 Feature Extraction: Wavelet Transforms

In this experiment the preprocessing of the raw data consists of three stages, of which the first one is feature extraction. Various different algorithms can be applied on the raw data set to perform feature extraction. Here, in this experiment we have used discrete wavelet transform algorithm.

Wavelet transform is basically a time-frequency domain technique and hence is found to have more advantages over other time domain techniques or frequency domain techniques, as unlike wavelet transform, these techniques lack the information about the other domain. Besides, frequency based techniques like Fourier transform, are not good at dealing EEG signals, as they are non-stationary signals. But Discrete Wavelet Transform (DWT) can combine information of both time and frequency domains and at a given instant of time, can also provide localized information related to frequency domain [12]. The energy distribution [14] equation for discrete wavelet transform is given as

$$\frac{1}{N}\sum_{t}|f(\mathrm{t})|^2=\frac{1}{N_J}\sum_{k}|a_J(k)|^2+\sum_{J=1}^{J}\left(\frac{1}{N_J}\sum_{k}|d_J(k)|^2\right) \tag{1}$$

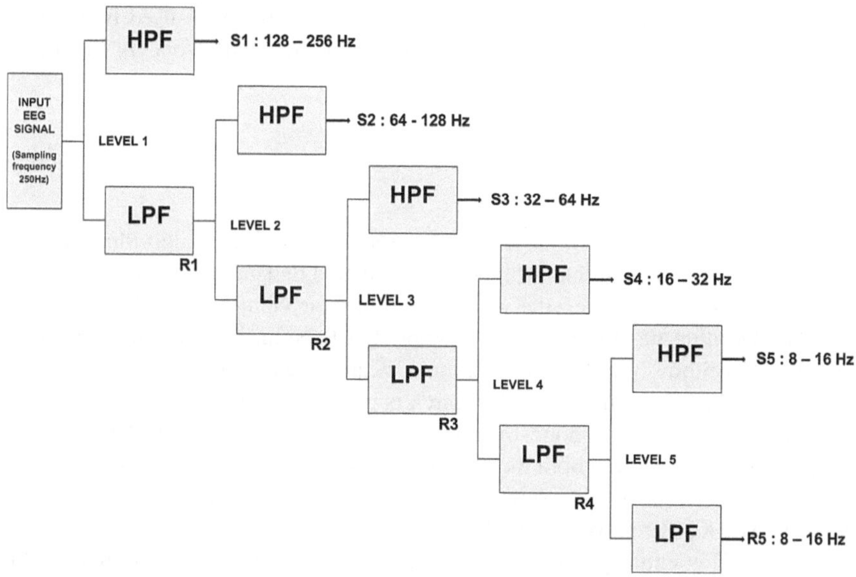

Fig. 2 The decomposition process in a discrete wavelet transform

Using (1) the features of power distribution of signals can be extracted. In DWT features are extracted by decomposing the input signals into two halves. This is done at every single level using two digital filters, a low pass and a high pass, as shown in Fig. 2. In this experiment a sampling frequency of 250 Hz is used and our required frequency band from which the signals are extracted, is of range 8–30 Hz. Hence, in order to achieve this particular frequency band, the signal has to be decomposes for 5 levels. As a single level passes by, the signal got divided into CA_i (coarse approximation) and DI_i (detailed information). The CA_i obtained from the low pass filter are further decomposed to get the subsequent levels. There are different mother (base) wavelets available, from which the Daubechies wavelet (db4) of fourth order is chosen. The outputs of levels 4 and 5 are selected post decomposition, as the desired frequency band (8–30 Hz) lies in these levels. The final dimensions of the feature vector is 7 electrodes × 35 features = 245 features.

2.4 Feature Selection

In the world of BCI, feature selection is considered to be an important step, especially while dealing with high dimensional features. The main goal is the reduction of the feature space dimension so as to get the most effective features. It so happens often that huge datasets contain a lot of irrelevant and ambiguous feature data that ultimately result in unnecessary computation complexity and lowering of classification accuracy. In this experiment, we have used two feature selection algorithms. The first one is ReliefF and the second one is Maximum Relevance Minimum Redundancy (MRMR).

2.4.1 ReliefF Feature Selector

The ReliefF algorithm [10] is a modification of the Relief algorithm. In this algorithm, the qualities of the attributes are estimated depending on how accurately they can distinguish between the values lying in their vicinity. In order to achieve that, the algorithm has to undergo two steps. Let R be an instance. In both of the steps the algorithms tries to find out the k nearest neighbors of R but the main difference is that, at the first step, the search is performed within the same class and in the next step, the it is done by taking all the classes into consideration. The nearest neighbors found in both the first and the second steps are known as nearHits and nearMisses respectively. Suppose, X is an attribute and its quality estimate is given by W [X]. Then W [X] will be updated depending on both nearHits and nearMisses. This entire process is repeated I times so that a good estimate of the weights can be obtained. The algorithm of the complete process is given below.

Algorithm 1 Relief(S,F,I,K)

S - Training Set, F - feature numbers,
I - Iteration times, K - number of nearest neighbors
Initialize all the weights W[X], to zero
For i= 1 to I **Do** begin
Randomly choose an instant Y in S
Find its K near hits and K near misses from each class different from which Y belongs to
For X=1 to F
W(X) = W(X) - Avg (diff (X,Y, near hit)) +Avg (diff (X,Y, near miss))
End
Return all weights W(X)

2.4.2 MRMR Feature Selector

In an unsupervised situation where the classifiers are not specified, minimal error requires the maximum statistical dependency of the target class *c* on the data distribution in the subspace R_m of m features. This scheme is called the maximal dependency (Max-Dependency). The most popular approach to realize Max-Dependency is maximal relevance (Max-Relevance). Some researchers have also employed minimum redundancy (Min-Redundancy) to reduce the redundancy of the feature set.

In terms of mutual information, feature selection algorithm aims to find a feature set S with m features $\{x_i\}$, which jointly have the largest dependency on the target class c,

$$maxD(S,c) = I(x_i, i = 1,\ldots,m; c) \tag{2}$$

Max-Relevance searches features satisfying (3), which approximates D(S, c) in (2) with the mean value of all mutual information values between individual features x_i and c:

$$\max D(S,c), D = \frac{1}{|S|}\sum_{x_i \in S} I(x_i; c) \tag{3}$$

It is likely that the features selected would have rich redundancy (dependency among the features are large). Thus, when two features are highly dependent on each other, the respective class discriminative power would not change much if one of them were removed. Therefore, Min-Redundancy is applied to select mutually exclusive features:

$$\max R(S), R = \frac{1}{|S|^2}\sum_{x_i, x_j \in S} I(x_i, x_j) \tag{4}$$

The criterion combining the above two constraints is called "minimal-redundancy-maximal-relevance" (*MRMR*).

Thus, the condition for any search algorithm applying this method is

$$\max\left[I(x_j; c) - \frac{1}{m-1}\sum_{x_i \in S_{m-1}} I(x_j; x_i)\right] \tag{5}$$

The computational complexity of this incremental search method is O(|S|.M), where S is the dimension of the feature set and M is the number of features. Also, a theorem exist which states that the first order incremental search, mRmR is equivalent to Max-Dependency.

2.5 Classifier: Distance Likelihood Ratio Test (DLRT)

The DLRT algorithm is a statistical tool for classification and quite efficiently used for BCI applications. This classification algorithm is a modification of the k-NN classifier and particularly suitable for the datasets whose feature distributions under the different classes are well defined and known to the user. If the probability distribution of the features is not well defined then application of DLRT may render a significant amount of error in the result. Non-parametric probability distributions are used in order to avoid such situations. This classifier first estimates the class conditional probability vector for the feature vector which is obtained from the feature extraction stage. The estimate is obtained using the following formula

$$\hat{p}(x|H_i) = frack\, N_i A(k, x) \tag{6}$$

where, k denotes the number of neighbors of x, N_i denotes the number of training points and A denotes the volume of the feature space containing the nearest neighbors (k). This equation is then used to calculate the likelihood ratio ($\hat{\lambda}$(x)) which is given by,

$$\log\left(\frac{nH_0}{nH_1}\left(\frac{\Delta k(0)}{\Delta k(1)}\right)D\right) = \log\left(\frac{nH_0}{nH_1}\right) + D\log(\log(\Delta k(0)) - \log(\Delta k(1)) \tag{7}$$

where $\log(\Delta_{k(1)})$ and $\log(\Delta_{k(0)})$ denote the log of the distances to the *k*th neighbor and D denotes the dimensionality of the feature vector. This estimated logarithmic ratio is then compared to the actual likelihood ratio λ(x) to verify whether the outputs of the DLRT are clustered mostly around the true ratio, i.e. if the algorithm has rendered satisfactory results [13].

3 Results and Discussions

In this study, the subject performed two separate cognitive tasks, viz, (i) spotting the difference from two sets of look-alike pictures, and (ii) solving a mathematical puzzle. EEG signals from 7 electrode locations are acquired for further analysis. Seven subjects performed the experiments in 3 sessions and each session consists of 10 sets of cognitive tasks (5 spotting the difference and 5 mathematical tasks). Wavelet Transform is used for feature extraction and the size of the original feature vector is 245. After introducing ReliefF and MRMR algorithm, the size of the feature vector is reduced to 5 best features. Following that, the reduced feature vector is fed as input to the DLRT classifier and the results are given below.

First, we illustrate the activity of the brain when the subject performs cognitive experiment. From Fig. 3 it is observed that the activation (as shown in red) occur mostly in the frontal (component 2, 3, 4 and 5), parietal (component 5 and 6) and temporal (component 7) regions. It is previously discussed, while performing any cognitive task, the activation of a human brain takes place mostly in the frontal, parietal and temporal lobes. This result, thus, validates our claim that the cognitive tasks are dominant in the three regions of the brain.

This study also reveals that the use of feature selection method in signal processing of EEG data, improves the results obtained. As it can be seen from Fig. 4 the accuracy increased almost 15 % on average after using feature selection. Same

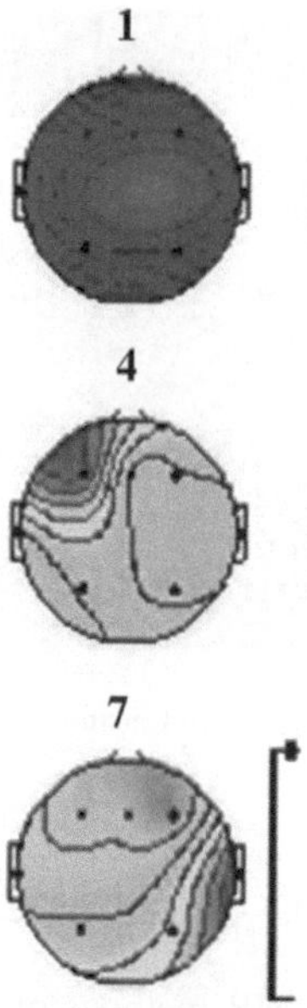

Fig. 3 An example of the activation map of the brain during the cognitive task performed by Subject *3*. *7* samples or components (denoted by numbers: *1*, *2*, …, *7* in the figure) are considered during the performance of the cognitive task. *Red marks* the maximum brain activation and *blue marks* the minimum brain activation

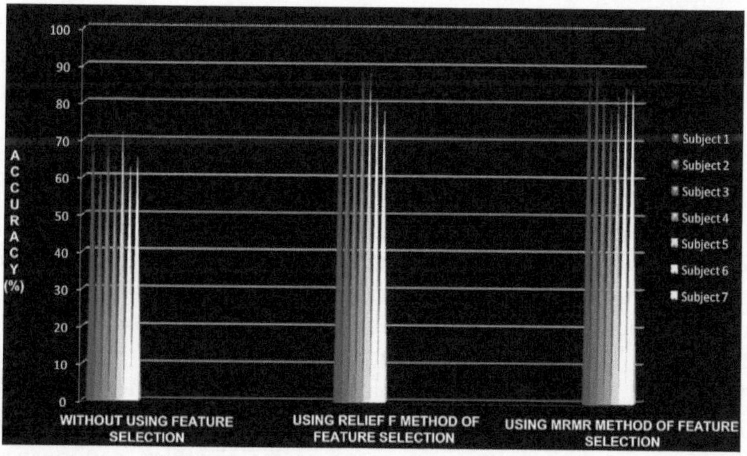

Fig. 4 Classification Accuracy (CA) of seven subjects without using any feature selection algorithm, ReliefF algorithm and MRMR algorithm

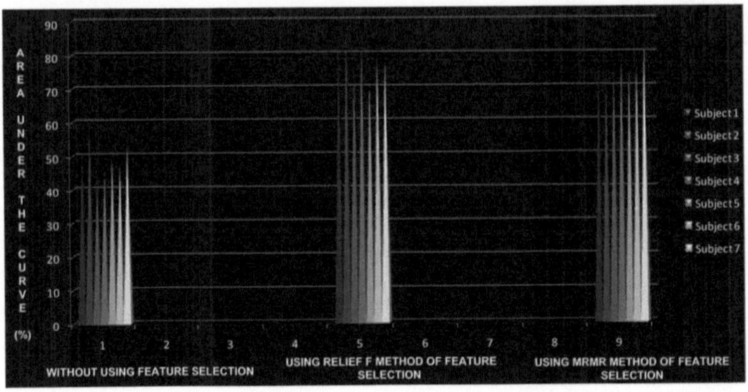

Fig. 5 Area under the curve (AUC) of seven subjects without using any feature selection algorithm, ReliefF algorithm and MRMR algorithm

is the result for area under the curve (AUC). In Fig. 5 it is evident that the AUC is almost 20 % less when feature selection is not used.

The performance measures of ReliefF and MRMR methods of feature selection are further validated by means of Friedmans Test [15]. We compare the performance, in terms of classification accuracy and AUC, of the two algorithms with other standard algorithms: Correlation Based Feature Selection (CFS), Principal Component Analysis (PCA) and Minimal Redundancy (MR) algorithms. Friedman test compares the relative performance of the two algorithms with 4 more feature selection algorithms. The null of the algorithm states that the rank of all the algorithms should be equal as all the algorithms. The Friedman statistic is given by Eq. (8).

$$\chi_F^2 = \frac{12N}{k(k+1)}\left[\sum_i R_i^2 - \frac{k(k+1)}{4}\right] \tag{8}$$

It is distributed accordingly to χ_F^2 with 1 degrees of freedom, where k is the number of algorithms to be compared and N is the number of parameters used for comparison. In this study, the mean of the two metric: classification accuracy (CA) and area under the curve (AUC), have been selected for evaluation, thus, k = 5 and N = 2 and Table 1 is the ranking table for all the algorithms.

From Table 1, the value of R_j is calculated, which is further used to obtain $\chi_F^2 = 8 > \chi_{5,0.95}^2$. Hence it can be seen that the proposition of null hypothesis that all the algorithms are equivalent is wrong and so the algorithm performances are determined from their ranks itself. It is clear from the table that the rank of MRMR is 1 and ReliefF is 2, claiming ReliefF and MRMR yields better results than its competitors. Also, MRMR works slightly better than ReliefF.

Table 1 Comparison and ranking of five feature selection algorithms

Algorithms	Mean CA (%)	Rank CA	Mean AUC (%)	Rank AUC	Average rank (R_j)
ReliefF	85.39	2	78.12	2	2
MRMR	85.77	1	80.60	1	1
CFS	81.23	3	77.56	3	3
PCA	63.45	5	60.00	5	5
MD	78.67	4	75.00	4	5

4 Conclusion

The work presented here aims to study the different brain processes during two different kinds of mental tasks, (i) while spotting the difference between two similar pictures, and (ii) while performing some a mathematical puzzle. For this purpose, we have used Wavelet Transforms for feature extraction and Distance Likelihood Ratio Test as classifier. We have also explored the change in performance of the classifier by introducing a feature selection step in between the feature extraction step and classification step. Here, we have used ReliefF and MRMR algorithm for this purpose. It is observed from the results that the accuracy of the classifiers increased by II. 15 % on inclusion of the feature selection step. From Friedman Test, it is observed that MRMR performs slightly better than ReliefF but both the algorithms are much higher as compared to other algorithms, which are, CFS, PCA and MD. From the brain activation maps, shown in results, it is noted that the parietal, frontal and temporal regions are most active.

Acknowledgments The authors would like to thank Council of Scientific and Industrial Research, India for their financial assistance.

References

1. Milner B, Squire LR, Kandel ER (1998) Cognitive neuroscience and the study of memory. Neuron 20:445–468
2. Davis CE, Hauf JD, Wu DQ, Everhart DE (2011) Brain functionwith complex decision making using electroencephalography. Int J Psychophysiol 79:175–183
3. Ramsey NF, van de Heuvel MP, Kho KH, Leijten FSS (2006) Towards human BCI applications based on cognitive brain systems: an investigation of neura signals recorded from the dorsolateral prefrontal cortex. IEEE Trans Neural Syst Rehabil Eng 14(2):214–217
4. Dauwels J, Vialatte F, Cichocki A (2010) Diagnosis of Alzheimer's disease from EEG signals: where are we standing? Curr Alzheimer Res 7(6):487–505
5. Giles GM, Radomski MV, Champagne T et al (2013) Cognition, cognitive rehabilitation, and occupational performance. Am J Occup Ther 67:S9–S31
6. Cososchi S, Strungaru R, Ungureanu A, Ungureanu M (2006) EEG features extraction for motor imagery. In: Proceedings of the 28th annual international conference IEEE engineering medicine and biology society EMBS '06, New York, USA, pp 1142–1145, 30 Aug–3 Sept, 2006

7. Schroder M, Bogdan M, Hinterberger T, Birbaumer N (2003) Automated EEG feature selection for brain computer interfaces. In: Proceedings of the 1st international IEEE EMBS conference on neural engineering, pp 626–629
8. Lotte F, Congedo M, L'ecuyer A, Lamarche F, Arnaldi B (2007) A review of classification Algorithms for EEG-based brain-computer interfaces. J Neural Eng 4(2):R1–R13
9. Koprinska I (2010) Feature selection for brain-computer interfaces. New Front Appl Data Min, Lect Notes Comput Sci 5669:106–117
10. Megchelenbrik W (2010) Relief-Based feature selection in bioinformatics: detecting functional specificity residues from multiple sequence Alignments. Radboud University, Nijmegen
11. Estevez PA, Tesmer M, Perez CA, Zurada JM (2009) Normalized mutual information feature selection. IEEE Trans Neural Netw 20(2):189–201
12. Darvishi S, Al-Ani A (2007) Brain-computer interface analysis using continuous wavelet transform and adaptive neuro-fuzzy classifier. In: Proceedings of the 29th international annual conference in IEEE engineering, medicine and biology society, pp 3220–3223
13. Remus JJ, Morton KD, Torrione PA, Tantum SL, Collins LM (2008) Comparison of a distance-based likelihood ratio test and k-nearest neighbor classification methods. In: IEEE workshop on machine learning for signal processing MLSP 2008, pp 362–367
14. Bhattacharyya S, Rakshit P, Konar A, Tibarewala DN, Janarthanan R (2013) Feature selection of motor imagery EEG signals using firefly temporal difference Q-Learning and support vector machine. In: Panigrahi B, Suganthan PN, Das S, Dash SS (eds) Swarm, evolutionary, and memetic, computing, lecture notes in computer science, vol 8298. Springer International Publishing, Switzerland, pp 534–545
15. Bhattacharyya S, Konar A, Tibarewala DN (2014) A differential evolution based energy trajectory planner for artificial limb control using motor imagery EEG signal. Biomed Signal Process Control 11:107–113

Generalised Orthogonal Partial Directed Coherence as a Measure of Neural Information Flow During Meditation

Laxmi Shaw, Subodh Mishra and Aurobinda Routray

Abstract Neural information flow in brain during meditation, can be addressed by brain connectivity studies. This work aims to obtain neural connectivity measures based on a strictly causal time varying Multi-Variate Auto-Regressive (MVAR) model, fitted to EEG signals obtained during meditation. The time varying Granger Causality based connectivity estimators as PDC (Partial Directed Coherence), g-PDC (generalized Partial Directed Coherence), OPDC (Orthogonalized Partial Directed Coherence) and g-OPDC (generalized Orthogonalized Partial Directed Coherence) are calculated using the adaptive autoregressive MVAR parameters. The MVAR model parameters have been estimated by Kalman Filter algorithm. In this work g-PDC and g-OPDC have been used to make the connectivity measures scale invariant. These connectivity estimators quantify the neural information flow between Electroencephalograph (EEG) channels. In addition g-OPDC is also immune to volume conduction artifact and gives better result compared to g-PDC. Finally, surrogate data statistics has been used to check the significance of the above connectivity estimators.

Keyword Coherence · EEG · Time varying auto regressive model · Connectivity measures · Kalman filter

L. Shaw (✉)
Silicon Institute of Technology, Bhubaneswar, India
e-mail: laxmishaw1983@gmail.com

S. Mishra
Pune, India

A. Routray
Indian Institute of Technology, Kharagpur, Kharagpur, India
e-mail: aroutray@ee.iitkgp.ernet.in

S. Gupta et al. (eds.), *Advancements of Medical Electronics*,
Lecture Notes in Bioengineering, DOI 10.1007/978-81-322-2256-9_13

1 Introduction

Frequency domain study of brain connectivity has assumed great importance in the recent years. There are many conventional frequency domain estimators of connectivity such as coherence and partial coherence for calculating coupling and direct coupling respectively [1]. Coherence and partial coherence being symmetric, give no information regarding the direction of information flow. The direction of neural information flow helps in determining causality.

A time series (e.g.: a single EEG channel data) can be said to cause another series when the information in the past of the former series can help in predicting the present value of the latter series. This is based on the famous Granger Causality principle which has been found to be very successful in econometric causality analysis.

Applications of Granger Causality (GC) are profoundly used in neuroscience. Granger Causality based brain connectivity measures are Directed Transfer Function (DTF), Partial Directed Coherence (PDC) and generalized-Partial Directed Coherence (g-PDC). They have also been established in [2–4], The generalized PDC (or g-PDC) is a scale invariant version of PDC and it is immune to static gain.

The connectivity measures discussed above are essentially derived from a strictly causal MVAR (Multi-Variate Auto-Regressive) model fitted to the multi-channel EEG data [1–4].

The estimators are generally defined for stationary, time invariant linear signals. This makes their application to EEG signals a challenging task because of EEG's nonlinearity and non-stationarity [5, 6]. To overcome this problem of non-stationarity, a strictly causal time varying MVAR model has been used in [7−9]. In this paper, we have considered only lagged effects; hence the proposed model is strictly causal MVAR model. The effects of instantaneous or zero lag causation can be modeled using an extended MVAR model [10]. PDC is non-zero only when a direct causal link between two channels exist, while DTF is non-zero if any causal pathway (both direct and indirect (or cascaded)) occurs between two channels. Mathematically, PDC is a result of the factorization of the partial coherence (PCoh) function [1]. DTF is a particularization of another causality measure i.e. Directed Coherence (DC) [1]. The PDC combines the qualities of partial coherence (PC) and direct coherence (DC) [1]. Major problem in estimating true connectivity measures are: (1) Effects of volume conduction and, (2) Different artifacts which includes electro galvanic signals (slow artifacts, movement artifacts and frequency artifacts) associated with brain signals. Volume conduction is the effect of a single source. The effects of volume conduction are seen in most of the electrodes [11]. Hence two or more electrodes may appear to be connected due to the underlying effect of volume conduction but such a connection is not a true indicator of interaction among electrodes.

An optimum estimator of brain connectivity should mitigate the effects of the volume conduction. But the MVAR model parameters that we measure are sensitive to volume conduction [12]. This problem can be overcome by connectivity analysis at the source level, which requires highly reliable source localization techniques [13, 14]. Spurious connectivity patterns which occur due to volume conduction can be overcome by orthogonalizing signal powers [15]. In the imaginary part of coherence between two channels is found to be free ofvolume conduction artifact [11]. A recent study combines orthogonalization and imaginary part of coherence to obtain an orthogonalized version of classical PDC called the Orthogonalized Partial Directed Coherence (OPDC). The OPDC is not affected by volume conduction artifacts. Hence it has the potential to be a true estimator of brain connectivity among EEG channels. Analogous to the definition of gPDC, the gOPDC can be formulated from OPDC [16].

This paper is organized as follows. In Sect. 1 the data collection method has been discussed. Then in Sect. 2 clear description of the EEG datasets and the preprocessing techniques are given. Section 3 throws some light on the MVAR model fitting on the EEG data and various connectivity estimators are briefly discussed. Section 4 then explains about the surrogate data method for statistical validation of obtained results. In Sect. 5 summarizes the results of this connectivity study. The paper concludes in Sect. 6.

2 Data Description and EEG Signal Preprocessing

EEG data has been collected from 23 meditating subjects. The data is collected at a sampling frequency of 256 Hz. To include all the five bands of brain waves, data from each channel is band pass filtered within 0–64 Hz. A notch filter has been used to remove the 50 Hz power-line interference frequency component from the EEG signal. The baseline has also been removed from the data. The five bands of brain waves are delta (0.4–4 Hz), theta (4–8 Hz), alpha (8–16 Hz), beta (16–32 Hz) and gamma (>32 Hz) bands [17]. Artifacts from the EEG data has been removed by using the technique of wavelet thresholding [18] using the 'db4' mother wavelet and scaling function. The db4 mother wavelet has been used owing to its structural similarity with the rhythmic EEG. The general block diagram of preprocessing technique has been given as in Fig. 1.

3 MVAR Model Fitting and Time Varying Connectivity Measures

Consider a multichannel EEG data with M channels and N data points. The strictly causal MVAR model that we fit into the above data set is of the form:

$$y(n) = \sum_{k=1}^{p} A(k)y(n-k) + w(n) \quad (1)$$

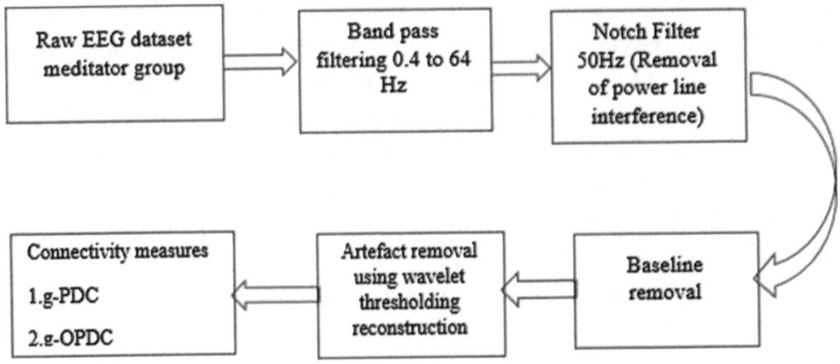

Fig. 1 General Block diagram of preprocessing techniques and followed by connectivity measures using g-PDC and g-OPDC

Here, $y(n) = [y_1(n), y_2(n),\ldots, y_M(n)]^T$ is a M × 1 vector of present values of the model, $y(n-k)) = [y_1(n-k), y_2(n-k),\ldots\ldots y_M(n-k)]^T$ is a M × 1 vector of values of y(n) at lag k, A(k) is a M × M coefficient matrix, p is the model order, which denotes the maximum lag of the model, $w(n) = [w_1(n), w_2(n), \ldots, w_M(n)]^T$ is a M × 1 vector of uncorrelated white noise, w(n) is also called innovation process. Since it is a matrix of uncorrelated noise, the correlation matrix given by $R_w = E[w(n)w^T(n-k)]$ is zero for all positive lags except for k = 0, where it equals the covariance matrix $\sum = \mathrm{cov}(w(n))$. $\sum$ is a diagonal matrix containing the variances of the innovation process as the diagonal elements. The diagonality of $\sum$ ensures that instantaneous effects are absent in the MVAR model described above [10].

$A(k)$, the coefficient matrix for a lag k is of the form:

$$A(k) = \begin{bmatrix} a_{11}(k) & \cdots & a_{1M}(k) \\ \vdots & \ddots & \vdots \\ a_{M1}(k) & \cdots & a_{MM}(k) \end{bmatrix} \tag{2}$$

The real parameters $a_{ij}(k)$ of A(k) matrix find the relation between time series i and j at a lag k. If $a_{ij}(k)$ is non-zero for at leas tone lag k, then series j is said to cause series i.

For all k varying from 1 to p we have different values of A(k). Hence, the total number of parameters to be estimated is M × Mp.

For reliable estimation of model parameters that give a good estimation of the actual data the number of data points MN must be significantly larger than number of parameters to be estimated [19], i.e.

$$MN \geq M^2 p \tag{3}$$

This implies N ≥ MP. The optimum model order p can be chosen using different information theory based criteria, the AIC (Akaike Information Criterion) and SBC (Schwarz Bayesian Information Criterion), to name a few. In [20] the SBC is found to outperform AIC for time series analysis. The model order must be high enough to account for all the delays and fluctuations in the original time series and low enough to allow authentic model identification from the measured data [19].

"Equation (1)" is the strictly causal MVAR model for time invariant systems. The time varying form of (1) can be written as:

$$y(n) = \sum_{k=1}^{p} A(n,k)y(n-k) + w(n) \tag{4}$$

Here in (4) the coefficients in the parameter matrix A(n, k) are time varying for all lag k. The time varying parameters take care of the non-linearity of the EEG channels. The time varying MVAR model given in (4) is fit into the data by using Adaptive Auto-Regressive (AAR) modeling algorithm. This algorithm uses Kalman filtering for estimating the time varying model parameters [20]. The kalman filtering based parameter estimation has been done using the 'mvaar' module of the BIOSIG toolbox. The model order p has been estimated by using the ARFIT module [21]. The model order is kept constant throughout the analysis. Optimal model order depends on the sampling rate and higher sampling rate often requires high model order.

The frequency domain representation of (4) is given as:

$$Y(f) = \sum_{k=1}^{p} (A(n,k)e^{-i2\pi fk})Y(f) + W(f) \tag{5}$$

Let

$$A(n,f) = \sum_{k=1}^{p} A(n,k)e^{-i2\pi fk} \tag{6}$$

$A(n, f)$ is a M × M matrix with each element $A_{ij}(n, f)$ given as

$$Aij(n,f) = \sum_{k=1}^{p} aij(n,k)e^{-i2\pi fk} \tag{7}$$

Using (5) and (6) we have:

$$Y(f) = A(n,f)Y(f) + W(f) \tag{8}$$

Taking $A(n, f)$, $Y(f)$ to the L.H.S., we have

$$(I - A(n,f))Y(f) = W(f) \tag{9}$$

Here I is a M × M identity matrix
Let

$$\hat{A}(n,f) = (I - A(n,f)) \tag{10}$$

Using (9) and (10), we have

$$\hat{A}(n,f)Y(f) = W(f) \tag{11}$$

The frequency domain MVAR equation in the form as in (11) can be used to define most of the strictly causal connectivity estimators [1].

We will directly proceed to write down the formulae of time varying PDC, g-PDC, OPDC and g-OPDC with the requisite explanation.

3.1 Time Varying PDC

The time varying version of PDC is given as:

$$\pi kl(n,f) = \frac{\left|\hat{A}kl(n,f)\right|}{\sqrt{\sum_{m=1}^{M}\left|\hat{A}ml(n,f)\right|^2}} \tag{12}$$

Here πkl measures the amount of time varying information flow from y_l *to* y_k through direct transfer path only relative to the total outflow leaving the structure at which y_k is measured [1].

Direct transfer path implies direct causality. The PDC measure does not take into consideration any cascaded path, thus being different from DTF. But this classical form of PDC is not scale invariant. It is affected by amplitude scaling which does not affect the causality structure [4]. To overcome this problem the g-PDC was developed.

3.2 Time Varying G-PDC

A scale invariant version of PDC can be given as [4]:

$$\dot{\pi} kl = \frac{\left|\gamma k^{-1}\hat{A}kl(n,f)\right|}{\sqrt{\sum_{m=1}^{M}\gamma m^{-2}\left|\hat{A}ml(n,f)\right|^2}} \tag{13}$$

Here γk^2 refers to the variance of the innovation processes $w_k(n)$. This is called the generalized PDC or simply g-PDC. The physical interpretation of g-PDC is same as that of PDC but the g-PDC is invariant to any amplitude scaling.

3.3 Time Varying Orthogonalized-PDC

The orthogonalized PDC is the result of a recent work [22]. The main concept behind OPDC and g-OPDC is instead of performing orthogonalization process at amplitude level, it is done at the level MVAR coefficients to mitigate the effect of volume conduction (or mutual sources). As given in [22] the time varying OPDC is defined as:

$$\alpha kl(n,f) = \frac{|Real(\hat{A}kl(n,f))|}{\sqrt{\sum_{m=1}^{M}|\hat{A}ml(n,f)|^2}} \frac{|imag(\hat{A}kl(n,f))|}{\sqrt{\sum_{m=1}^{M}|\hat{A}ml(n,f)|^2}} \tag{14}$$

here $k \neq l$.

The physical interpretation is same as classical PDC except that the OPDC does not take into consideration the effect of mutual sources.

3.4 Time Varying Generalized−Orthogonalized-PDC

The time varying g-OPDC is the scale invariant version of OPDC [16]. It is defined as:

$$\dot{\alpha} kl(n,f) = \frac{1}{\gamma k^2} \frac{|Real(\hat{A}kl(n,f))|}{\sqrt{\sum_{m=1}^{M}\frac{1}{\gamma m^2}|\hat{A}ml(n,f)|^2}} \frac{|imag(\hat{A}kl(n,f))|}{\sqrt{\sum_{m=1}^{M}\frac{1}{\gamma m^2}|\hat{A}ml(n,f)|^2}} \tag{15}$$

here $k \neq l$.

Interpretation of this formula is same as OPDC except that this one is scale invariant.

4 Statistical Assessment and Its Significance for connectivity Measures

The PDC measures calculated in this paper have a nonlinear relation to the time series data from which they are derived [23] hence the probability distributions of their estimators are not well defined, making the statistical testing of significance very difficult. In this paper we have used the surrogate data method [24].

4.1 Surrogate Data Analysis

Surrogate data is the random data generated, keeping the mean, variance and autocorrelation function as same as the original data and this techniques of time series can be used to test for nonlinear dynamics [25]. The data points in all the channels are randomly permutated to remove any causal ordering. Then a time varying MVAR model is fit to this shuffled data and the connectivity measures are calculated from it. This process is undertaken several times to create an empirical distribution for connectivity measures. The estimators calculated from the surrogate data set serve the null hypothesis which assumes that there is no causal relationship between the channels of the data set. Using this new distribution we can assess the significance of causal measures calculated from actual data. The method has been validated in [2] and found to be effective.

Hence, the basic problem in our analysis is to check the null hypothesis H_0:

$$H_0 : \pi_{kl}(n,f) = 0 \tag{16}$$

This hypothesis if rejected implies the existence of direct information flow from y_l *to* y_k. And this happens when $a_{kl}(k)$ is non-zero at least for one k in [1, p].

5 Results and Discussion

To study intra and inter hemispheric interaction, eight electrodes have been selected out of 64 [24]. The electrodes are F3, F4, Fz, C3, C4, P3, P4 and Pz. Here these eight electrodes represent the midline hemisphere of the brain. The result of g-PDC and g-OPDC for one of the meditators are given below. Figure 2 is the output for g-PDC estimator which shows the amount of information flow out of an electrode to another electrode in the direction of the arrow shown in the figure. The magnitude of information flow for each corresponding time frequency point is shown by the color in the plot. Red stands for high value of information flow while blue stands for negligible information flow. The x axis of each subplot is samples and the y axis is the frequency axis. For each time frequency pair we find the amount of information flow. From the figure it is clearly visible that g-PDC is not a symmetric measure which clearly implies

$$\text{g-PDC}_{ij} \neq \text{g-PDC}_{ji} \tag{17}$$

The figures are testimony to the directional nature of g-PDC and other PDC based measures (Fig. 3).

According to the g-PDC figure, we see some information flow from F4 to almost all electrodes except Fz. Similarly we find significant information flow from F4 to almost all other electrodes. But the g-OPDC measure is not as colored as the g-PDC

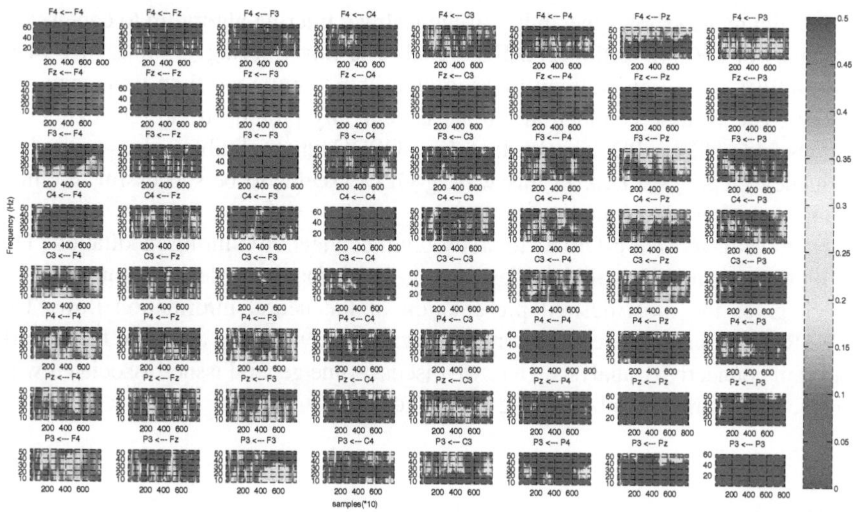

Fig. 2 Time varying connectivity analysis of the selected 8 scalp EEG electrodes: g-PDC measure

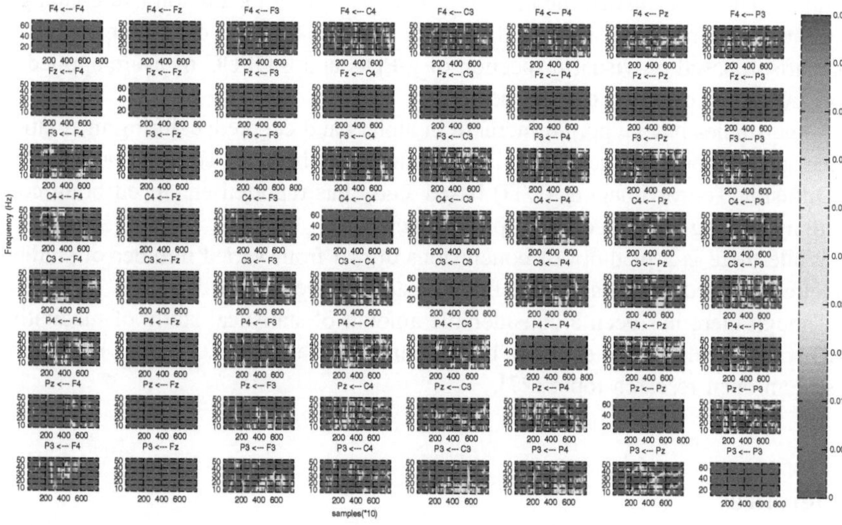

Fig. 3 Time varying connectivity analysis of the selected 8 scalp EEG electrodes: g-OPDC measure

because the g-OPDC does not take into consideration the effect of volume conduction and for this reason g-OPDC has lighter shades as compared to g-PDC plots. All the interpretations written above for g-PDC also hold good for g-OPDC. Also the magnitude of g-OPDC is significantly lower compared to g-PDC, which is evident from the color bars present in both the figures.

The figure also shows the ability of the Kalman filtering algorithm to capture the dynamics of the non-stationary EEG signal. The time varying versions of g-PDC and g-OPDC based on Kalman filtering algorithm show good time frequency resolution. The volume conduction artifact is generally present in the low frequency regions and unlike g-PDC we see that the g-OPDC has negligible values in the low frequency range ($\sim$3 Hz). In this way, g-OPDC seems to be successful in eliminating the effect of volume conduction.

Here we have used g-OPDC to study brain connectivity during meditation. The g-OPDC measure takes time series scaling and information leakage into account which gives the most desired presentation of the neural information flow. The MVAR model based coherence connectivity estimator offers a common framework for brain connectivity analysis. Here we discuss some general issues associated with connectivity comparisonbetween g-PDC and g-OPDC.

6 Conclusion

This paper measures the brain connectivity during meditation. Not many studies in literature have investigated the brain connectivity during meditation using Granger causality measures using PDC and its other variants on EEG data. Our study is a comparison based on visual inspection of g-PDC and g-OPDC measure applied to the EEG signal captured during meditation.

Some studies have reported increase in alpha band coherence during meditation while some reported reduced functional connectivity between cortical sources [26]. A Diffusion Tensor Imaging (DTI) based study has reported enhanced brain connectivity in long term meditation practitioners.

While there is a plethora of conclusions drawn from varied number of studies, the effect of mediation on human brain is still not well defined.

Though there has been a considerable amount of statistical study of meditative EEG waves, the study of effective brain connectivity using EEG during meditation is still an open research topic [27].

References

1. Laskovski A (ed) (2011) Multivariate frequency domain analysis of causal interactions in physiological time series. Biomedical Engineering trends in electronics, communication and software, Intech, pp 403–428
2. Kaminiski M, Ding M, Truccolo WA, Bressler SL (2001) Evaluating causal relationship in neural systems: granger causality, DTF and statistical assessment of significance. Biol Cybern 85:145–157 (Springer verlag)
3. Baccala LA, Sameshima K (2001) Partial directed coherence: a new concept in neural structure determination. Biol Cybern 84:463–474 (Springer verlag)

4. Baccala LA, Sameshima K (2007) Generalized partial directed coherence. In: 15th International IEEE conference on digital signal processing, pp 163–166
5. Rankine L, Stevenson N, Mesbah M, Boashash B (2007) A non-stationary model of new born EEG. IEEE Trans Bio Med Eng 54(1):19–28
6. Ting CM, Salleh SH, Zainuddin ZM, Bahar A (2011) Spectral estimation of non stationarity of EEG using particle filtering with application to event related desynchronization. IEEE Trans Bio Eng 58(2):321–331
7. Hesse W, Moller E, Arnold M, Schack B (2003) The use of time-variant EEG Granger causality for inspecting directed interdependencies of neural assemblies. J Neurosci Methods 123:27–44
8. Astolfi L, Cincotti F, Mattia D, Fallani F, Tocci A, Colosimo A, Salinari S, Marciani MG, Hesse W, Witte H, Ursino M, Zavaglia M, Babiloni F (2008) Tracking the time varying cortical connectivity patterns by adaptive MV estimation. IEEE Trans Biomed Eng 55(3):902–913
9. Sommerlade L, Henschel K, Wohlmuth J, Jachan M, Amtage F, Hellwig B, Lückin C, Timmer J, Schelte B (2009) Time variant estimation of directed influences during parkinsonian tremor. J Physiol 103(6):348–352
10. Faes L, Nollo G (2010) Extended causal modelling to assess PDC in multiple time series with significant instantaneous interaction. Biol Cybern Biol Cybern 103(5):387–400
11. Nolte G, Bai O, Wheaton L, Mari Z, Vorbach S, Hallett M (2004) Identifying true brain interaction from EEG data using the imaginary part of coherency. Clin Neurophysiol 115:2292–2307
12. Gomez G (2010) Brain connectivity analysis with EEG. Doctoral dissertation, Department of Signal Processing, Tampere University of Technology
13. Brookes MV, Woolrich MJ, Luckhoo M, Price H, Hale D, Stephenson JR, Barnes MC, Smith GR, Morris SM, Peter G (2011) Investigating the physiological basis of resting state networks using MEG. Proc Natl Acad Sci USA 108(40):16783–16788
14. Palva S, Kulashekhar S, Hmalinen, Palva JM (2010) Neural Synchrony reveals working memory networks and predicts individual memory capacity. Proc Natl Acad Sci USA 107:7580–7585
15. Hipp JF, Hawellek DJ, Corbetta M, Siegel M, Engel AK (2012) Large scale cortical correlation structure of spontaneous oscillatory activity. Nat Neurosci 15:884–890
16. Omidvarnia A, Azemi G, Boashash B, O'Toole JM, Colditz P, Vanhatalo S (2014) Measuring time varying information flow in scalp EEG signals: orthogonalized partial directed Coherence. IEEE Trans Biomed Eng 61(3):680–693
17. Murugappan M, Nagarajan R, Yaccob S Discrete wavelet transform based selection of salient EEG frequency band for assessing human emotion. http://cdn.intechopen.com/pdfs.wm/19508.pdf
18. Kumar P, Arumuganathan R, Sivakumar K, Vimal C (2008) A wavelet based statistical method for denoising of ocular artifacts: artifact in EEG signals. Int J Comput Sci Netw Secur 8(9):87–92
19. Hytti H, Takalo R, Ihalainen H (2006) Tutorial on multivariate autoregressive modeling. J Clin Monit Comput 20(2):101–108
20. Koehler AB, Murphree ES (1988) A comparison of akaike and Schwarz criteria for selecting model order. Appl Stat 37(2):187–195
21. Arnold M, Miltner W, Witte H, Bauer R, Braun C (1998) Adaptive AR modeling of non-stationary time series by means of kalman filtering. IEEE Trans Biomed Eng 45(5):553–562
22. Schneider T, Neumaier A (2001) Algorithm 808: ARFIT-a Matlab package for the estimation of parameters and eigenmodes of multivariate autoregressive models. ACM-Trans Math Softw 27:58–65
23. Omidvarnia A, Azemi G, Boashash B, Toole J, Colditz P, Vanhatalo S (2012) Orthogonalized partial directed coherence for functional connectivity analysis of newborn. Neural Inf Proc EEG 7664:683–691
24. Theiler J, Eubank S, Longtin A, Galdrikian B, Farmer J (1992) Testing of non-linearity in time series: the method of surrogate data. Physica D 58:77–94

25. Theiler J, Eubank S, Longtin A, Galdrikian B, Farmer D (1992) Testing for nonlinearity in time series: the method of surrogate data. Physica D 58(92):77–94
26. Hebert R, Lehmann D, Tan G, Travis F, Alexander A (2005) Enhanced EEG alpha time-domain phase synchrony during transcendental meditation: implications for cortical integration theory. J Signal Process 85(11):2213–2232
27. Tang Y, Rothbart M, Posner MI (2012) Neural correlates of establishing, maintaining, and switching brain states. Trends Cogn sci 16(6):330–337

An Approach for Identification Using Knuckle and Fingerprint Biometrics Employing Wavelet Based Image Fusion and SIFT Feature Detection

Aritra Dey, Akash Pal, Aroma Mukherjee and Karabi Ganguly Bhattacharjee

Abstract Identification is an essential part of our lives. Identification of authentic candidate is essential in E-commerce, in keeping track of criminals, in airport and railway surveillance and many more aspects of the modern world. In this paper an approach has been proposed combining the knuckle and fingerprint biometrics for the purpose of identification of a person using SIFT (Scale Invariant Feature Transform).

Keywords SIFT · DWT · Knuckle-fingerprint

1 Introduction

In the field of identification, biometric identification techniques such as Fingerprint recognition, Facial recognition, Voice pattern, Handwritten Signature, Retina recognition, Iris recognition are well established. However, none of the biometrics is giving cent percent accuracy. Multi biometric systems [1] remove some of the drawbacks of the uni-modal biometric systems by acquiring multiple sources of information together in a group, which has richer detail. Utilization of these

A. Dey (✉)
Department of Electrical Engineering, JIS College of Engineering, Kalyani, India
e-mail: aritradey33@gmail.com

A. Pal
Department of Computer Science and Engineering, JIS College of Engineering, Kalyani, India
e-mail: akashpa93@gmail.com

A. Mukherjee · K.G. Bhattacharjee
Department of Biomedical Engineering, Kalyani, India
e-mail: aroma.mukherjee@yahoo.com

K.G. Bhattacharjee
e-mail: kg73_kin@yahoo.co.in

S. Gupta et al. (eds.), *Advancements of Medical Electronics*,
Lecture Notes in Bioengineering, DOI 10.1007/978-81-322-2256-9_14

biometric systems depends on more than one physiological or behavioral characteristic for enrollment and verification/identification [2]. Multi-resolution approach in wavelet is well suited to manage the different image resolutions. Many research works have studied on multi-resolution representation of signals and have established that multi-resolution information for a number of image processing applications including the image fusion. Wavelet coefficients coming from different images can be appropriately combined to obtain new coefficients, so that the information in source images is collected appropriately. The discrete wavelet transform (DWT) allows the image decomposition in different kinds of coefficients preserving the image information. A wavelet-based image fusion method is therefore required to identify the most important information in the input images and to transfer it into the fused image [3].

Zhang et al. proposed a new kind of biometric identifier, called Finger-Knuckle-Print (FKP) an alternate for personal identity authentication. In those works [4–8], they used different techniques for Feature Detection and matching. In general, most of the previous SIFT FKP based authentication systems claim that they have achieved greater than 98 % accuracy. In all the previous works only knuckle has been used. This work focuses in exploring new possibilities of identification by combing the knuckle and finger print.

2 Wavelet Decomposition for Images

This process basically takes two images of knuckle and finger print [9] after wavelet decomposition is performed on both images followed by a fusion of decomposition of two images which produces fused image of low resolution. It has the capability to provide good localization for both frequencies and space domains. The wavelet based image fusion would be applied to two dimensional multispectral knuckle and fingerprint at each level [10].

3 SIFT Descriptor

The scale invariant feature transform, called SIFT [11, 12] descriptor, has been proposed by and proved to be invariant to image rotation, scaling, translation, partly illumination changes. Following are the major stages of computation used to generate the set of image features.

3.1 Scale Space Extrema Detection

The first stage of computation is to create a scale space of images. This is done by constructing a set of progressively Gaussian blurred images with increasing values of sigma. Then the difference between pairs of Gaussian is taken to obtain a Difference of Gaussian (DOG) which is similar to the function Laplacian of Gaussian (LOG) to obtain potential locations for finding features. The image is then sub-sampled (i.e. 1/4th resolution of lower octave) to obtain the next octave and the same process is repeated to obtain DOG pyramid (Fig. 1).

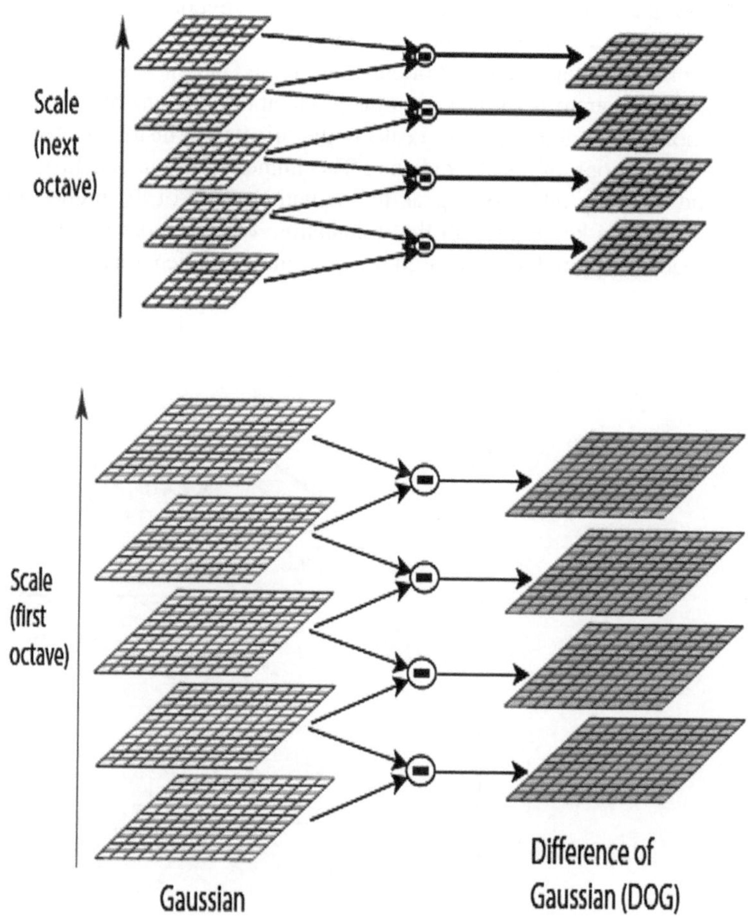

Fig. 1 Operations within same octaves (same scale)

3.2 Keypoint Localization

Accurately locates the feature keypoints by comparing a pixel (X) with 26 pixels in current scale and adjacent scales (Green Circles). The pixel (X) is selected if it is larger/smaller than all 26 pixels. There are still a lot of points; some of them are not good enough. The locations of keypoints may be not accurate. Eliminating edge points, keypoints are selected from the extrema based on measures of their stability.

3.3 Orientation Assignment

This step assigns orientation to each keypoint, the keypoint descriptor can be represented relative to this orientation and therefore achieve invariance to image rotation. It computes magnitude and orientation on the Gaussian smoothed images. An orientation histogram is formed from the gradient orientations of sample points within a region around the keypoint. Peaks in the orientation histogram correspond to dominant directions of local gradients. The highest peak in the histogram is detected, and then any other local peak that is within 80 % of the highest peak is used to also create a keypoint with that orientation. One or more orientations are assigned to each keypoint location based on local image gradient directions (Fig. 2).

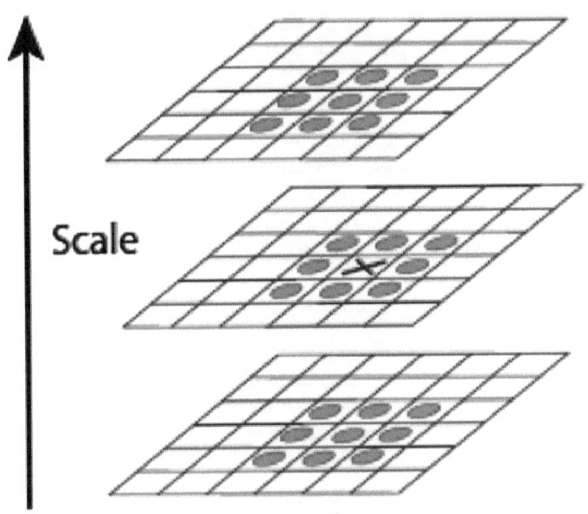

Fig. 2 Operations between different octaves (different scale)

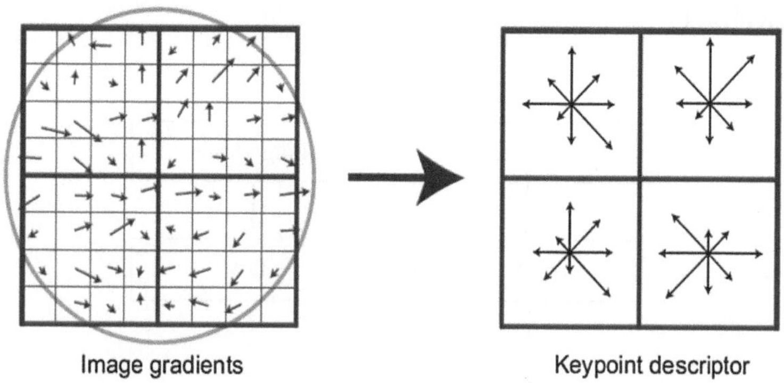

Fig. 3 The keypoint descriptor

3.4 Keypoint Descriptor

This step describes the keypoint as a high dimensional vector. The local image gradients are measured at the selected scale in the region around each keypoint. It computes relative orientation and magnitude in a 16 × 16 neighborhood around each keypoint. It forms weighted histogram (8 bin) for 4 × 4 regions. Finally it concatenates 16 histograms in one long vector of 128 dimensions.

These are transformed into a representation that allows for significant levels of local shape distortion and change in illumination. This approach has been named the Scale Invariant Feature Transform (SIFT), as it transforms image data into scale invariant coordinates relative to local features. An important aspect of this approach is that it generates large number of features that densely cover the image over the full range of scales and locations (Fig. 3).

4 Proposed Method

1. Two images (i.e. knuckle and fingerprint) of same size are read and 1-level wavelet decomposition is performed for both images.
2. Fuse synthesized images.
3. Perform SIFT on the fused image.
4. Save extracted corners saved as a feature point for tracking and recognizing objects in the database for matching.

5 Flowchart

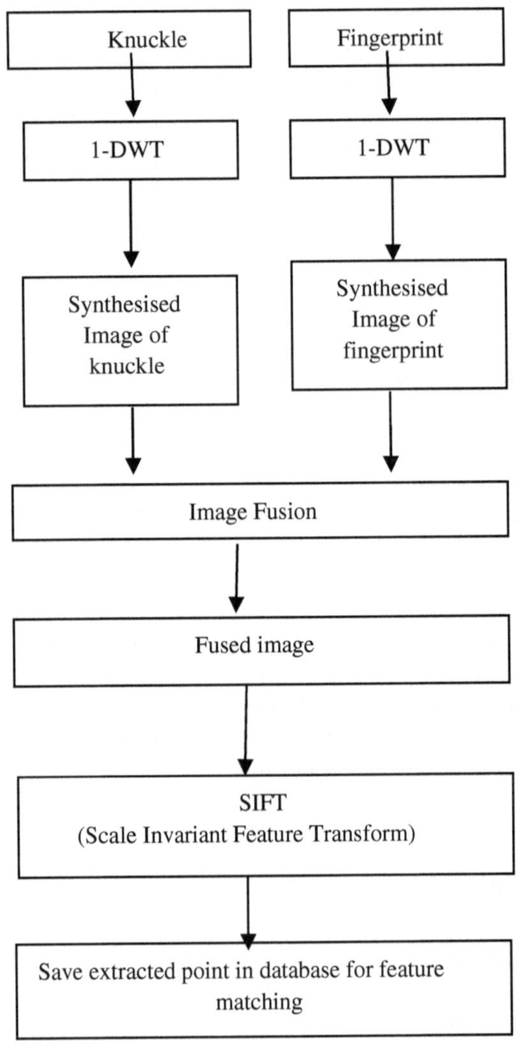

6 Steps of Matching

1. Matching function reads two images, finds their SIFT features, and displays lines connecting the matched Keypoints. A match is accepted only if its distance is less than distance Ratio times the distance to the second closest match. It returns the number of matches displayed.

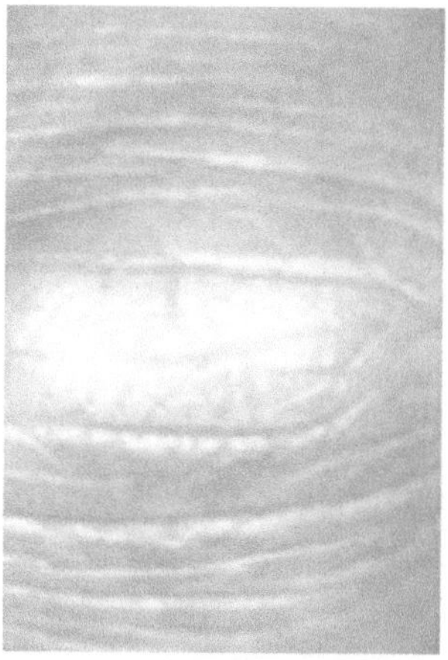

Fig. 4 Knuckle

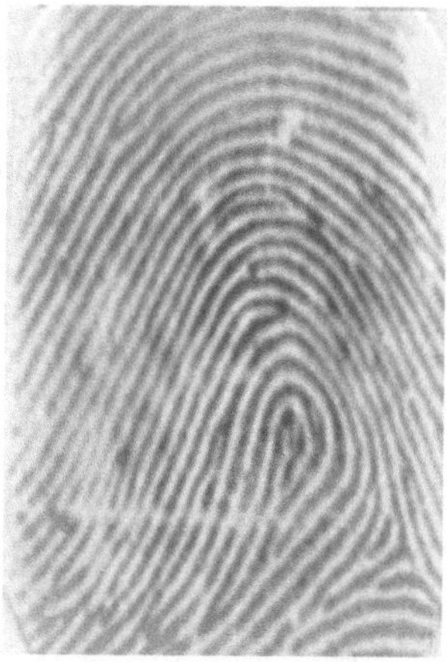

Fig. 5 Fingerprint

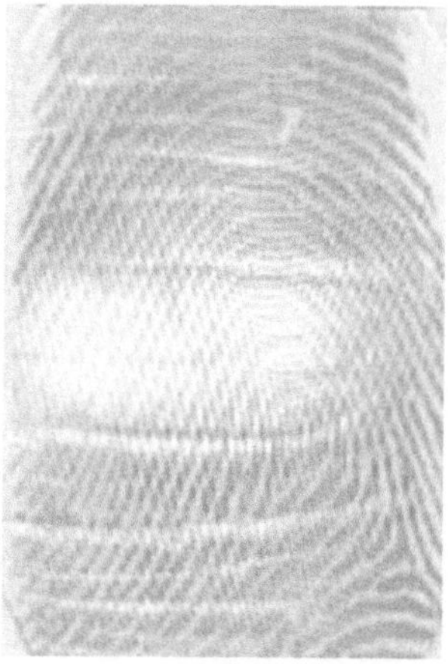

Fig. 6 Synthesized image

Fig. 7 Extracted SIFT points

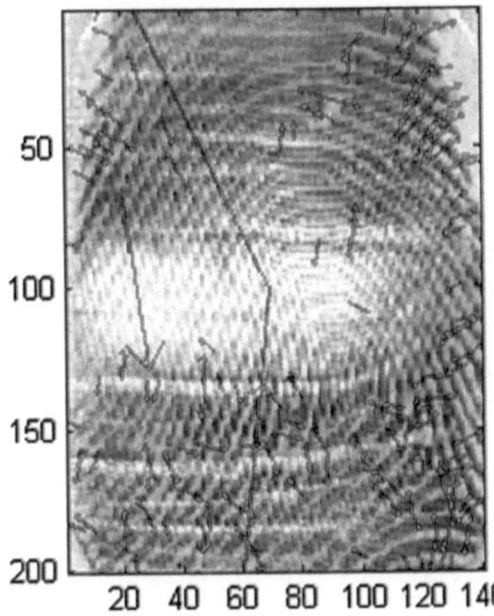

2. Matching function finds SIFT (Scale Invariant Fourier Transform) Keypoints for each image.
3. Assume some distance ratio for example suppose distance ratio = 0.4 it means that it only keep matches in which the ratio is less than distance Ratio = 0.5. Now for each descriptor in the first image, it selects its match to second image.
4. Then it creates a new image showing the two images side by side with lines joining the accepted matches (Figs. 4, 5, 6 and 7).

7 Matching

See (Figs. 8, 9, 10 and 11).

Fig. 8 Normal matching

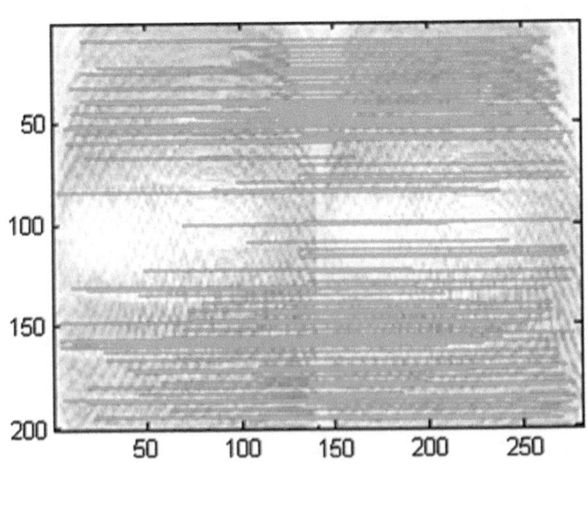

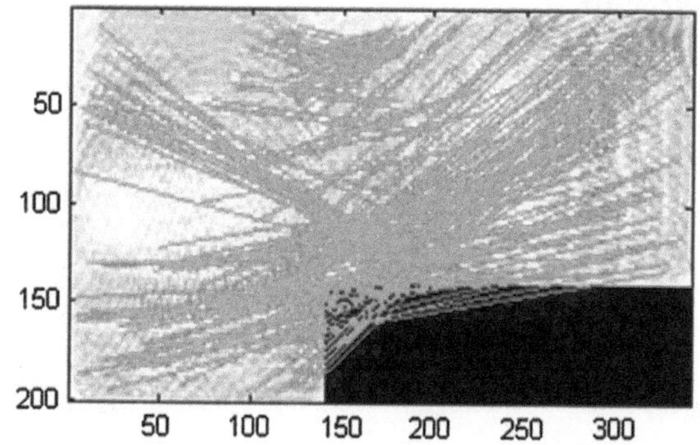

Fig. 9 Image rotated by 90°

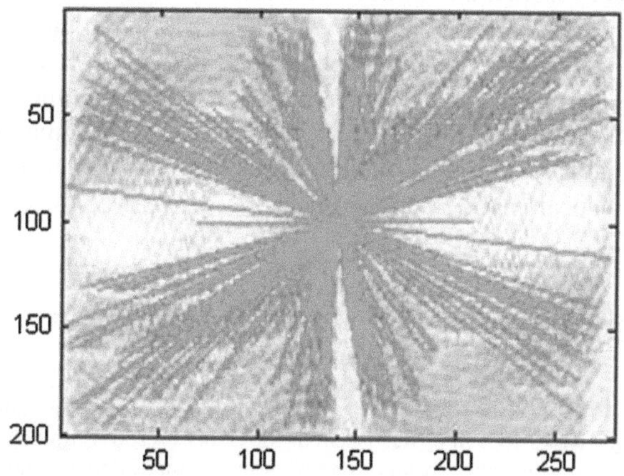

Fig. 10 Image rotated by 180°

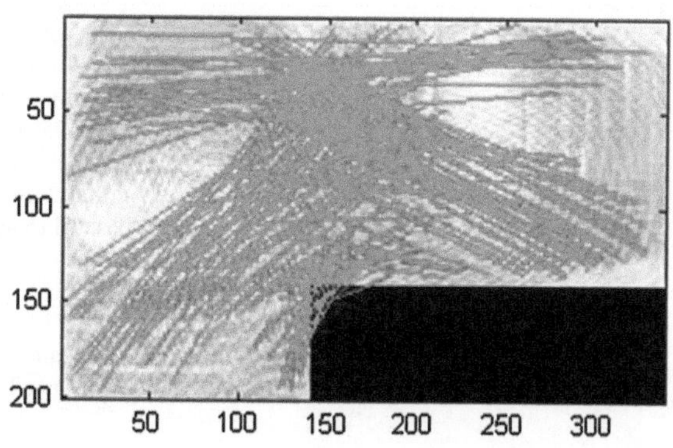

Fig. 11 Image rotated by 270°

8 Result

Number of keypoints found in the fused image is 193

Image transformation	Match found
0°	193
90°	191
180°	188
270°	186

9 Conclusion

The total number of keypoints extracted can be stored. These points can be saved in the database for identification and for future image processing operations like tracking or recognition of objects. Future work would involve in establishing a database of knuckle-fingerprint including plotting the ROC Receiver Operating Characteristic (ROC) which would project the comparison of true vs. false positive rate, at various threshold settings. Accuracy can also be measured in terms of Area under the ROC Curve (AUC).

References

1. Jain AK, Ross A (2004) Multibiometric systems. Commun ACM 47(1):34–40
2. Kisku RD, Sing KJ, Tistarelli M, Gupta P (2009) Multisensor biometric evidence fusion for person authentication using wavelet decomposition and monotonic-decreasing graph. In: 2009 seventh international conference on advances in pattern recognition, 2009 IEEE
3. Dey N, Das S, Rakshit P (2011) A novel approach of obtaining features using wavelet based image fusion and Harris corner detection. Int J Mod Eng Res 1(2):396–399
4. Zhang L, Zhang L, Zhang D (2009) Finger-knuckle-print: a new biometric identifier. In: 2009 16th IEEE international conference on image processing (ICIP), Cairo, 7–10 Nov 2009, 1981–1984
5. Zhang L, Zhang L, Zhang D (2009) Finger-knuckle-print verification based on band-limited phase-only correlation (Lecture Notes). Computer analysis of images and patterns, pp 141–148
6. Zhang L, Zhang L, Zhang D, Zhu HL (2010) Online finger-knuckle-print verification for personal authentication. Pattern Recogn 43(7):2560–2571
7. Zhang L, Zhang L, Zhang D (2010) Monogeniccode: a novel fast feature coding algorithm with applications to finger-knuckle-print recognition. In: 2010 international workshop on emerging techniques and challenges for hand-based biometrics (ETCHB), Istanbul, 22 Aug 2010, pp 1–4
8. Zhang L, Zhang L, Zhang D, Zhu H (2011) Ensemble of local and global information for finger-knuckle-print recognition. Pattern Recogn 44(9):1990–1998
9. Shameem Sulthana ES, Kanmani S (2014) Implementation and evaluation of SIFT descriptors based finger-knuckle-print authentication system. Indian J Sci Technol 7(3):374–382

10. Karanwal S, Kumar D, Maurya R (2010) Fusion of fingerprint and face by using DWT and SIFT. Int J Comput Appl 2(5):0975–8887
11. Lowe DG (2004) Distinctive image features from scale invariant keypoints. Int J Comput Vis 60(2):91–110
12. Fergus R, Perona P, Zisserman A (2003) Object class recognition by unsupervised scale invariant learning. In: IEEE conference on computer vision and pattern recognition, Madison, Wisconsin, pp 264–271

Development of a Multidrug Transporter Deleted Yeast-Based Highly Sensitive Fluorescent Biosensor to Determine the (Anti)Androgenic Endocrine Disruptors from Environment

Shamba Chatterjee and Sayanta Pal Chowdhury

Abstract A competent and consistent androgen receptor transactivation assay has been developed using pleiotropic drug resistance transporters *Pdr5, Snq2 and Yor1* deleted yeast strain, *Saccharomyces cerevisiae,* intended to express the human androgen receptor and androgen response element (probasin promoter) driving the expression of green fluorescent protein to determine endocrine disruptors from pulp and paper mill effluents (PPME). Stimulation of cells by known androgens, correlated with androgenic activities as measured by other reported bioassay systems. This yeast-based assay system when applied to evaluate anti-androgenic activities, the known anti-androgens effectively inhibited fluorescence reporter gene induction by dihydrotestosterone. The specificity of the assay was experienced by incubating the recombinant yeast cells with supraphysiological concentrations of non-androgenic steroidal compounds and none of them yielded considerable response. Further, the assay was used to analyze the extracted PPME from different mills confirmed strong androgenic activities. In conclusion, these results support the earlier report by us that PPME are rich in androgenic compounds and the employed detection system provides novel high throughput fluorescence based biosensor system for successful well sensitive (picogram level) screening of (anti)androgenic chemicals from various environmental sources.

Keywords Endocrine disruptors · Androgen · *Saccharomyces cerevisiae* · Fluorescence based biosensor

1 Introduction

Environmental levels of an array of endocrine-disrupting chemicals have been escalating and thus affecting human health by interacting with hormone receptors followed by the disruption in normal endocrine activity [1–5]. Such compounds are

---

S. Chatterjee (✉) · S.P. Chowdhury
Department of Biotechnology, Haldia Institute of Technology, Haldia, India
e-mail: shambac@gmail.com

S. Gupta et al. (eds.), *Advancements of Medical Electronics*,
Lecture Notes in Bioengineering, DOI 10.1007/978-81-322-2256-9_15

suspected to raise developmental, reproductive and tumorigenic effects to wildlife and humans even at very low concentrations. Amongst all endocrine disruptors, environmental estrogens are the most introspected arena [6]. Whereas, reports on successful, robust and highly sensitive androgen assay with short time period are available in a scanty mode. As we know, androgens are basically essential for the masculinization of male genitalia in utero, the augmentation of secondary sex characteristics in males, and the continuation of male sexual function in adults. After entering in the target cell, it binds to androgen receptor (AR) driving a ligand-dependent transcriptional activity. Finally, AR enters into the nucleus and binds successively to the regulatory region of the target gene as a homodimer that in turn modulates transcriptional activity [7, 8]. Earlier report shows a crucial involvement of the xeno-androgens through androgen receptor in development of abnormal sex related disorders that can be eventually inclined towards testicular cancer [9]. Several assays have been devised for the detection of environmental endocrine disruptors (EDCs), using both mammalian and yeast cells. But, yeast-based assay systems achieved its popularity following its low cost, easy handling approach, complete absence of crosstalk (lack of known endogenous receptors), and use of media that are devoid of steroids [10–18].

The major aim of this introspection was initially, to develop a yeast *yEGFP* reporter-based, rapid, efficient, highly sensitive short timed and economic assay system, appropriate for screening several samples concurrently for their androgenic or anti-androgenic activities. To fulfill this notion, the pleiotropic drug resistance transporters deleted yeast strain (well versed to facilitate efflux of organic compounds including steroids and other chemotherapeutics), was transformed with human AR (hAR) and androgen response element (ARE) and minimal yeast-based promoter (CYC1) driving *yEGFP* reporter gene in the corresponding yeast-based expression vectors. The recombinant *Saccharomyces cerevisiae* strain was then employed to analyze the androgenicity of different steroids and synthetic chemicals.

The assay was also utilized to test the pulp and paper mill effluents for their in vitro androgenicity. Earlier, one Swedish group has published a report on the analysis of PPME using recombinant yeast-based reporter system [19]. Later on, we have already reported a recombinant yeast-based rapid and sensitive β-galactosidase biosensor assay [20]. Here, the developed *yEGFP* reporter based assay is much more sensitive than the earlier framed one and can thus be utilized to screen (anti) androgenic chemicals from various environmental sources.

2 Materials and Methods

2.1 Chemicals

D-glucose and yeast nitrogen base without amino acids and without ammonium sulphate were acquired from Himedia (Mumbai, India). Trizol reagent, L-leucin, tryptophan and uracil were from Sigma (St. Louis, MO, USA). Ammonium sulphate, chloroform, isoamyl alcohol, isopropanol, ethanol absolute and dimethyl sulfoxide (DMSO) were obtained from Merck (Merck, Mumbai, India). DNase I, Ribonuclease inhibitor and restriction endonucleases were all procured from Promega (Madison, WI, USA). Testosterone (T), Dihydrotestosterone (DHT), estrogen (E), progesterone (Prog), all trans retinoic acid (RA), dexamethasone (Dex), hydroxy flutamide (HF), cyproterone acetate (CPA), spironolactone, p,p′-DDE and vinclozolin were kindly provided by Prof. Ilpo Huhtaniemi, Imperial College, London, UK.

2.2 PPME Sample Collection and Preparation

PPME effluents were collected from the outlets of five different pulp and paper industries of northern India. These samples (2 L) were extracted instantaneously after collection by employing solid phase extraction process. They were first filtered through 0.1 μm glass fibre filters (Type GMF5, Rankem, Mumbai, India), followed by an acidification process with concentrated sulphuric acid to pH 2.0, and finally separated into two 1 L samples. Then, 1 L of the sample was extracted utilizing reverse phase C18 solid phase extraction columns (RP-C18 SPE, Rankem, Mumbai, India) and dissolved in 1,000 μL of DMSO (a concentration factor of 1,000). Next, a 1:100 dilution in medium for the above extracted samples yielded in the highest test concentration of 1 mL eq./well.

2.3 Plasmids and Yeast Strains

The pRS425-Leu2-ARS yeast expression vector (kindly provided by Dr. A. Bachhawat, IMTECH, Chandigarh, India) was used to generate hAR construct. Yeast compatible hAR construct, pRS425-Leu2-ARS-hAR was made as per the procedure mentioned earlier by Chatterjee et al. [20]. The high copy *Escherichia coli*/yeast shuttle vector pYEX-BX (Clontech, Palo Alto, USA) was employed to construct the green-fluorescent protein reporter plasmid, pARE-CYC1-GFP. Incorporation of the CYC1 minimal promoter and yeast-enhanced *yEGFP* gene of *Aequorea victoria* in plasmid pERE-CYC1-GFP was elaborated earlier [21]. Next, the estrogen receptor-responsive element cassette (ERE) was primarily reinstated by using the *Xba*I and *Bam*HI restriction sites and end-filled with Klenow polymerase. Finally, a 0.3 kb fragment of probasin promoter was utilized as androgen response element (ARE) which was digested with *Bgl*II and *Hin*dIII from pGL3-Probasin-Luciferase expression vector, endfilled with Klenow polymerase and both the blunt ended products were ligated using T4 DNA Ligase to form pARE-CYC1-GFP (ARE-GFP). The recombinant plasmids were mapped by restriction analysis and confirmed by sequencing.

Table 1 Haploid yeast strains

Haploid yeast strains	Genotype	Source
FY1679-28C	*MATa, ura 3-52, trp 1Δ63, leu 2Δ1, his3Δ200, GAL2+*	Kolaczkowski et al. (1998)
FYAK26/8-10B1	*MATa, ura 3-52, trp 1Δ63, leu 2Δ1, his3Δ200, GAL2+, pdr5-Δ1::hisG, snq2::hisG, yor1-1::hisG*	Kolaczkowski et al. (1998)
SCY1	FYAK26/8-10B1	This study
	[pARE-CYC1-GFP]	
SCY2	FYAK26/8-10B1	This study
	[pARE-CYC1-GFP]	
	[pRS425-Leu2-ARS-hAR]	

The yeast strain *Saccharomyces cerevisiae* FYAK26/8-10B1 (*MATa, ura 3-52, trp 1Δ63, leu 2Δ1, his3Δ200, GAL2+, pdr5-Δ1::hisG, snq2::hisG, yor1-1::hisG*) was a kind gift from Dr. M. Ghislain (Table 1).

2.4 Transformation and Culture of Recombinant Yeast Cells

Plasmids harboring hAR and ARE-*lacZ* were co-transformed into *Saccaromyces cerevisiae,* FYAK26/8-10B1 using the lithium acetate protocol [20]. Transformants were grown in minimal YNB media consisting of 0.67 % yeast nitrogen base (YNB) with $(NH_4)_2SO_4$ and 0.5 % d-glucose adjusted to pH 6.4 and supplemented with the suitable amino acids without tryptophan, leucine and uracil. Glycerol stocks of the recombinant strains were prepared and stored at −80 °C.

2.5 Isolation of RNA and RT-PCR

Total RNA was isolated according to the method described by Ausubel et al. [22]. The RNA pellet thus obtained was resuspended in 50 µL of DEPC treated water and stored at −80 °C for future use.

RT-PCR was carried out with 1 µg of total RNA as template by using the MMLV reverse transcriptase and oligo-dT primers (Promega, Madison, WI, USA). Oligonucleotide primers were designed from areas conserved in the published sequences of hAR cDNA sequences as follows: sense, 5′-ACCATGTTTTGCCCATTGAC-3′; antisense, 5′-GCTGTACATCCGGGACTTGT-3′. The PCR was performed using 28 PCR cycles with Taq polymerase (94 °C for 30 s, 50 °C for 75 s and 72 °C for 90 s and finally at 72 °C for 10 min for one cycle). The amplified DNA samples were run on a 1.5 % agarose gel and bands were visualized with ethidium bromide.

2.6 Western Blot

Western blot analysis was carried out according to the protocol depicted by Laemmli [23]. In brief, the proteins thus obtained from yeast biosensors with/without pRS425-Leu2-ARS-hAR expression vectors were loaded on a 10 % SDS polyacrylamide gel. Finally, proteins were transferred to a nitrocellulose membrane and it was incubated overnight with a polyclonal antibody for hAR (1:200) in the presence of blocking buffer. This was further incubated with alkaline phosphatase labeled secondary antibody (1:1,000). Color development was achieved in BCIP/NBT solution. The extracted protein samples from LNCaP cells were utilized as positive control.

2.7 Confocal Microscopy

Transformed yeast strains were cultured with exposure to testosterone for 16.5 h and 1 ml of the culture was washed with MilliQ followed by formaldehyde fixation. Next, cells were harvested and washed twice with PBS + BSA (1 mg/ml) + 0.1 % (v/v) Triton X-100. Then, the pellet was resuspended in 50 μl PBS/BSA buffer. Finally, cells were washed twice with the same buffer without Triton X-100 and were examined under Olympus Fluoview FV-1,000 (Olympus, Japan) with a 60× oil-immersion objective.

2.8 Biosensor Assay Conditions and Fluorescence Monitoring

For yeast cells with receptor/reporter construct, quantitative analysis of growth phenotypes and fluorescence development have been carried out with logarithmic growing cells (diluted to a start OD_{600} of 0.4) equivalent to 3.25×10^6 cells/ml (Pharmacia Ultrospec 2000 Spectrophotometer). Recombinant yeast cells were treated with different concentrations of standards (testosterone, dihydrotestosterone), non-androgenic steroids (estradiol, progesterone, retinoic acid, dexamethasone), test chemicals (hydroxyflutamide, cyproterone acetate, vinclozolin, *p,p'*-DDE, PPME) in the presence or absence of dihydrotestosterone depending on experiments. Test compounds were dissolved in DMSO and added to the test cultures of co-transformed yeast strains in a total volume of 200 μl followed by the incubation at 30 °C. The growth was obtained by OD_{600} measurements in 15 min intervals in 96-well plates using a microplate reader (Tecan, Spectrofluor Plus).

For fluorescence measurements, the excitation wavelength was accustomed to 485 nm and emission was detected at 535 nm in 15 min intervals for 16.5 h. Dose–response curves were achieved as follows:

FLUO/OD = $(FLUOci - FLUO_{Bci})/(ODci - OD_{Bci})$, where the fluorescence at the test concentration (ci) minus the corresponding values for the blank control (Bci) is divided by growth determined as OD 600 (corrected for blanks).

2.9 Statistical Analysis

The statistical analysis of the results obtained was carried out using the Student's *T*-test. The acceptance level was set at $p < 0.05$.

3 Results and Discussion

3.1 Establishment of Recombinant Yeast Strain and Level of Expression of AR

This paper portrays the development of a recombinant yeast-based highly sensitive (picogram level) fluorescent bioassay for screening different environmental pollutants from PPME. To achieve this purpose, *Saccaromyces cerevisiae* FYAK26/8-10B1 was co-transformed with pRS425-Leu2-ARS-hAR and ARE-GFP. The recombinant FYAK26/8-10B1 strain carrying *yEGFP* reporter plasmid under the control of CYC1 promoter, the androgen-dependent interaction between hAR and ARE can be sensed by the expression of reporter gene. Expression of the transformed hAR cDNA was confirmed by transcriptomic analysis by isolation of total RNA and RT-PCR amplification (Fig. 1).

Figure 2 demonstrates the Western blot of positive yeast cells that contain a single copy of the ARE-GFP and pRS425-Leu2-ARS-hAR receptor expression constructs. An immunoreactive protein band with apparent molecular weight of 110 kDa was detected in transformant tested indicating the exact translation of the AR protein and was at par with the band pattern from LNCaP cells, utilized as positive control. The band was absent in untransformed yeast cells as negative control.

Successful expression of yEGFP in recombinant FYAK26/8-10B1 strain has been confirmed further by confocal microscopy (Fig. 3).

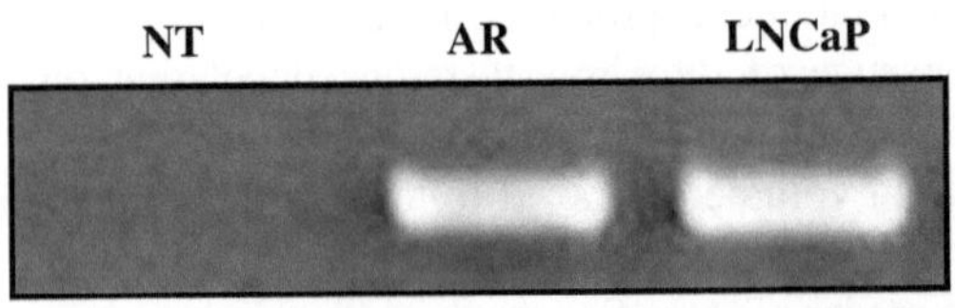

Fig. 1 RT-PCR analysis of androgen receptor mRNA expression in yeast cells. The total RNA isolated from the yeast cells and LNCaP cell line, reverse transcribed and cDNA obtained was subjected to PCR. *NT* Nontransformed yeast cells; *AR* yeast cells transformed with androgen receptor and *LNCaP* LNCaP cell line expression as positive control

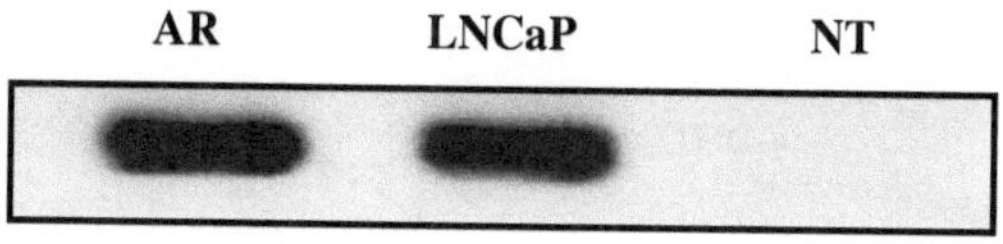

Fig. 2 Western blot analysis for the expression of androgen receptor in transformed yeast cells. In the *upper panel*, the first lane represents *AR* protein in transformed cells (*AR*), the second lane is from *LNCaP* cells used as positive control (*LNCaP*) and third lane is nontransformed yeast cell extract used as negative control (NT)

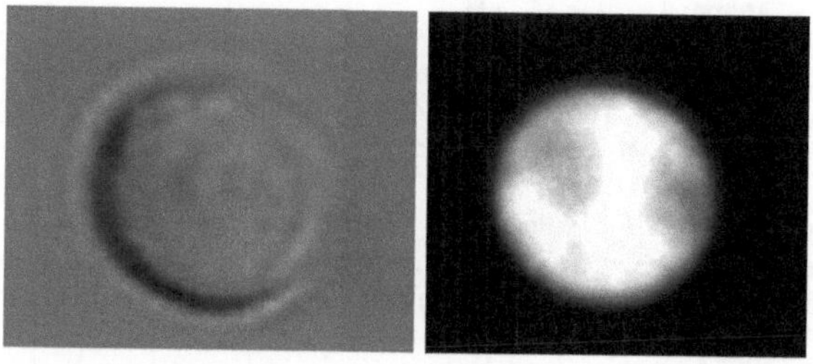

Fig. 3 Confocal microscopy analysis of *yEGFP* expression in co-transformed yeast cells exposed to testosterone

3.2 Dose-Response Curves Obtained with the YEGFP-Based Yeast Androgen Bioassay

Androgen-response capability analysis of the transformed FYAK26/8-10B1 strain was performed with increasing concentrations of androgens (Fig. 4). The *yEGFP* activity was induced by testosterone (T) and dihydrotestosterone (DHT) in a dose dependent manner with the saturation of *yEGFP* induction at around 60 pg/ml in both the situations. The developed assay unveiled a half maximal effect (EC_{50}) in the transformed yeast strain at 55 pg/ml for T and 43 pg/ml for DHT, which is well consistent with the relative AR binding affinity and agonist activity of these androgens in yeast-based assays [20, 24–26] and mammalian cells [27, 28].

3.3 Dose-Dependent Ligand Specificity in the Transformed Yeast Cells

Specificity with steroidal and non-steroidal hormones was analyzed by the induction of *yEGFP* activity in the recombinant yeast strain. Recombinant FYAK26/8-10B1 strain was incubated with increasing concentrations of T, DHT, 17β-estradiol (E), progesterone (Prog), all-trans-retinoic acid (RA) and dexamethasone (Dex) followed by the measurement of their *yEGFP* activity (Fig. 5). T and DHT notably induced

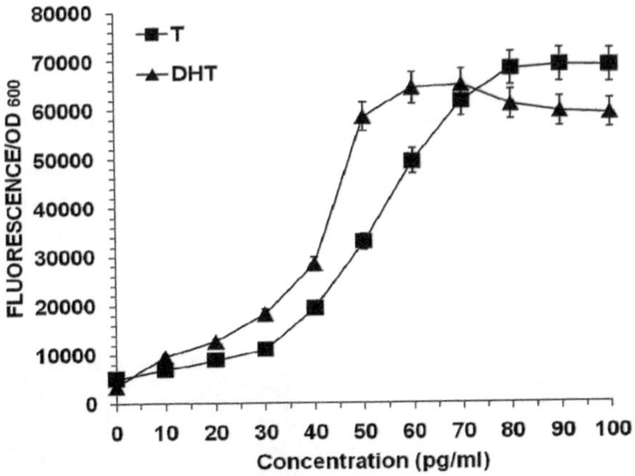

Fig. 4 Dose-dependent *yEGFP* induction by increasing concentrations of T and DHT in co-transformed yeast cells. The *yEGFP* activity in the absence of ligand was set as zero. The values represent the mean ± S.E.M. of four independent experiments each performed in quadruplicates

Fig. 5 Determination of ligand specificity in recombinant yeast strains in response to increasing concentrations of androgenic and non-androgenic steroids. The values represent the mean ± S.E.M. of four independent experiments each performed in quadruplicates

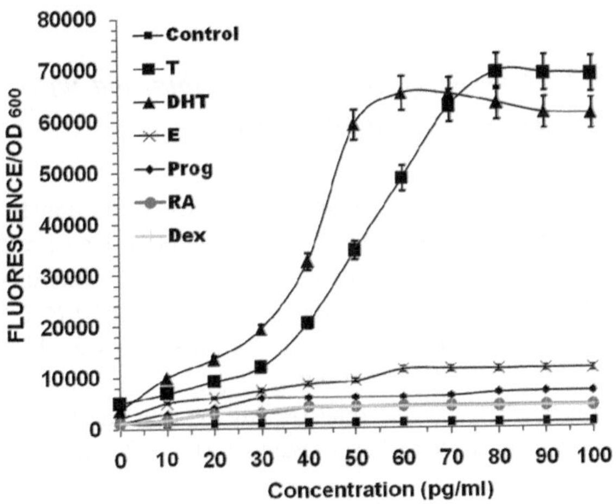

yEGFP activity at a concentration of 50 pg/ml, while the other hormonal compounds failed to induce noteworthy enzymatic activity in the same range.

Only 17β-estradiol yielded a minor agonistic response. However, this compound is already known to exert androgenic effects [24, 26]. Dexamethasone was also depicted as an androgen receptor agonist earlier, showing androgenic activity in few mammalian assays, but gave a negative result in our *yEGFP*-based yeast androgen bioassay.

3.4 Anti-androgenic Activity in Transformed Yeast Strain

The specificity of the new yeast androgen bioassay was further demonstrated by the ability of anti-androgens to suppress the induction of yEGFP. Figure 6 shows the anti-androgenic activity of the known antagonist flutamide and the other compounds like cyproterone acetate, spironolactone, *p,p′*-DDE, and vinclozolin. The antagonistic properties were examined by co-exposure with a concentration of DHT that induced a submaximal response (50 nM). None of these five compounds were able to show an agonistic response, but Fig. 6 clearly shows that all three were able to inhibit the response induced by DHT. Their IC_{50} values were estimated approximately as 5.8 μM for flutamide, 6.78 μM for cyproterone acetate (CPA), 9.93 μM for *p,p′*-DDE and 33.79 μM for vinclozolin ($p < 0.05$) using our yeast-based yEGFP assay.

3.5 Detection of Androgenic Activity of PPME

Recombinant FYAK26/8-10B1 cells were incubated on 96-well plates applied with 1:100 dilution of solid phase extracted water sample of PPME at 30 °C and fluorescence was measured in a time-dependent manner (16–24 h) (Fig. 7). Our results

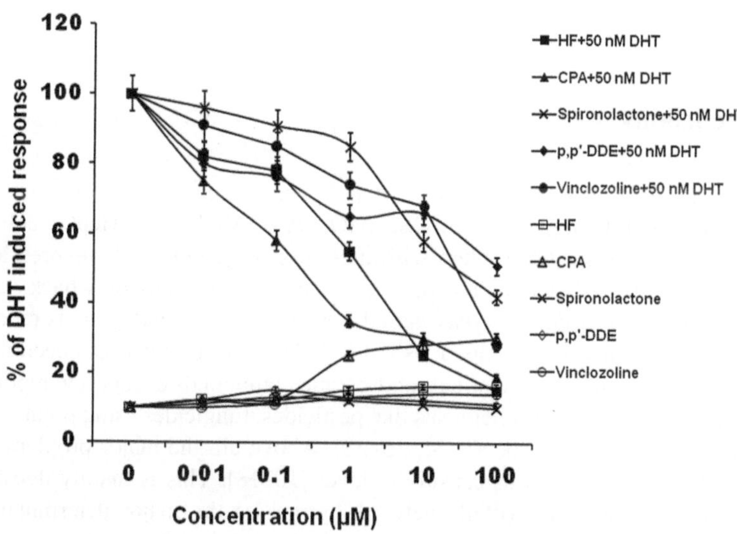

Fig. 6 Anti-androgenic activities of some androgen receptor antagonists in recombinant yeast cells. The cells were treated either with increasing concentrations of test compounds alone or in the presence of 50 nM DHT. The *yEGFP* activities were compared with 50 nM DHT only, which was set as 100. The values represent the mean ± S.E.M. of four independent experiments each performed in quadruplicates

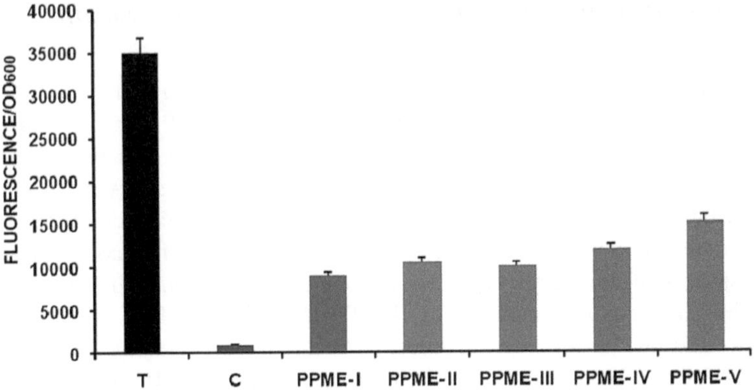

Fig. 7 Detection of androgenic effects of solid phase extracted effluents from five different PPME samples (*I–V*) with respect to *yEGFP* induction by the developed recombinant yeast cells employing 96-well plate format. The *yEGFP* activities were evaluated with that obtained with control (vehicle treated). The values represent the mean ± S.E.M. of four independent experiments each performed in quadruplicates

showed detectable levels of androgen receptor mediated *yEGFP* activity at levels significantly greater than the background ($p < 0.05$). The data suggests that the solid phase extracted water samples of all the five PPME samples comprise potential androgen mimicking compounds and they are in the range of 0.001–1.5 μg/L eq. of T.

4 Conclusions

Earlier transactivation assays with AR and androgen-responsive reporter genes have been carried out utilizing DU-145, COS, CHO, CV-1, MCF-7 and HepG2 cell lines [5, 27, 29–32]. The major problem with these assay systems is the expression of other non-specific endogenous receptors [5]. To overcome this drawback several attempts have been made with yeast-based assays devoid of endogenous receptors for screening of estrogenic compounds [21, 33–35], but not much has been unravelled for screening androgenic or anti-androgenic compounds except few reports on screening different synthetic chemicals like pesticides, fungicides using β-gal or GFP as reporter genes [20, 25, 26, 33, 34, 36–38]. Above all, the major problem associated with these reports is its sensitivity level [10–18]. This is mainly due to the pleiotropic drug resistance (PDR) network comprises the major determinants of multiple drug resistance in *S. cerevisiae.* Thus, induction by a drug or toxin mediates the up-regulation of gene expressions that encode transporters resulting in drug extrusion, followed by multidrug resistance. So to enhance the sensitivity of our *yEGFP*-based assay (picogram level), the hAR and *yEGFP* constructs have been expressed in a yeast strain deleted in the multidrug ABC transporter genes.

The present report thus unveils a better model in sensitivity than the other as reported earlier [24, 25, 36–38]. To the best of our knowledge, this is the only report, other than Svenson and Allard [19] and our self report (Chatterjee et al. [20]), where androgenicity of chemicals in PPME has been reported employing a recombinant yeast-based model with picogram level sensitivity. As the initial application of yeast-based androgen bioassay, we analyzed some PPME samples in the northern region of India. Our data revealed detectable levels of androgen receptor mediated transcriptional activity at levels significantly greater than the background recorded in the effluents ($p < 0.05$) indicating its androgenic activity. Above all, the assay seems to be more robust and more specific for detecting compounds with a pure androgenic mode of action.

Acknowledgments SC would like to thank CSIR and DAAD for providing research fellowship in order to carry out this work. We thank Dr. Anand Bachhawat for the vectors, Dr. M. Ghislain for the the FYAK strain and Prof. Ilpo Huhtaniemi for the kind gift of all steroidal and non-steroidal test chemicals. We thank Prof. P. Roy, Prof. C.B. Majumder and Prof. M. Höfer for their valuable technical suggestions whenever needed.

References

1. Pike MC, Spicer DV, Dahmoush L, Press MF (1993) Estrogens, progestogens, normal breast cell proliferation, and breast cancer risk. Epidemiol Rev 15:17–35
2. Routledge EJ, Parker J, Odum J, Ashby J, Sumpter JP (1998) Some alkyl hydroxy benzoate preservatives (parabens) are estrogenic. Toxicol Appl Pharmacol 153:12–19
3. Skakkebaek NE, Jørgensen N, Main KM, Rajpert-DeMeyts E, Leffers H, Andersson A, Juul A, Carlsen E, Mortensen GK, Jensen TK, Toppari J (2006) Is human fecundity declining? Int J Androl 29:2–11
4. Kelce WR, Wilson EM (2001) Environmental anti-androgens: developmental effects, molecular mechanisms and clinical implications. J Mol Med 75:198–207
5. Roy P, Salminen H, Koskimies P, Simola J, Smeds A, Sáuco P, Huhtaniemi IT (2004) Screening of some anti-androgenic endocrine disruptors using a recombinant cell-based in vitro bioassay. J Steroid Biochem Mol Biol 88:157–166
6. Jobling S, Reynolds T, White R, Parker MG, Sumpter JP (1995) A variety of environmentally persistent chemicals, including some phthalate plasticizers, are weakly estrogenic. Environ Health Perspect 103:582–587
7. Keller ET, Ershler WB, Chang C (1996) The androgen receptor: a mediator of diverse responses. Front Biosci 1:59–71
8. Blankvoort BM, de Groene EM, van Meeteren-Kreikamp AP, Witkamp RF, Rodenburg RJ, Aarts JM (2001) Development of an androgen receptor gene assay (AR-LUX) utilizing a human cell line with an endogenously regulated androgen receptor. Anal Biochem 298:93–102
9. Henley DV, Lipson N, Korach KS, Bloch CA (2007) Prepubertal gynecomastia linked to lavender and tea tree oils. N Engl J Med 356:479–485
10. Bovee TF, Lommerse JP, Peijnenburg AA, Fernandes EA, Nielen MW (2008) A new highly androgen specific yeast biosensor, enabling optimisation of (Q)SAR model approaches. J Steroid Biochem Mol Biol 108:121–131
11. Bovee TF, Schoonen WG, Hamers AR, Bento MJ, Peijnenburg AA (2008) Screening of synthetic and plant-derived compounds for (anti)estrogenic and (anti)androgenic activities. Anal Bioanal Chem 390:1111–1119

12. Rijk JC, Bovee TF, Wang S, Van Poucke C, Van Peteghem C, Nielen MW (2009) Detection of anabolic steroids in dietary supplements: the added value of an androgen yeast bioassay in parallel with a liquid chromatography-tandem mass spectrometry screening method. Anal Chim Acta 637:305–314
13. Bovee TF, Thevis M, Hamers AR, Peijnenburg AA, Nielen MW, Schoonen WG (2010) SERMs and SARMs: detection of their activities with yeast based bioassays. J Steroid Biochem Mol Biol 118:85–92
14. Plotan M, Elliott CT, Scippo ML, Muller M, Antignac JP, Malone E, Bovee TF, Mitchell S, Connolly L (2011) The application of reporter gene assays for the detection of endocrine disruptors in sport supplements. Anal Chim Acta 700:34–40
15. Rijk JC, Ashwin H, van Kuijk SJ, Groot MJ, Heskamp HH, Bovee TF, Nielen MW (2011) Bioassay based screening of steroid derivatives in animal feed and supplements. Anal Chim Acta 700:183–188
16. Rijk JC, Bovee TF, Peijnenburg AA, Groot MJ, Rietjens IM, Nielen MW (2012) Bovine liver slices: A multifunctional in vitro model to study the prohormone dehydroepiandrosterone (DHEA). Toxicol In Vitro 26:1014–1021
17. Reitsma M, Bovee TF, Peijnenburg AA, Hendriksen PJ, Hoogenboom RL, Rijk JC (2013) Endocrine-disrupting effects of thioxanthone photoinitiators. Toxicol Sci 132:64–74
18. de Rijke E, Essers ML, Rijk JC, Thevis M, Bovee TF, van Ginkel LA, Sterk SS (2013) Selective androgen receptor modulators: in vitro and in vivo metabolism and analysis. Food Addit Contam Part A Chem Anal Control Expo Risk Assess 30:1517–1526
19. Svenson A, Allard AS (2004) In vitro androgenicity in pulp and paper mill effluents. Environ Toxicol 19:510–517
20. Chatterjee S, Majumder CB, Roy P (2007) Development of a yeast-based assay to detrmeine the (anti)androgenic contaminants from pulp and paper mill effluents in India. Environ Toxicol Pharmacol 24:114–121
21. Sievernich A, Wildt L, Lichtenberg-Frate H (2004) In vitro bioactivity of 17alpha-estradiol. J Steroid Biochem Mol Biol 92:455–463
22. Ausubel FM, Brent R, Kingston RE, Moore DD, Seidman JG, Smith JA, Struhl K (1995) Current protocols in molecular biology. Greene Publishing Associates and Wiley-Interscience, New York
23. Laemmli UK (1970) Cleavage of structural proteins during the assembly of the head of bacteriophage T4. Nature 227:680–685
24. Gaido KW, Leonard LS, Lovell S, Gould JC, Babai D, Portier CJ, McDonnell DP (1997) Evaluation of chemicals with endocrine modulating activity in yeast-based steroid hormone receptor gene transcription assay. Toxicol Appl Pharmacol 143:205–212
25. Lee HJ, Lee YS, Kwon HB, Lee K (2003) Novel yeast bioassay system for detection of androgenic and antiandrogenic compounds. Toxicol In Vitro 17:237–244
26. Michelini A, Leskinen P, Virta M, Karp M, Roda A (2005) A new recombinant cell-based bioluminescent assay for sensitive androgen-like compound detection. Biosens Bioelectrons 20:2261–2267
27. Kemppainen JA, Langley E, Wong CI, Bobseine K, Kelce WR, Wilson EM (1999) Distinguishing androgen receptor agonists and antagonists: distinct mechanisms of activation by medroxyprogesterone acetate and dihydrotestosterone. Mol Endocrinol 13:440–454
28. Terouanne B, Tahiri B, Georget V, Belon C, Poujol N, Avances C Jr, Orio F, Balaguer P, Sultan C (2000) A stable prostatic bioluminescent cell line to investigate androgen and antiandrogen effect. Mol Cell Endocrinol 160:39–49
29. Lobaccaro JM, Poujol N, Terouanne B, Georget V, Fabre S, Lumbroso S, Sultan C (1999) Transcriptional interferences between normal or mutant androgen receptors and the activator protein 1-dissection of the androgen receptor functional domains. Endocrinology 140:350–357
30. Vinggaard AM, Joergensen EC, Larsen JC (1999) Rapid and sensitive reporter gene assays for detection of antiandrogenic and estrogenic effects of environmental chemicals. Toxicol Appl Pharmacol 155:150–160

31. Xu LC, Sun H, Chen JF, Bian Q, Qian J, Song L, Wang XR (2005) Evaluation of androgen receptor transcriptional activities of bisphenol A, octylphenol and nonylphenol in vitro. Toxicology 216:196–203
32. Sun H, Xu X-L, Xu L-C, Song L, Hong X, Chen J-F, Cui L-B, Wong X-R (2007) Antiandrogenic activity of pyrethroid pesticides and there metabolite in reporter gene assay. Chemosphere 66:474–479
33. Bovee TF, Helsdingen RJ, Koks PD, Kuiper HA, Hoogenboom RL, Keijer J (2004) Development of a rapid yeast estrogen bioassay, based on the expression of green fluorescent protein. Gene 325:187–200
34. Bovee TF, Helsdingen RJ, Rietjens IM, Keijer J, Hoogenboom RL (2004) Rapid yeast estrogen bioassays stably expressing human estrogen receptors α and β, and green fluorescent protein: a comparison of different compounds with both receptor types. J Steroid Biochem Mol Biol 91:99–107
35. Wozei E, Hermanowicz SW, Holman H-YN (2006) Developing a biosensor for estrogens in water samples: study of the real-time response of live cells of the estrogen sensitive yeast strain RMY/ER-ERE using fluorescence microscopy. Biosens Bioelectron 21:1654–1658
36. Sohoni P, Sumpter JP (1998) Several environmental estrogens are also antiandrogens. J Endocrinol 158:327–339
37. Beck V, Pfitscher A, Jungbauer A (2005) GFP-reporter for a high throughput assay to monitor estrogenic compounds. J Biochem Biophys Methods 64:19–37
38. Bovee TFH, Helsdingen RJR, Hamers ARM, van Duursen MBM, Nielen MWF, Hoogenboom RLAP (2007) A new highly specific and robust yeast androgen bioassay for the detection of agonists and antagonists. Anal Bioanal Chem 389:1549–1558

Simulation of ICA-PI Controller of DC Motor in Surgical Robots for Biomedical Application

Milan Sasmal and Rajat Bhattacharjee

Abstract The medical device industry is considered a hub of innovation with constantly improving designs and processes, the possibilities seem endless. Over the past 20 years, dc motors have played its role in the development of medical instruments like surgical robots, plate readers, liquid and specimen handling systems, chromatography, and In Vitro diagnostic machines. Regardless of what ever the application may be, all devices in the medical industry should have high Reliability and flexibility. This paper presents a new controller model using Imperialist Competitive Algorithm (ICA) for efficient search and optimization of Proportional Integral controller parameter in order to achieve a better close loop control of the dc motor in bio-medical application. This control method was simulated using MATLAB/SIMULINK to control dc motor. From the simulation, it was found that the proposed method gives better performance with improved response in any user defined function as the settling time, overshoot and speed error are reduced.

Keywords Surgical · Robot · DC motor · Speed control · Controller · Optimization · Imperialist competitive algorithm (ICA)

1 Introduction

Now a day's high performance motor drives are very essential for medical applications especially in surgical robots, liquid and specimen handling system. A high performance motor drive system must have good dynamic speed command tracking and load regulating response. DC motors provide excellent control of speed for

M. Sasmal (✉) · R. Bhattacharjee
Electrical Engineering Department, JIS College of Engineering, Kalyani, Nadia, India
e-mail: milan164sasmal@yahoo.com

R. Bhattacharjee
e-mail: rajat_b@hotmail.com

S. Gupta et al. (eds.), *Advancements of Medical Electronics*,
Lecture Notes in Bioengineering, DOI 10.1007/978-81-322-2256-9_16

acceleration and deceleration. Separately excited DC motor has better voltage variation, so it is used for speed and torque control applications.

DC drives, because of their simplicity, ease of application, reliability and favorable cost have long been a backbone of industrial applications as well as medical applications as compared to AC drive. Motion of the drive can be controlled smoothly by adjusting proper parameters of the speed controller. Here the controller used is basically a Proportional Integral type which removes the delay and provides fast control. Modeling of separately excited DC motor is done by using MATLAB 7.0 version 2007. The controller value is optimized [1] by using an artificial intelligence technique called Imperialist Competitive Algorithm (ICA) [2–7].

Furthermore, the motor is able to run smoothly at an extended speed range by the help of reference speed signal. There is no need to hardware control, only by changing the reference speed signal we can control the device. Thus we can get a computer controlled automated system. Specifically, we aim to maximize the smoothness and the speed with which the tasks are performed. Increased smoothness reduces damage to human tissue and the robotic mechanism, and increased speed reduces operation time.

2 Surgical Robot

In 1979, the Robot Institute of America, an industrial trade group, defined a robot [8] as “a reprogrammable, multi- functional manipulator designed to move materials, parts, tools, or other specialized devices through various programmed motions for the performance of a variety of tasks.” Such a definition leaves out tools with a single task (e.g., stapler), anything that cannot move (e.g., image analysis algorithms), and nonprogrammable mechanisms (e.g., purely manual laparoscopic tools). As a result, robots are generally indicated for tasks requiring programmable motions, particularly where those motions should be quick, strong, precise, accurate, untiring, and/or via complex articulations. The greatest impact of medical robots has been in surgeries, both radiosurgery and tissue manipulation in the operating room, which are improved by precise and accurate motions of the necessary tools. Through robot assistance, surgical outcomes can be improved, patient trauma can be reduced, and hospital stays can be shortened, though the effects of robot assistance.

Surgical robots cover the following multiple next-generation systems on the future of the medical field [8]

- The expected benefit of robot assistance in orthopedics is accurate and precise bone resection
- General Laparoscopy
- Percutaneous
- Steerable Catheters
- Radiosurgery

- Assistive and Rehabilitation Systems
- Prosthetics and Exoskeletons
- Emergency Response—Few medical robot systems are suitable for use outside of the operating room, despite significant research funding on medical devices for disaster response and battlefield medicine.

3 Circuit Model and Procedure

The block diagram of the model is shown in Fig. 1. DC motor is supplied from a converter and the feedback of the motor is fed to a comparator after a filtration. The error signal generated from comparator is fed to the controller [9] and the controller output to the motor.

In the first circuit model, the dc motor is driven by a group of four mosfets as shown in Fcircuit is modified as shown in Fig.2, two mosfets are for two quadrants and rest two for two other quadrants i.e. four quadrant operation. Again these mosfets are controlled by the gate signal generating from Pulse Width Modulation block (PWM). To design a close loop control, Speed feedback is taken from the measuring terminal of the motor through de-multiplexer, then is fed to the speed controller (PI controller), again the output of PI controller [1] and armature current feedback signal is fed to the current controller. The output of the current controller is feed to the PWM as input. The PWM transfers the reference signal to the corresponding pulses which will be fed to the two mosfet and the inverted pulses of the PWM output is feed to other two mosfets. The mosfets will be turned on-off according to the pulses.

For optimization the previous circuit is modified as shown in Fig. 3. To calculate the instantaneous error, an error signal is generated by a comparison between the motor speed and reference speed signal by the comparator. The error signal is fed to the controller block and the output of which is fed to the MATLAB base workspace for calculation of optimization. During the optimization the controller parameter is again fed from base workspace to the controller of the model again in each step of optimization. After a complete process of the ICA [6], the final accurate value of the parameters are fed to the controller and system runs with this (Table 1).

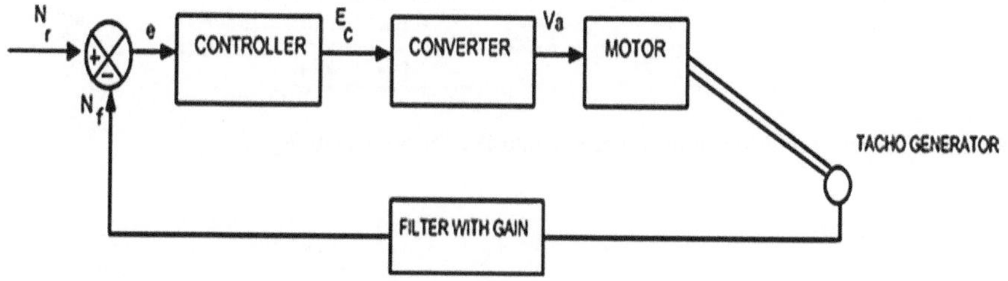

Fig. 1 Block diagram of the model

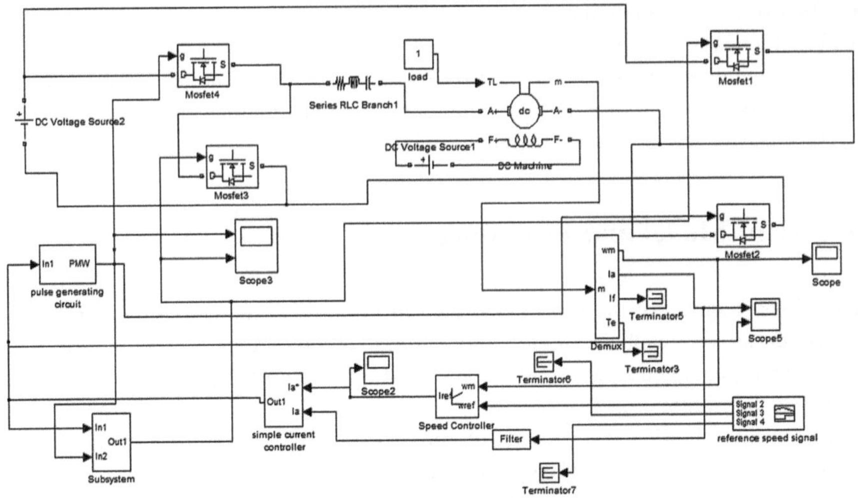

Fig. 2 Circuit diagram of speed controller before optimization

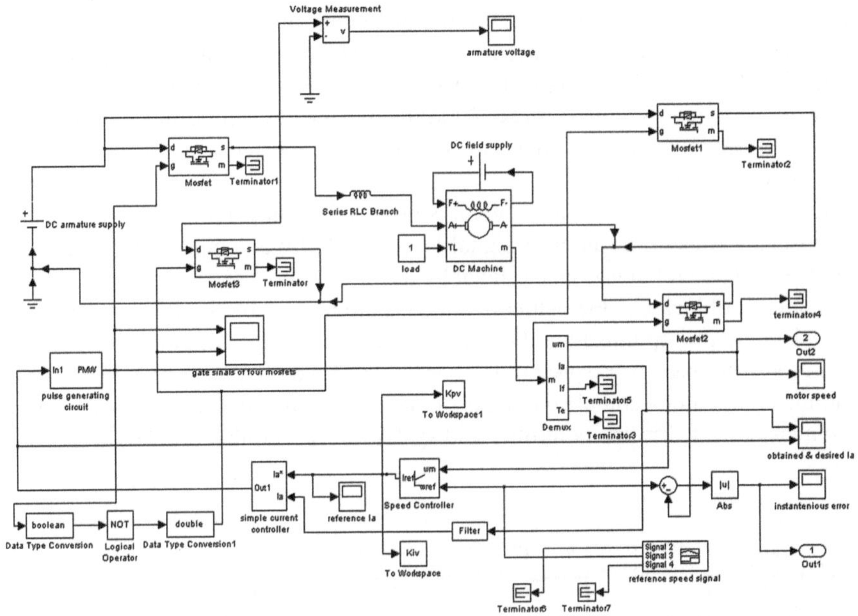

Fig. 3 Complete circuit diagram of speed controller for optimization

Table 1 DC motor parameter

Armature resistance	0.232 ohm
Armature inductance	0.000557 H
Field resistance	78.91 ohm
Inertia	0.2053 * e^{-5} kg.m^2
Load torque	1 (constant)
Coulomb friction	5.282 N.m
Initial speed	0 rad/s

4 ICA to Calculate the Controller Parameter

The Imperialist Competitive algorithm is a technique of Artificial Intelligence for optimization, inspired by the imperialistic competition. Like other evolutionary ones, the algorithm starts with an initial population. Population individuals called country are in two types: colonies and imperialists that all together form some empires. Imperialistic competition among these empires forms the basis of the evolutionary algorithm. During this competition, weak empires collapse and powerful ones take possession of their colonies. Imperialistic competition hopefully converges to a state in which there exist only one empire and its colonies are in the same position and have the same cost as the imperialist. Applying this algorithm to some of benchmark cost functions and in a expert system based on MATLAB shows its ability to drive an computer operated surgical robot and many more things which are used in medical application (Fig. 4).

5 Results and Discussion

In our experiment, our goal is to drive the system in a pre-defined motion which can be used in a surgical robot. To achieve the user defined motion i.e. speed we use a reference speed signal (speed signal 2) which is shown in Fig. 5, where we assume that the drive has no initial speed and starts to accelerate up to 2 s and reaches a speed of 100 revolution per minute, after that it maintain a constant speed and after 3 s it starts to decelerate and again reaches a speed of 100 revolution per minute in opposite direction and again maintain a constant speed between 7 and 8 s.

The output results of the motor corresponding to the reference speed signal 2 before optimizing the value of PI controller, the output follows the reference speed signal but the output consist with a large error including a settling time of 0.3 s approximate and speed error of 4.5 % (Fig. 6).

After optimizing the value of PI controller, the output results of the motor shown in Fig. 7 also follows the corresponding reference speed signal 2 where error is minimized and settling time decreases to 0.035 s approximate with a very small speed error.

Fig. 4 Flowchart of ICA

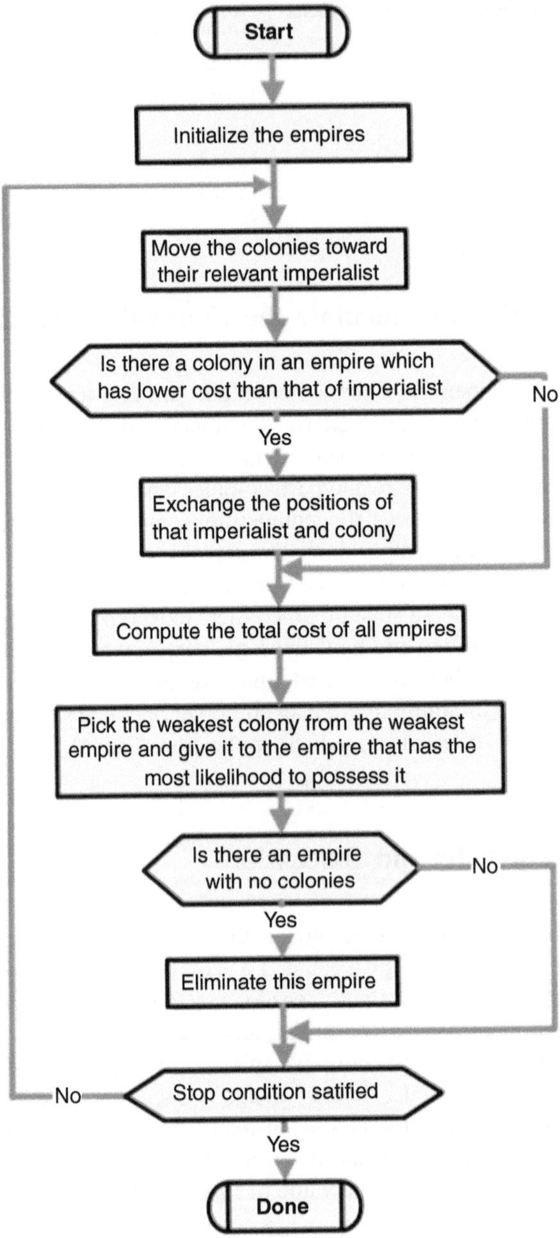

An another user defined motion is taken to drive the system, where reference speed signal 3 is used and is shown in Fig. 8, where we assume that the drive has no initial speed and has stepped accelerated motion from 0 to 100 revolution per

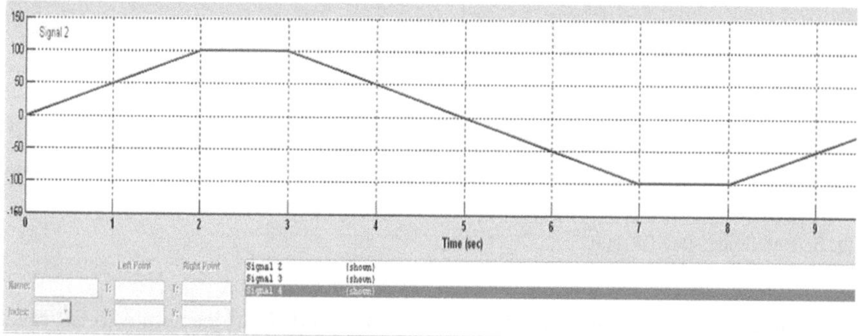

Fig. 5 Reference speed signal 2

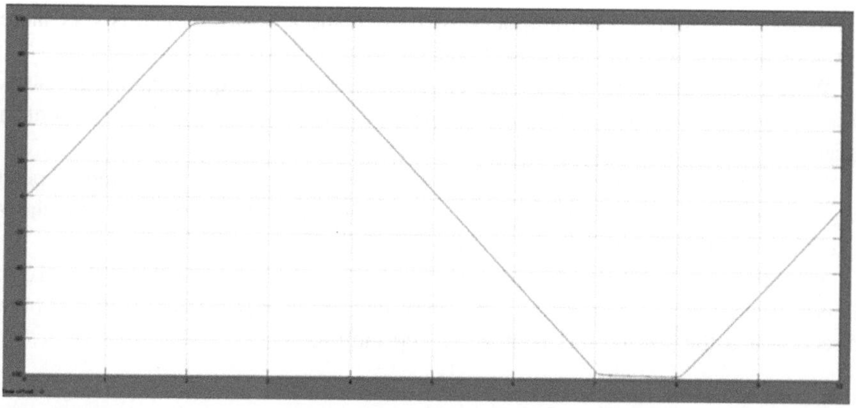

Fig. 6 Speed-time characteristics of the motor for reference sig 2 before optimization

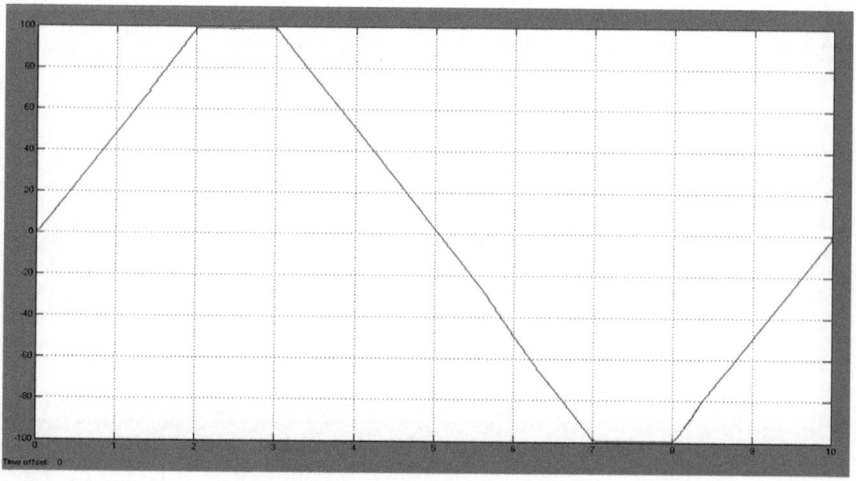

Fig. 7 Speed-time characteristics of the motor for reference sig 2 after optimization

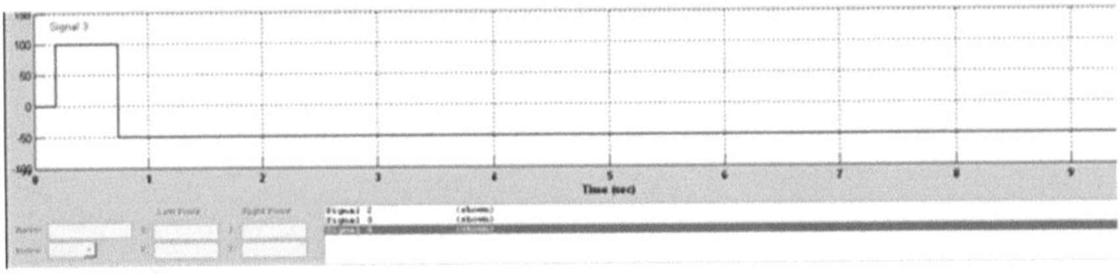

Fig. 8 Reference speed signal 3

minute and also has a constant speed for a while and then decelerate to 50 revolution per minute in reverse direction.

The output results of the motor corresponding to the reference speed signal 3 before optimizing the value of PI controller, output shown in Fig. 9 follows the reference speed signal 3 but output consist a overshoot of 3 % which is very large for sophisticated medical instrument.

Without optimizing the value of PI controller the motor output response involves a huge steady state error including long settling time, rise time with over shoot in large magnitude, shown in Fig. 9.

Figure 10 is the output results of the motor corresponding to the reference speed signal 3 after a complete optimizing the value of PI controller, where the overshoot is completely minimized with a percent of 0.12.

The reference input speed time curve (Figs. 5 and 8) is taken from 0 to 10 s. Between the two reference speed time curve first one is smooth controlling in both direction and the second one (Fig. 8) is step motion.

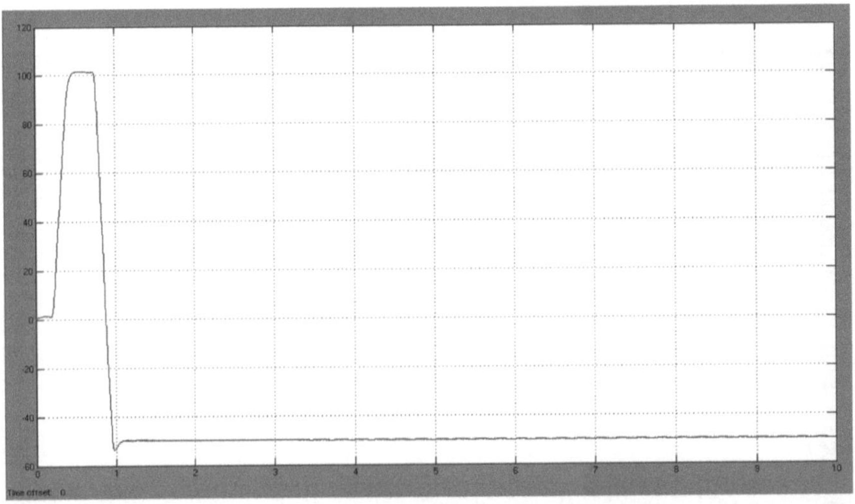

Fig. 9 Speed-time characteristics of the motor for reference sig 3 before optimization

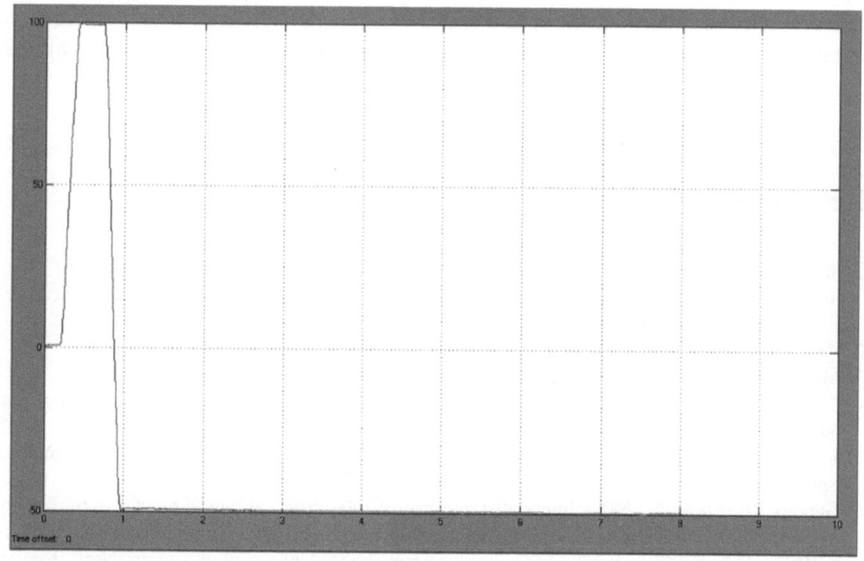

Fig. 10 Speed-time characteristics of the motor for reference sig 3 after optimization

Table 2 Performance criteria of the drive

Parameter values and their response	Before optimization	After optimization
Kp	9.971	6.174
Ki	3.854	0.021
Settling time (s)	0.28	0.035
Overshoot (%)	3	0.12
Speed error (%)	4.5	1.14

After optimization the over shoot has completely eliminated. Settling time and rise time has reduced to great extent. Steady state error has been reduced which is shown graphically (Table 2).

6 Conclusion

As the Surgical robot and medical appliances are very sophisticated tools so we need very smooth and fast control with high reliability. This paper presents a new design of controller for surgical robot and other medical appliances using ICA based optimization to optimize the controller parameters. The comparison between PI controller without and with ICA implemented has been made. From the result it is shown that ICA method can be used to optimize PI controller is more efficient and give better performance for motion system in biomedical appliances with reduced settling time, overshoot and speed error of the system. Further research is intended to be focused in the application of the proposed method for designing the controller to control multiple surgical devices in biomedical application.

References

1. Safura Hashim NL, Yahya A, Andromeda T, Abdul Kadir MR, Mahmud N, Samion S (2012) Simulation of PSO-PI controller of dc motor in micro-EDM system for biomedical application. ELSEVIER Procedia Eng 41:805–811
2. Haupt RL, Haupt SE (2004) Practical genetic algorithms, 2nd edn. Wiley, Hoboken
3. Melanie M (1999) An introduction to genetic algorithms. MIT Press, Massachusetts
4. Dorigo M, Blum C (2005) Ant colony optimization theory: a survey. Theoret Comput Sci 344:243–278
5. Johnston RL, Cartwright HM (2004) Applications of evolutionary computation in chemistry. Springer, Berlin
6. Gargari EA, Lucas C (2007) Imperialist competitive algorithm: an algorithm for optimization inspired by imperialistic competition: control and intelligent processing center of excellence (CIPCE). School of Electrical and Computer Engineering, University of Tehran, North Kargar Street, Tehran, Iran
7. Varol HA, Bingul ZA (2004) New PID tuning technique using ant algorithm, In: Proceeding of the 2004 American control conference, Boston, Massachusetts, June 30–July 2 (2004)
8. Beasley RA (2012) Medical robots—current systems and research directions: Hindawi Publishing Corporation. J Robot 2012(401613):14. doi:10.1155/2012/401613
9. Nasri M, Maghfoori M (2007) Pso—based optimum design of pid controller for a linear brushless dc motor. World Acad Sci Eng Tech 26:211–215

Development of a Wireless Attendant Calling System for Improved Patient Care

Debeshi Dutta, Biswajeet Champaty, Indranil Banerjee, Kunal Pal and D.N. Tibarewala

Abstract The present proposal revolves around the fabrication of a finger movement based wearable wireless attendant calling system. The system comprised of a flex sensor and a hall-effect sensor coupled with Arduino UNO and worked synchronously with patient hand movement. The concurrent activation of both the sensors enables the conveyance of patient location (ward and bed numbers) to the nurse station through Xbee protocol and a one-way SMS correspondence to a preloaded mobile number through GSM protocol. The device is capable of handling multiple patient requests at a time with minute time interval. A graphical user interface in MATLAB program monitors the patient status at the nursing station. The proposed device is expected to improve the quality of the patient care.

Keywords Wearable · SMS · Wireless · Attendant calling system

1 Introduction

There is an acute shortage of nursing staff across the globe. The condition is worse in the developing and under-developed countries. As per the Nursing Council of India (NCI), only 40 % of the registered nursing staffs are actively working. This has resulted in the decrease in the nurse: patient ratio. The suggested ratio is 1:1, 1:3 and 1:6 in the critical care unit, intermediate care unit and general ward, respec-

D. Dutta · B. Champaty · I. Banerjee (✉) · K. Pal (✉)
Department of Biotechnology and Medical Engineering,
NIT-Rourkela, Rourkela, Odisha 769008, India
e-mail: indraniliit@gmail.com

K. Pal
e-mail: pal.kunal@yahoo.com

D.N. Tibarewala
School of Bioscience and Engineering, Jadavpur University,
Kolkata 700032, India

S. Gupta et al. (eds.), *Advancements of Medical Electronics*,
Lecture Notes in Bioengineering, DOI 10.1007/978-81-322-2256-9_17

tively. Since the duty of the nursing staff occurs in 3 shifts, the shortage of the nursing staff looks grimmer. As per the recent report, in many public hospitals the nurse: patient ratio hovers around 1:60 during the evening and the night shifts. This puts a stress on the on-duty nurses. The stress and the work pressure have been reported to be the major deterrent for the nurses to join as a nursing staff. The aforementioned factors hamper the patient care to a great extent. Keeping this in mind, in the current study we propose to develop a low cost attendant calling system which can help reducing the work pressure for the nurses. Generally, the nurses have to be regularly at round to get an insight about the need of the patients. We have devised a prototype finger-movement based attendant calling system. The prototype developed is a wearable device which can wirelessly transmit signal to the nursing station and can inform the healthcare staffs about the need of the patient [1–3].

2 Materials

Xbee-S2 shield (Digi International, USA), flex sensor (Sparkfun, USA), hall-effect sensor (Evelta Electronics, India), GSM-GPRS shield (Seedstudio, China), Arduino UNO (Arduino, Italy) microcontroller boards, 16x2 LCD (Sunrom Technologies, India) and Emic 2 Text-to-speech module (Parallax Inc., USA) are the major components used in this study.

3 Methodology

The developed device consists of two units: (a) transmission unit and (b) receiver unit. The construction of the complete device has been described below.

3.1 Transmission Unit

The transmission unit consists of three input interfaces, namely, flex sensor, hall-effect sensor and a push button. The signals from the input interfaces are fed into an Arduino UNO microcontroller for decision making and triggering of the communication protocol. Two types of communication protocol were used in this study: one being the GSM protocol to send SMS to the nursing staff on activation and the other being Xbee based wireless transmission of the alert signal to the nursing station. The transmitter unit was a wearable device [3, 4]. The setup was assembled on a wearable hand-glove. The circuit was powered with a 9 V rechargeable battery.

3.2 Receiver Unit

The receiver unit consists of an Xbee receiver connected to Arduino UNO microcontroller. The microcontroller was interfaced with a laptop. A GUI (graphical user interface) based MATLAB program was made to monitor the status. The microcontroller was also connected to a LCD panel and a text-to-speech module [1, 3, 4]. The schematic diagram and the gist of functioning of the proposed device has been shown in Fig. 1.

3.3 Testing of the Proposed Device

The device has been developed for application under hospital environment. Two modules of the transmission units were developed to mimic a two bed situation in a ward.

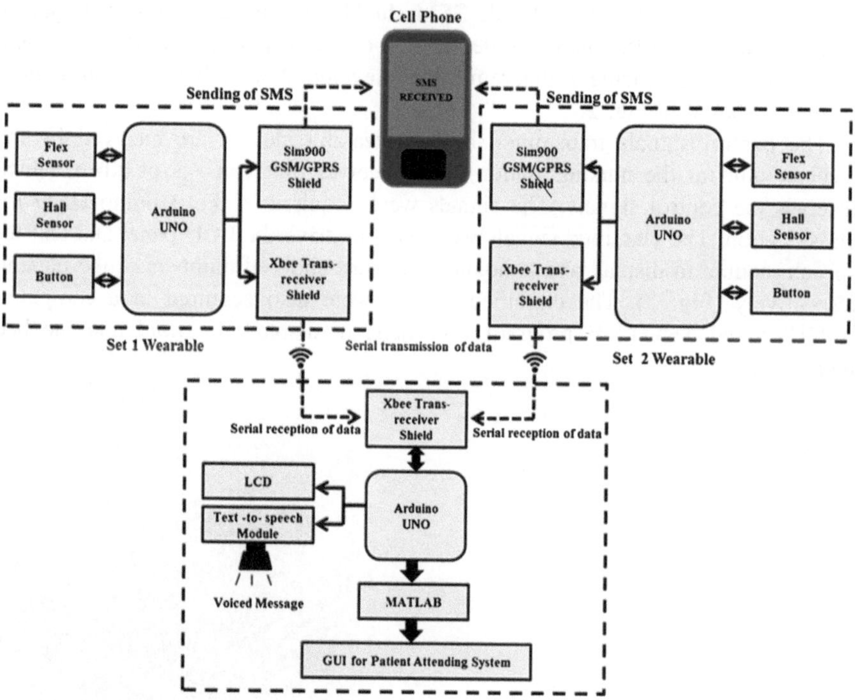

Fig. 1 Schematic diagram and functioning of the proposed device

4 Results and Discussions

An easy to use wearable attendant calling system was devised based on the movement of the finger. The device consists of two sub-units, namely, transmission unit and receiver unit. The transmission unit consists of two sensors (flex sensor and Hall-effect (HE) sensor). Flex sensor is a resistive sensor, which changes its resistance when it is bent. The sensor was attached over the index finger region of a hand-glove such that when the index finger is flexed there is a change in resistance [5, 6]. The change in the resistance was monitored in the microcontroller, which generated control signals when the resistance increased beyond a threshold level. But there are chances of accidental switching of the device when the patient is carrying out other day-to-day activities. Hence, a magnet was attached towards the arm-end of the glove [5, 7, 8].

The magnet was used to activate the HE sensor. The HE sensor was used as a switching device. The program, monitoring the change in the resistance of the flex sensor, was modulated to generate the control signals only when the outputs from both the sensors were in a high state. The control signals were used to drive Xbee and GSM shields. The Xbee shield wirelessly transmitted the control signals to the nursing station. On the other hand, GSM shield sent an alert SMS to a specific mobile. The alert SMS contained the location of the patient, i.e. ward number and bed number. The schematic diagram of the developed wearable attendant calling system is shown in Fig. 2.

The control signals transmitted by the wearable device are received by the receiver unit (at the nursing station). The receiver unit consists of Xbee, which receives the control signals. The signals were acquired in the Arduino UNO for classification. The classified signals were used to drive the LCD panel and text-to-speech module to display and announce the ward and bed numbers of the patient, respectively (Fig. 3). The classified signals were also acquired in a computer. A GUI based MATLAB program was made to display the ward and the bed numbers [9].

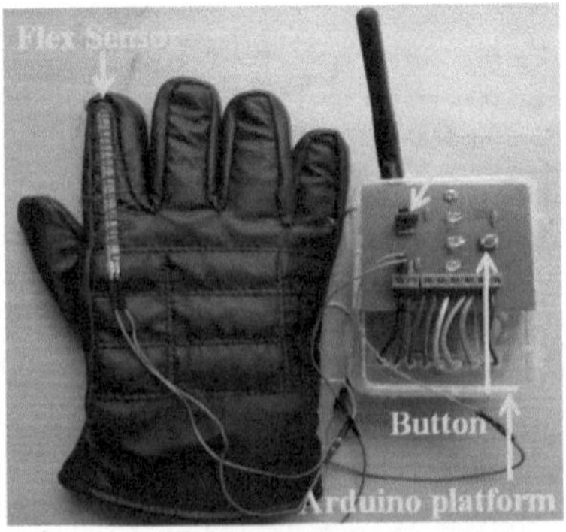

Fig. 2 The developed wearable attendant calling system (transmitter unit)

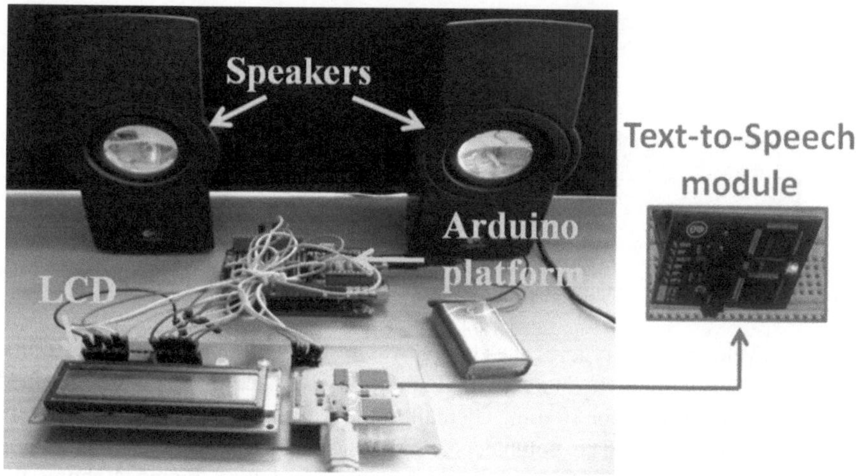

Fig. 3 Receiver unit connected with the LCD and the text-to-speech modules

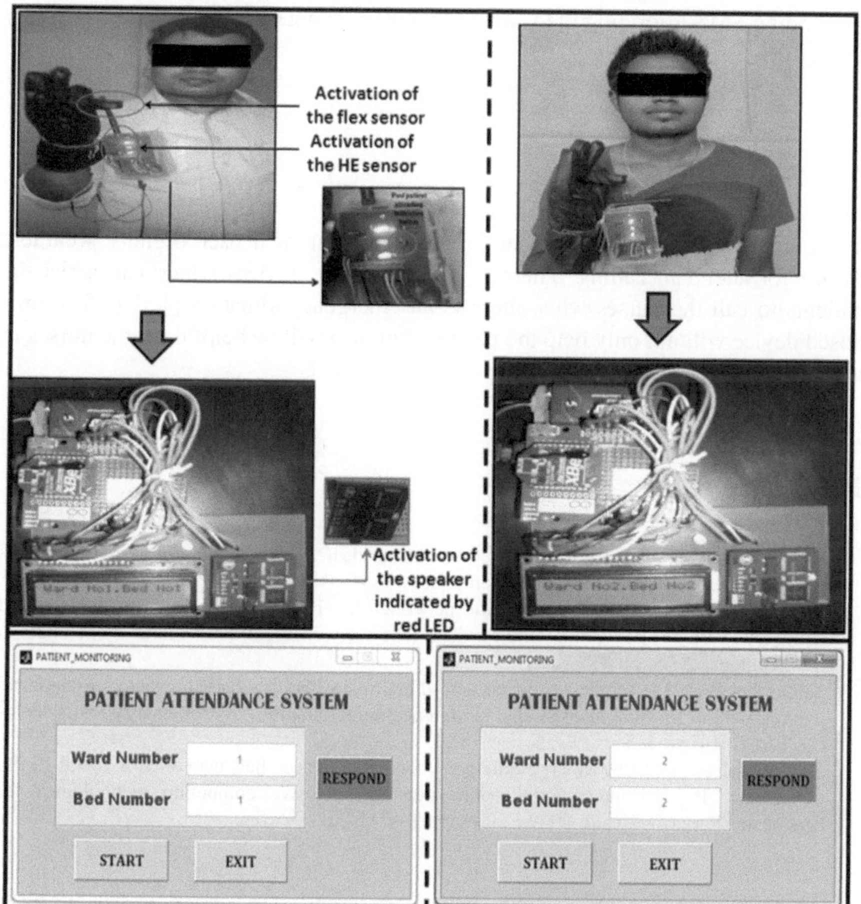

Fig. 4 Volunteers activating the prototype attendant calling units

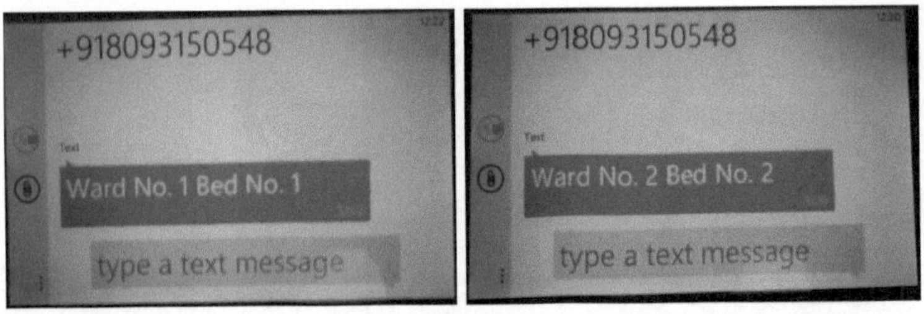

Fig. 5 SMS sent by the wearable device to the specified mobile number

The developed prototype of the device was tested using two transmitter units. Two volunteers were invited to participate in the study. The volunteers were trained on the transmitter units for 10 min. Thereafter, the wearable transmission units were put on the volunteers. The volunteers were advised to activate the transmission units. The volunteers were able to easily activate the device. Figure 4 shows the activation of the LCD, the text-to-speech and MATLAB based GUI display modules at the nursing station. Figure 5 shows the activation of the GSM module for sending SMS informing the ward and the bed number to the healthcare provider [3, 7, 10].

5 Conclusion

The current study describes about the development of a user friendly wearable device for attendant calling. There is a need for the devices which can assist the patients to call the nurses when there is an emergency situation [1, 10]. The proposed device will not only help the patients but also will be helpful for the nurses in their day-to-day activities of the professional life [3].

References

1. George WR et al (1991) Communication system for deaf, deaf-blind, or non-vocal individuals using instrumented glove. Google Patents (ed)
2. Jafari R et al (2005) Wireless sensor networks for health monitoring. In: Mobile and ubiquitous systems: networking and services. The second annual international conference on MobiQuitous, 2005, pp 479–481
3. Sahoo S et al (2014) Wireless transmission of alarm signals from baby incubators to neonatal nursing station. In: First international conference on automation, control, energy and systems (ACES), 2014, pp 1–5
4. Borges LM et al (2009) Smart-clothing wireless flex sensor belt network for foetal health monitoring. In: 3rd international conference on pervasive computing technologies for healthcare (PervasiveHealth 2009), 2009, pp 1–4

5. Simone L et al (2004) A low cost method to measure finger flexion in individuals with reduced hand and finger range of motion. In: 26th Annual international conference of the IEEE engineering in medicine and biology society IEMBS'04, 2004, pp 4791–4794
6. Saggio G (2012) Mechanical model of flex sensors used to sense finger movements. Sens Actuators, A 185:53–58
7. Connolly J et al (2012) A new method to determine joint range of movement and stiffness in Rheumatoid Arthritic Patients. In: Annual international conference of the IEEE engineering in medicine and biology society (EMBC), 2012, pp 6386–6389
8. Grimaud JJG et al (1999) Tactile feedback mechanism for a data processing system. Google Patents (ed)
9. Kim JH et al (2005) Hand gesture recognition system using fuzzy algorithm and RDBMS for post PC. In: Fuzzy systems and knowledge discovery. Springer (ed), pp 170–175
10. Zimmerman TG et al (1987) A hand gesture interface device. In: ACM SIGCHI Bulletin pp 189–192

A Review on Visual Brain Computer Interface

Deepak Kapgate and Dhananjay Kalbande

Abstract The primary aim of brain computer interface (BCI) is to establish communication link between computers and severely disable people those are partially or totally paralyzed (locked in state) due to neurological disorders. Recently much attention is focused on visual BCI (V-BCI) because of its low cost and minimal user training time. Characteristics of visual stimuli plays important role in V-BCI as it decides the strength of visual signals in brain and system communication rate. Recently new concept called Hybrid V-BCI emerge based on concept of combining advantages of one or more basic V-BCIs to further enhance ITR and accuracy of system. In this paper we review properties of brain signals involved in V-BCI, stimulation methods used, hybrid V-BCI and discuss current challenges in V-BCI with their possible solutions.

Keywords Visual BCI · Stimulation methods · Visual signals · Hybrid V-BCIs · Information transfer rate · Repetitive visual stimuli

1 Introduction

Brain computer interface (here after as BCI) also referred as brain machine interface allow users to communicate with computer or external devices without any muscular movement and provide interface to allow cerebral activity alone to convey message and commands to computers [1]. Out of all neuroimaging methods available for monitoring brain activity electroencephalography (EEG) is most widely used and best neuroimaging technique due to its high portability, relative

D. Kapgate (✉)
Nagpur University, Nagpur, India
e-mail: deepakkapgate32@gmail.com

D. Kalbande
Department of C.T., S.P.I.T., Mumbai, India
e-mail: drkalbande@spit.ac.in

S. Gupta et al. (eds.), *Advancements of Medical Electronics*,
Lecture Notes in Bioengineering, DOI 10.1007/978-81-322-2256-9_18

low cost, high temporal resolution and few risks to users. EEG is direct cerebro-electrical activity measurement technique, having approximate ~ 0.05 s temporal resolution and ~ 10 mm spatial resolution [2].

There are four conventional non-invasive EEG based BCI paradigms based on type of brain potential they are using for command generation as visual evoked potentials (VEPs) based BCI, Slow cortical potentials (SCP) based BCI, P300 evoked potentials (P300) based BCI and sensorimotor rhythms (ERD/ERS) based BCI [3]. Conventional EEG based BCI types e.g. VEPs based BCI uses only one brain signal for command generation but recently this concept overcome by researchers as potential BCI applications go far beyond clinical settings leads to generation of new subtypes of BCIs as active BCI [4], Reactive BCI [5], Passive BCI [6], Emotional BCI [7], Collaborative BCI [8], Visual BCI [9], Auditory BCI [10]. There is no general consensus about whether the new BCI subtypes conform to original BCI definition.

Active BCI is a BCI which derives its output as control commands from brain activity which is directly consciously controlled by user, independent from external events called active BCI [4]. Reactive BCI is based on external stimuli generated brain activity to generate its output, which is indirectly modulated by the user for controlling an application [5]. Passive BCI derives its output from arbitrary brain activity without the purpose of voluntary control [6]. Emotional BCI is BCI which generates its output as control commands from brain activity which can provide significant insight into user's emotional state [7]. Collaborative BCI is BCI which generates its output by integrating brain activity information from multiple users [8]. Visual BCI (here after referred as V-BCI) is based on external visual stimuli generated brain signal modulations in visual cortex to generate its control commands [9]. Auditory BCI is based on endogenous potentials linked to reaction to external auditory stimuli to generate its control commands.

Recent study shows that many classical V-BCI paradigms demonstrate the promising prospect of real life BCI application. First Farwell and Donchin [11] demonstrated 6×6 matrix visual speller system based on P300 evoked potential in 1988. In first decade of new century number of research groups working on V-BCI and V-BCI focused scientific publications increased tremendously [12]. This is due to V-BCI systems provide high communication speed and classification accuracy up to approximately 95 %, with minimal or none user training.

In this review is focused only on V-BCI, Sect. 2 explains different brain signals involved in V-BCI, Sect. 3 discusses stimulation methods used in V-BCI. Section 4 discusses Hybrid V-BCI; Sect. 5 explains challenges and solutions of current V-BCI systems.

2 Brian Signals in V-BCI

The brain signals commonly used in V-BCI are as follows –

2.1 Brain Signals Modulated by Exogenous Stimuli

(1) *Visual evoked potential* (*VEP*)—The subtypes of VEPs include in V-BCI are transient—VEPs (TVEPs), steady state visual evoked potentials (SSVEPs) and Motion onset SSVEPs (M-SSVEPs). VEPs are generated by external visual stimulus at visual cortex. Theses brain activity modulations are relatively easy to detect as amplitude of VEPs depends on external stimuli [13]. Transient—VEPs occur with lower visual stimulation frequency (<6 Hz) while SSVEPs occurs with higher visual stimulation frequency (>6 Hz) [14]. SSVEP based V-BCI further classified as (i) time modulated VEP (t-VEP based V-BCI) where the external flash visual sequences of different targets are orthogonal in time [2]. (ii) Frequency modulated VEP (f-VEP based V-BCI) where each target flashed at unique frequency [15]. Middendorf et al. [16] first developed frequency modulated VEP based BCI with higher transfer rates. (iii) Pseudorandom code modulated VEP (c-VEP) based BCI where pseudorandom sequence determine the duration of ON/OFF state of target flash [17, 18]. Sutter [17] first demonstrated m-sequence c-VEP based BCI with very high communication rate of 10–12 words/min. Recently Kimura et al. [18] use FSK-modulated visual stimuli to achieve higher ITR.

(2) *Flash TVEPs*—It present a series of positive and negative peaks. The most prominently peaks are negative N1 and N2 peaks at around 40 and 90 ms and positive P1 and P2 peaks at around 60 and 120 ms respectively. Lee et al. [19] uses flicker stimulus to generate Flash VEPs to generate N1, P1, N2, P2 responses. Further Hong et al. [20] uses moving line stimulus to implement speller BCI with negative (N2) and positive (P2) potentials. Yuan et al. [21] developed collaborative BCI with visual Go/noGO stimulus to generate N2 signals in multiple subjects. Steady state motion VEP based BCI rely on paradigm where human perception of motion oscillates in two opposite direction [22]. Recently [23] demonstrated hybrid V-BCI by combining P300 + SSVEP signals to generate BCI with higher accuracy and improved information transfer rate (ITR).

2.2 Endogenously Modulated Brain Signals

(1) *P300 evoked Potentials*—Infrequent visual, auditory or somatosensory stimuli evokes endogenic responses about 300 ms after attending oddball stimulus called P300 potentials. According to [24] the accuracy and ITR of P300 based BCI is largely based on classifier using by BCI. Researchers developed oddball paradigm using P300 evoked potentials for large applications like visual speller, smart home control, mental prosthesis control, entertainment and so on [2].

(2) *N400 event related potentials* (*ERPs*)—These ERPs are related to cognitive responses to meaningful words in sentence. It is negative Peaks evoked about 400 ms post-stimulus onset. Principe [25] developed N400 ERP based BCI that use users thoughts to move a ball on computer screen.

(3) *N170 Event Related Potentials (ERPs)*—These are evoked (negative peaks) 130–200 ms after stimulus presentation. N170 ERPs represent neural response to faces (generally stimuli consist of images of faces). Zhang et al. [26] developed hybrid BCI combining N170 and P300 ERPs sensitive to configure processing of human faces.

Studies shown that hybrid V-BCI i.e. combining brain signals evoked due to visual stimuli and pseudo random code modulated VEP (c-VEP) based BCI are suitable for real life applications due to their higher communication speed (higher ITR), no/minimal user training time and also user acceptability as compared to other BCI types.

3 Stimulation Methods Used in V-BCI

In all V-BCI research there are three types of repetitive visual stimuli present as:

(A) *Light Stimuli*—LEDs, fluorescent lights and Xe-lights are used to generate light stimuli which are modulated at specified frequency. V-BCI system using flickering LED stimuli achieved a higher information transfer rate approximately about 68 bis/min [27]. The important factor of light stimuli that affect visual signal modulation are intensity of light stimulus, light luminance, background luminance, illumination sequence and frequency of illumination. First SSVEP based BCI that uses fluorescent light to render brain stimuli was presented in 1996 [28].
(B) *Single graphic stimuli*—Mostly computer screens with single graphic in form of squares, rectangle and arrows [29] that appears from and disappear into background used in single graphic stimuli. The performance of such BCIs is mainly depending on color of stimulus and frequency of stimulus (stimulus rate); also these parameter affects user safety and comfort fatigues and commercial acceptability of V-BCI.
(C) *Patter reversal stimuli*—Mostly used in transient—VEP based BCI research. It consists of graphical patterns that are alterated in certain oscillations e.g. checkerboards [30], line boxes, moving lines and so on.

Direct comparison of V-BCI performance based on different stimuli is difficult as there are number of other factors that affect BCI performance. Unfortunately most of articles fail to mention BCI performance (ITR) or did an offline analysis [31]. So in this review paper we are doing comparison based on frequency of visual stimulus that are broadly classifies into three frequently bands as Low (1–12 Hz), Medium (12–30 Hz) and high (30–60 Hz) bands. Table 1 demonstrates the comparison as shown.

4 Hybrid Visual BCI

In recent BCI research term evolved called “Hybrid BCI” where two or more conventional BCIs are combined in either simultaneous or sequential system

Table 1 Different stimulation methods used in visual BCI

Frequency band	Device	Visual stimuli type	Response	Color	References	Bit rate (bits/min)
Low (1–12 Hz)	LED	Light stimuli	SSVEP	–	[44]	–
				Green	[45]	–
	LCD /CRT	Single graphic stimuli	SSVEP	White/black	[46]	–
				Green	[29]	–
		Row-column paradigm	P300	White/black	[11]	–
		Single character paradigm			[47]	–
		Pattern reversal stimuli (checker board)			[48]	–
		Pattern reversal stimuli (checker board)	SSVEP		[49]	
					[50]	10.3
	LCD	Flashing stimuli	FVEPs	–	[19]	23.06
		Moving line	N2, P2	Red, green, blue, purple, yellow	[20]	–
	CRT	Visual oddball Protocol	P300	–	[51]	29.35
		Meaningful words	N400	–	[25]	
Medium (12–30 Hz)	LED	Light stimuli	SSVEP	Red	[52]	
	Fluorescent light			–	[28]	
	CRT	Single graphic stimuli	SSVEP	White/black	[53]	
				–	[54]	7.5
	LCD	Pattern reversal stimuli	SSVEP	White/black	[55]	
High (30–60 Hz)	LED	Light stimuli	SSVEP	–	[56]	
	Xe-light		SSVEP	–	[57]	

(continued)

Table 1 (continued)

Frequency band	Device	Visual stimuli type	Response	Color	References	Bit rate (bits/min)
Low + Medium (6–30 Hz)	LED	Flicker stimuli	SSVEP	White	[15]	–
				Red	[58]	27–31.5
				Green	[59]	51.47
					[27]	68
	CRT/LCD	Single graphic stimuli	SSVEP	White/black	[60]	21–58
	LCD	Flickering stimuli	C-VEP	–	[61]	
	TFT	Hybrid	SSVEP + P300	–	[62]	
	LCD	Pattern reversal stimuli	SSVEP	Red, green, yellow	[63]	–
		Pattern reversal stimuli	SSVEP	White/black	[64]	
		Pattern reversal stimuli	mVEPs (motion-onset VEPs)		[65]	42.1
Medium + High (12–60 Hz)	LED	Light stimuli	SSVEP	White	[66]	
	CRT	Single graphic stimuli		–	[37]	
Low + High	CRT	Single graphic stimuli	SSVEP		[67]	
Low + Medium + High (6–60 Hz)	LED	Light stimuli	SSVEP	Red	[68]	

organization [32]. The main goal of hybrid BCI is to improve accuracy, ITR, and error minimization by combining advantages of two conventional BCIs.

Here in this review we are discussing hybrid visual BCI where one or more brain signals involved in hybrid BCI is evoked through external repetitive visual stimuli (visual signal) i.e. out of all types of brain signals used in hybrid BCI at least one of the brain signal is used in visual BCI systems. In this review we are further dividing hybrid V-BCI in Partial Hybrid V-BCI and Complete Hybrid V-BCI.

(A) *Partial Hybrid V-BCI* are where brain signals to be processed consisting of at least one of signal as visual signal and at least one is non-visual signal e.g. SSVEP + ERD based BCI, P300 + ERD based BCI and so on.

(1) *SSVEP + ERD based hybrid V-BCI*—Both sequential and simultaneous system organizations were possible in SSVEP/ERD based BCI. Pfurtscheller et al. [33] used SSVEP + ERD hybrid V-BCI in sequential combination for orthosis control application. Result shown that false positive rate reduced by more than 50 % in hybrid BCI as compared to independent conventional BCI.

(2) *P300 + ERD based Hybrid V-BCI*—Both system organizations as sequential and simultaneous possible. In such BCI P300 potentials are suitable for discrete control commands and motion imaginary based signals (ERD) are suitable for continuous control commands. P300 + ERD based BCI found suitable for several applications as wheelchair control, robotic control decision applications [34], virtual environment and so on [32].

Another types of partial hybrid V-BCI proposed are NIRS + SSVEP based V-BVI as brain switch to ON/OFF SSVEP BCI [35] for orthosis control applications another type is ECG (heart rate) + SSVEP based sequential hybrid V-BVI where changes of heart rate measured in RRT were used to ON/OFF SSVEP operated prosthetic hand (Table 2).

(B) *Complete Hybrid V-BCI* are BCI where all of brain signals involved in hybrid V-BVI are visual signals. Studies on SSVEP + P300 hybrid V-BCI revealed that SSVEP signals are suitable for continuous control commands and P300 signals suitable for discrete control commands. Yin et al. [23] showed simultaneous processing in SSVEP + P300 hybrid V-BCI gives higher ITR and accuracy with minimum user training. Table 3 gives brief overview of complete hybrid V-BCI systems with their performance rates.

5 Challenges and Solutions

(A) *System Communication Rate (ITR) improvement*—Regarding to V-BCI the system communication rate (ITR) depends broadly on two aspects as enhancement of signal to noise ratio and separability of multiple classes.

(B) *Enhancement in SNR*—the ratio between related signal and nose is improved by decreasing other nuisance level and increasing stimulation relates signal level. Spatial filtering based signal processing methods are good in improving SNR because they combines signals from different channels as signals recorded from

Table 2 Partial hybrid visual BCI

Hybrid type	System organization	Device	Frequency band (Hz)	Visual stimuli type	References	Bit rate (bits/min)	Accuracy (%)
SSVEP + ERD	Sequential	LED	8 and 13 (Low + Medium)	Light stimuli	[33]	–	80
	Sequential	LED	15, 17 and 19 (Medium)	Light stimuli	[69]	–	98.1
	Simultaneous	LED	8 and 13 (Low + Medium)	Light stimuli	[70]	6.3	95.6
	Simultaneous	LED	8 and 13 (Low + Medium)	Light stimuli	[71]	–	81.0
P300 + ERD	Sequential	LCD	4.5	Single graphic stimuli	[72]	–	>80
	Sequential	LCD	7.6	Single graphic stimuli	[73]	22.7	90.1
	Simultaneous	LCD/ CRT	6.6	Single graphic stimuli	[34]	–	82
ECG (Heart rate) + SSVEP	Sequential	LED	6.3 and 17.3	Light stimuli	[74]	–	80

Table 3 Complete hybrid visual BCI

Hybrid type	System organization	Device	Frequency band (Hz)	Visual stimuli type	References	Bit rate (bits/min)	Accuracy (%)
P300 + SSVEP	Sequential	LCD/CRT	18	Single graphic stimuli	[75]	19.05	88.15
	Simultaneous	LED	8.18, 8.97, 9.98, 11.23, 12.85 and 14.99	Light stimuli	[23]	56.44	93.85
N170 + P300	Sequential	LCD/CRT	–	Facial images stimuli	[26]	38.7	88.7
P300 + M-VEP (Motion onset VEP)	Sequential	LCD/CRT	4.9	Light stimuli	[76]	26.7	96
P300 + ErrP (error potential)	Simultaneous	LCD/CRT	4.0	Single graphic stimuli	[77]	–	90

multiple channel are less affected by other nuisance signals. In [36] MCC pre-processing methods is used to maximize ration between visual signals and background signals with little higher computation time, other also proposed as CCA [37], ICA [38], PCA [39] etc. Also user efforts to maintain attention on stimuli affect visual signal strength [40]. User attention distraction from stimuli can deteriorate SNR. One possible solution to this is make visual stimuli move along with controlled elements.

(C) *Focused research towards Independent BCI* Mostly V-BCI are dependent in nature, as they require gaze movement in stimuli direction which may not suitable for people with severe disability in which user can not reliably control gaze [41]. Solution to this is to develop visual stimuli without users gaze.

(D) *Non-linearity and Non stationarity of EEG signals* EEG signals are non-linear in nature which may deteriorate BCI performance (classification accuracy). A solution is use of non-linear dynamic methods rather than linear methods for EEG signal characterization. Diverse behavioral and mental states of human mind leads to non-stationarity in EEG signals. Adaptive classification methods can address this problem as they automatically update classifier during online session.

(E) *Reduction in complexity of target detection* to address challenges like decrease in frequency resolution due to one frequency per target, target detection time, user habituation, probable solution could be use of different relative phases of stimuli rather than different frequencies, dual frequency stimulation to solve frequency resolution problem; use of machine learning methods in single trail classification, adaptive methods and optimization of stimulus coding method to minimize target detection time [42] and consideration of issues like attention blinks, repetition blindness, target to target interval to overcome user habituation problem.

(F) *Reduction of user fatigue* Temporary inability of user to respond optical stimuli due to use of system from longer period called mental fatigue. A lot of solutions are proposed towards this as use of displays that do not produce negative side effects, use of improved software's and dry electrodes, optimization of physical properties of stimulus e.g. Image based stimulus [43] but still this challenge need to solve effectively [12, 42] in context of V-BCI.

6 Conclusion

BCI is attractive for disable people suffering from disorders like amyotrophic lateral sclerosis, brain stem stroke or spinal cord injury as it enable them to perform many daily life activities independently which improve their quality of life, making them more independent, increasing productivity at the same time reducing the cost of intensive care. This review provides a background for finding new methodologies and paradigms to improve V-BCIs performance further. The future work in V-BCI area will be focused on design and development of efficient Hybrid V-BCI also integrating V-BCI with multimodal interfaces. Further work needed to make current V-BCI technologies effective for real life applications. Collaborative efforts from engineers, researchers, neuroscience experts and users will be needed to achieve this goal.

References

1. Wolpaw J, Birbaumer N, McFarland DJ (2002) Brain-computer inter-faces for communication and control. Clin Neurophysiol 113:767–791
2. Nicolas-Alonso LF et al (2012) Brain computer interfaces, a review. Sensors 12:1211–1279. doi:10.3390/s120201211
3. Pasqualotto E, Federici S, Belardinelli MO (2012) Toward functioning and usable brain-computer interfaces (BCIs): a literature review. Disabil Rehabil Assist Technol 7:89–103
4. Yongwook C et al (2012) Toward brain-actuated humanoid robots: asynchronous direct control using an EEG-based BCI. IEEE Trans Rob 28(5):1131–1144
5. Zander TO et al (2009) Detecting affective covert user states with passive brain-computer interfaces. In: 3rd IEEE international conference on affective computing and intelligent interaction and workshops, 2009
6. Zander TO, Kothe C (2011) Towards passive brain–computer interfaces: applying brain–computer interface technology to human–machine systems in general. J Neural Eng 8 (2):025005
7. Molina GG et al (2013) Emotional brain–computer interfaces. Int J Auton Adapt Commun Syst 6(1):9–25
8. Wang Y, Jung T-P (2011) A collaborative brain–computer interface for improving human performance. PLoS ONE 6(5):e20422
9. Cheng M, Gao X, Gao S, Xu D (2002) Design and implementation of a brain–computer interface with high transfer rates. IEEE Trans Biomed Eng 49(10):1181–1186
10. Hill NJ et al (2005) An auditory paradigm for brain–computer interfaces. In: Advances in neural information processing systems. MIT Press, Cambridge, pp 569–576
11. Farwell LA, Donchin E (1988) Talking off the top of your head-toward a mental prosthesis utilizing event-related brain potentials. Electroencephalogr Clin Neurophysiol 70(6):510–523
12. Rezaiey RF et al (2012) P300 brain computer interface: current challenges and emerging trends. J Front Neuroeng 5:14
13. Yijun W et al (2006) A practical VEP-based brain-computer interface. IEEE Trans Neural Syst Rehabil Eng 14:234–240
14. Xiaorong G et al (2003) A BCI-based environmental controller for the motion-disabled. IEEE Trans Neural Syst Rehabil Eng 11:137–140
15. Wu ZH, Yao DH (2008) Frequency detection with stability coefficient for steady-state visual evoked poten-tial (SSVEP)-based BCIs. J Neural Eng 5(1):36–43
16. Middendorf M, McMillan G, Calhoun G, Jones KS (2000) Brain–computer interfaces based on the steady-state visual-evoked response. IEEE Trans Rehabil Eng 8(2):211–213
17. Sutter EE (1984) The visual evoked response as a communication channel. IEEE Trans Biomed Eng 31(8):583
18. Kimura Y, Tanaka T, Higashi H, Morikawa N (2013) SSVEP-Based brain–computer interfaces using FSK-modulated visual stimuli. IEEE Trans Biomed Eng 60(10):2831–2838
19. Lee PL, Hsieh JC, Wu CH, Shyu KK, Chen SS, Yeh TC, Wu YT (2006) The brain computer interface using flash visual evoked potential and independent component analysis. Ann Biomed Eng 34(10):1641–1654
20. Hong B, Guo F, Liu T, Gao X, Gao S (2009) N200-speller using motion-onset visual response. Clin Neurophysiol 120(9):1658–1666

21. Yuan P et al (2013) A collaborative brain–computer interface for accelerating human decision making. In: Stephanidis C, Antona M (eds) UAHCI/HCII, Part I, LNCS 8009. Springer, Berlin, pp 672–681
22. Xie J et al (2012) Steady-state motion visual evoked potentials produced by oscillating Newton's rings: implications for brain-computer interfaces. J PLoS ONE 7(6):e39707
23. Yin E, Zhou Z, Jiang J, Chen F, Liu Y, Hu D (2014) A speedy hybrid BCI spelling approach combining P300 and SSVEP. IEEE Trans Biomed Eng 61(2):473–483
24. Rivet B et al (2009) Algorithm to enhance evoked potentials: application to brain-computer interface. IEEE Trans Biomed Eng 56:2035–2043
25. Principe JC (2013) The cortical mouse: a piece of forgotten history in noninvasive brain–computer interfaces. IEEE Pulse 4(4):26–29
26. Zhang Y, Zhao QB, Jin J, Wang XY, Cichocki A (2012) A novel BCI based on ERP components sensitive to configural processing of human faces. J Neural Eng 9(2):026018
27. Gao X, Xu D, Cheng M, Gao S (2003) A BCI-based environmental controller for the motion-disabled. IEEE Trans Neural Syst Rehabil Eng 11(2):137–140
28. Calhoun GL, McMillan GR EEG-based control for human-computer interaction. In: Proceedings of the 3rd annual symposium on human interaction with complex systems (HICS'96), pp 4–9. Dayton, Ohio, USA, Aug 1996
29. Beverina F, Palmas G, Silvoni S, Piccione F, Giove S (2003) User adaptive BCIs: SSVEP and P300 based interfaces. Psychol J 1:331–354
30. Kluge T, Hartmann M (2007) Phase coherent detection of steady-state evoked potentials: experimental results and application to brain-computer interfaces. In: Proceedings of the 3rd international IEEE EMBS conference on neural engineering, pp 425–429, May 2007
31. Zhu D et al (2010) A survey of stimulation methods used in SSVEP-based BCIs. J Comput Intell Neurosci, Article ID 702357
32. Pfurtscheller G et al (2010) The hybrid BCI. Front Neurosci 4:30
33. Pfurtscheller G, Solis-Escalante T, Ortner R, Linortner P, Muller-Putz GR (2010) Self-paced operation of an SSVEP-based orthosis with and without an imagery-based "brain switch": a feasibility study towards a hybrid BCI. IEEE Trans Neural Syst Rehabil Eng 18(4):409–414
34. Riechmann H, Hachmeister N, Ritter H, Finke A (2011) Asynchronous, parallel on-line classification of P300 and ERD for an efficient hybrid BCI. In: Proceedings of the 5th international IEEE/EMBS conference on neural engineering (NER'11), pp 412–415, May 2011
35. Coyle SM et al (2007) Brain–computer interface using a simplified functional near-infrared spectroscopy system. J Neural Eng 4:219–226
36. Garcia-Molina G, Zhu DH, Abtahi S (2010) Phase detection in a visual-evoked-potential based brain computer interface. In: Proceedings of 18th European signal processing conference, pp 949–953
37. Lin Z, Zhang C, Wu W, Gao X (2007) Frequency recognition based on canonical correlation analysis for SSVEP-based BCIs. IEEE Trans Biomed Eng 54(6):1172–1176
38. Kun L et al Single trial independent component analysis for P300 BCI system. In: Proceedings of the 31th annual international conference of the IEEE engineering in medicine and biology society (EMBCS'09), pp 4035–4038. Minneapolis, MN, USA, Sept 2009
39. Pouryazdian S, Erfanian A (2009) Detection of steady-state visual evoked potentials for brain-computer interfaces using PCA and high-order statistics. Proc World Cong Med Phys Biomed Eng 25:480–483
40. Muller MM, Malinowski P, Gruber T, Hillyard SA (2003) Sustained division of the attentional spotlight. Nature 424(6946):309–312
41. Allison BZ et al (2008) Towards an independent brain-computer interface using steady state visual evoked potentials. Clin Neurophysiol 119(2):399–408
42. Gao S et al (2014) Visual and auditory brain–computer interfaces. IEEE Trans Biomed Eng 61 (5):1436–1447
43. Rakotomamonjy A, Guigue V (2008) BCI competition III: dataset II-ensemble of SVMs for BCI P300 speller. IEEE Trans Biomed Eng 55:1147–1154

44. Maggi L, Parini S, Piccini L, Panfili G, Andreoni G A four command BCI system based on the SSVEP protocol. In: Proceedings of the 28th annual international conference of the IEEE engineering in medicine and biology society (EMBC'06), pp 1264–1267. New York, NY, USA, Aug 2006
45. Piccini L, Parini S, Maggi L, Andreoni G A wearable home BCI system: preliminary results with SSVEP protocol. In: Proceedings of the 27th annual international conference of the IEEE engineering in medicine and biology society (EMBC'05), vol 7, pp 5384–5387. Shanghai, China, Sept 2005
46. Wang Y, Gao X, Hong B, Jia C, Gao S (2008) Brain-computer interfaces based on visual evoked potentials: feasibility of practical system designs. IEEE Eng Med Biol Mag 27(5):64–71
47. Guger C et al (2009) How many people are able to control a P300-based brain-computer interface (BCI)? Neurosci Lett 462:94–98
48. Townsend G et al (2010) A novel P300-based brain-computer interface stimulus presentation paradigm : moving beyond rows and columns. Clin Neurophysiol 121:1109–1120
49. Trejo LJ, Rosipal R, Matthews B (2006) Brain-computer interfaces for 1-D and 2-D cursor control: designs using volitional control of the EEG spectrum or steady-state visual evoked potentials. IEEE Trans Neural Syst Rehabil Eng 14(2):225–229
50. Lalor EC, Kelly SP, Finucane C et al (2005) Steady-state VEP-based brain-computer interface control in an immersive 3D gaming environment. EURASIP J Appl Sig Process 2005 (19):3156–3164
51. Lenhardt A, Kaper M, Ritter HJ (2008) An adaptive P300-based online brain–computer interface. IEEE Trans Neural Syst Rehabil Eng 16(2):121–130
52. Leow RS, Ibrahim F, Moghavvemi M Development of a steady state visual evoked potential (SSVEP)-based brain computer interface (BCI) system. In: Proceedings of the international conference on intelligent and advanced systems (ICIAS'07), pp 321–324. Kuala Lumpur, Malaysia, Nov 2007
53. Cecotti H, Graeser A Convolutional neural network with embedded fourier transform for EEG classification. In: Proceedings of the 19th international conference on pattern recognition (ICPR'08), pp. 1–4. Tampa, Fla, USA, Dec 2008
54. Kelly SP, Lalor E, Reilly RB, Foxe JJ Independent brain computer interface control using visual spatial attention-dependent modulations of parieto-occipital alpha. In: Proceedings of the 2nd international IEEE EMBS conference on neural engineering, pp 667–670. Arlington, Va, USA, March 2005
55. Kelly SP, Lalor E, Finucane C, Reilly RB (2004) A comparison of covert and overt attention as a control option in a steady-state visual eyoked potential-based brain computer interface. In: Proceedings of the 26th annual international conference of the IEEE engineering in medicine and biology society (EMBC'04), vol 2, pp 4725–4728. San Francisco, Calif, USA, Sept 2004
56. Garcia Molina G (2008) High frequency SSVEPs for BCI applications. In: Brain-computer interfaces for HCI and games
57. Huang M, Wu P, Liu Y, Bi L, Chen H (2008) Application and contrast in brain-computer interface between Hilbert-Huang transform and wavelet transform. In: Proceedings of the 9th international conference for young computer scientists (ICYCS'08), pp 1706–1710, Nov 2008
58. Muller GR et al (2008) Control of an electrical prosthesis with an SSVEP-based BCI. IEEE Trans Biomed Eng 55(1):361–364
59. Parini S, Maggi L, Turconi AC, Andreoni G (2009) A robust and self-paced BCI system based on a four class SSVEP paradigm: algorithms and protocols for a high- transfer-rate direct brain communication. Comput Intell Neurosci 2009:11 pp, Article ID 864564
60. Zhang Y, Xu P, Liu T, Hu J, Zhang R, Yao D (2012) Multiple frequencies sequential coding for SSVEP-based brain–computer interface. PLoS One 7(3):e29519
61. Vasquez PM, Bakardjian H, Vallverdu M, Cichocki A (2008) Fast multi-command SSVEP brain machine interface without training. In: Proceedings of the 18th international conference on artificial neural networks (ICANN'08), pp 300–307, Sept 2008

62. Xu M, Qi H, Wan B, Yin T, Liu Z, Ming D (2013) A hybrid BCI speller paradigm combining P300 potential and the SSVEP blocking feature. J Neural Eng 10:026001
63. Cheng M, Gao X, Gao S, Xu D Multiple color stimulus induced steady state visual evoked potentials. In: Proceedings of the 23rd annual international conference of the IEEE engineering in medicine and biology society (EMBC'01), vol 2, pp 1012–1014. Istanbul, Turkey, Oct 2001
64. Martinez P, Bakardjian H, Cichocki A (2008) Multi command real-time brain machine interface using SSVEP: feasibility study for occipital and forehead sensor locations. In: Advances in cognitive neurodynamics, pp 783–786
65. Liu T, Goldberg L, Gao S, Hong B (2010) An online brain–computer interface using non-flashing visual evoked potentials. J Neural Eng 7:036003
66. Wang Y, Wang R, Gao X, Gao S (2005) Brain-computer interface based on the high-frequency steady-state visual evoked potential. In: Proceedings of the 1st international conference on neural interface and control, pp 37–39, May 2005
67. Sami S, Nielsen KD (2004) Communication speed enhancement for visual based brain computer interfaces. In: Proceedings of the 9th annual conference of the international FES Society
68. Ruen SL, Ibrahim F, Moghavvemi M (2007) Assessment of steady-state visual evoked potential for brain computer communication. In: Proceedings of the 3rd Kuala Lumpur international conference on biomedical engineering, pp 352–354
69. Savic A, Kisic U, Popovic M (2012) Toward a hybrid BCI for grasp rehabilitation. In: Proceedings of the 5th European conference of the international federation for medical and biological engineering, pp 806–809
70. Brunner C, Allison BZ, Altstätter C, Neuper C (2011) A comparison of three brain-computer interfaces based on event related de-synchronization, steady state visual evoked potentials, or a hybrid approach using both signals. J Neural Eng 8(2), Article ID 025010
71. Allison BZ, Brunner C, Kaiser V, Müller-Putz GR, Neuper C, Pfurtscheller G (2010) Toward a hybrid brain-computer interface based on imagined movement and visual attention. J Neural Eng 7(2), Article ID 026007
72. Yuanqing L et al (2010) An EEG-based BCI system for 2-D cursor control by combining Mu/Beta rhythm and P300 potential. IEEE Trans Biomed Eng 57(10):2495–2505
73. Yu T (2013) A hybrid brain-computer interface-based mail client. J Comput Math Methods Med 2013, Article ID 750934
74. Scherer R, Müller-Putz GR, Pfurtscheller G (2007) Self initiation of EEG-based brain-computer communication using the heart rate response. J Neural Eng 4(4):L23–L29
75. Panicker RC, Puthusserypady S, Sun Y (2011) An asynchronous P300 BCI with SSVEP-based control state detection. IEEE Trans Biomed Eng 58(6):1781–1788
76. Jin J et al (2012) A combined brain computer interface based on P300 potentials and motion onset visual evoked potentials. J Neurosci Methods 205:265–276
77. Dal Seno B et al (2010) Online detection of P300 and error potentials in a BCI speller. Comput Intell Neurosci, Article ID 307254

Design of Lead-Lag Based Internal Model Controller for Binary Distillation Column

Rakesh Kumar Mishra and Tarun Kumar Dan

Abstract Lead-Lag based Internal Model Control method is proposed based on Internal Model Control (IMC) strategy. In this paper, we have designed the Lead-Lag based Internal Model Control for binary distillation column for SISO process (considering only bottom product). The transfer function has been taken from Wood and Berry model. We have find the composition control and disturbance rejection using Lead-Lag based IMC and comparing with the response of simple Internal Model Controller.

Keywords SISO · Lead-lag · Internal model control · Wood and berry · Distillation column

1 Introduction

For purification of final products distillation process is used in chemical and petroleum industries. Distillation columns are used to enhance mass transfer or for transferring heat energy. Binary distillation column consists of a vertical column, where trays are used to increase the efficiency of separation [1]. Condenser and reboiler is the main part of distillation column. Here condenser is use to condense the vapour and a reboiler is used to provide heat for the necessary vaporization. A reflux drum is used to hold the condensed vapour to recycle the liquid reflux to back from top of the column [2]. Figure 1 has shown the distillation column.

Internal Model Control and Model Predictive Control have successful controller design technique for industrial system. For SISO system, IMC design procedure is

R.K. Mishra (✉) · T.K. Dan
Department of Electronics and Communication Engineering,
National Institute of Technology, Rourkela, India
e-mail: rakeshbbu@gmail.com

T.K. Dan
e-mail: tkdan@nitrkl.ac.in

S. Gupta et al. (eds.), *Advancements of Medical Electronics*,
Lecture Notes in Bioengineering, DOI 10.1007/978-81-322-2256-9_19

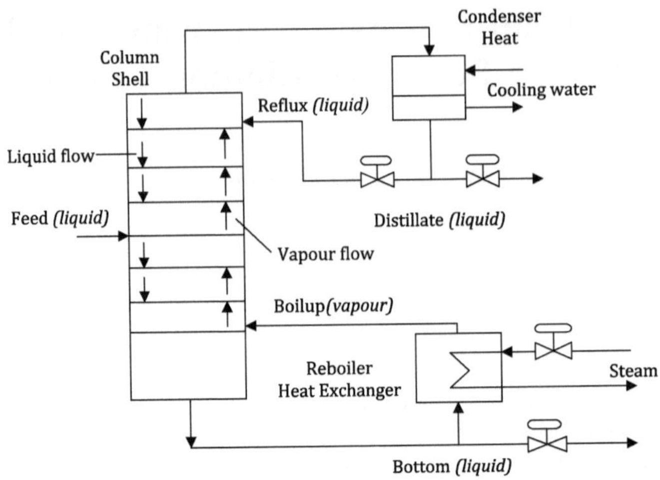

Fig. 1 Distillation column

similar to open loop controller design and there are number of advantages such as easier to tune, good set point tracking, compensation for disturbance and model uncertainty [3, 4].

2 Internal Model Control Algorithm

The Internal Model control algorithm is given below the Fig. 2 indicate the block diagram of internal model control [5, 6].

Where Q(s) is the primary controller (IMC) transfer function, $G_p(s)$ is the process transfer function, $G_m(s)$ is the process model transfer function, r(s) is set point, e(s) is error, c(s) is manipulated variable, d(s) is disturbance, $y_m(s)$ is model output and y(s) is controlled variable (process output).

Internal Model Control perform faster in dynamics stage compare to statics one. Hence

$$\mathrm{Q(s)} = \frac{1}{G_P(s)} \tag{1}$$

But Eq. (1) valid for stable system without delay, so we can design IMC for time delay system using lag-lead network.

The controller design procedure has been generalized to the following step [3].

1. First we have identified the process model into good stuff and bad stuff by using all pass formulation or using simple factorization.

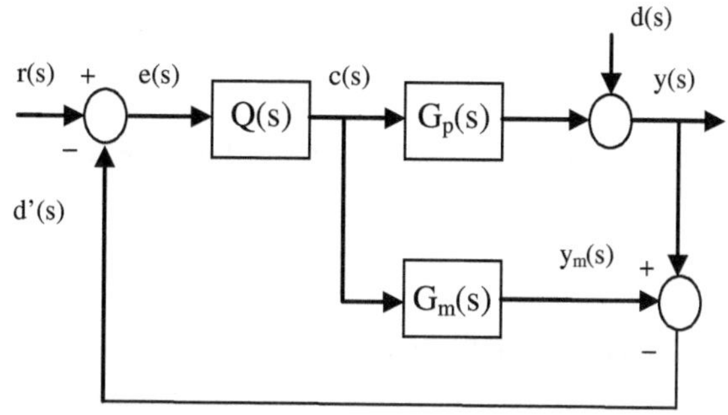

Fig. 2 Internal model control structure

2. Invert the (good stuff) invertible portion of the process model and to make proper add the filter.
3. After adding the filter we add lag-lead network which is a form of $\alpha s + 1/\beta s + 1$.

$$Q(s) = \frac{f(s)}{G_{m\_}(s)} T(s) \tag{2}$$

Different strategy in which Internal Model Control work which we given below.

When Model Perfect, No Disturbances [$G_p(s) = G_m(s)$ and $d(s) = 0$]:

$$y(s) = G_p(s)\, Q(s)\, r(s) \tag{3}$$

When model is Perfect and Disturbance Effect the process:

$$\begin{aligned} y(s) = [&G_p(s)\, Q(s)\, r(s) \\ &+ \{1 - G_m(s)Q(s)\}d(s)] \end{aligned} \tag{4}$$

With only Disturbance and Disturbance Rejection:

$$\begin{aligned} y(s) &= [1 - G_m(s)\ Q(s)]d(s) \\ d(s) &= G_d(s)\, L(s) \end{aligned} \tag{5}$$

L(s) is load disturbance.
For IMC design we have taken filter

$$f(s) = \frac{1}{(1 + \lambda s)^n} \tag{6}$$

For disturbance rejection we have use the Lag-Lead network which is in form of

$$T(s) = \frac{(\alpha s + 1)}{(\beta s + 1)} \tag{7}$$

where α and β are time constant and used as a tuning parameter for lead-lag based Internal Model Controller.

3 Distillation Column

Distillation column has two output top X_D and bottom X_B composition and four inputs such as two disturbance feed flow (F) and feed composition (X_F) and two manipulated variable reflux (L) and bottom product flow (B/S). The process model is described by first order transfer function with dead time (8) having gain constant and time constant for process channel [7, 8] (Fig. 3).

$$G = \frac{K_m e^{-\tau s}}{Ts + 1} \tag{8}$$

K_m represents the process gain, T is the time constant and τ is the dead time. The process model is a nonlinear and represented by using several parameters given below in Table 1 [9, 10].

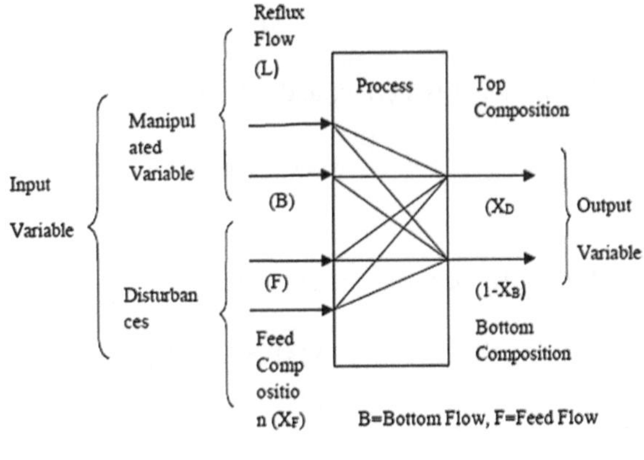

Fig. 3 Distillation process block diagram

Table 1 Model parameters

Channel (bottom product)	K_m (process gain)	T (time in min.)	τ (dead time)
$S{-}x_B$	−19.4	14.4	3
$X_F{-}x_B$	4.9	13.2	3.4

4 Internal Model Control for Distillation Column

We have taking wood and berry 2 × 2 process for distillation column and considering only bottom product (X_B). We have found several response using Lead-Lag based IMC controller. The several responses for different IMC strategies are given below.

4.1 When Process is Perfect and No Distrubance

When process is Perfect and no disturbance, it mines process model is equal to actual process and there is no mismatch. Hence manipulated and control variable response given below.

Figure 4 shows the rate of stem flow (s), which indicates that by using lead-lag based internal model controller response, is accurate, good set point tracking and less settling time as compare to generalize internal model controller response. The lead-lag based internal model controller response approach to critical damped but generalizes internal model controller approach to over damp.

Figure 5 shows that controlled variable response X_B (bottom product) of distillation column. It indicated that with lead-lag based internal model controller output response is more accurate good set point tracking and less settling time as compare to generalize internal model controller.

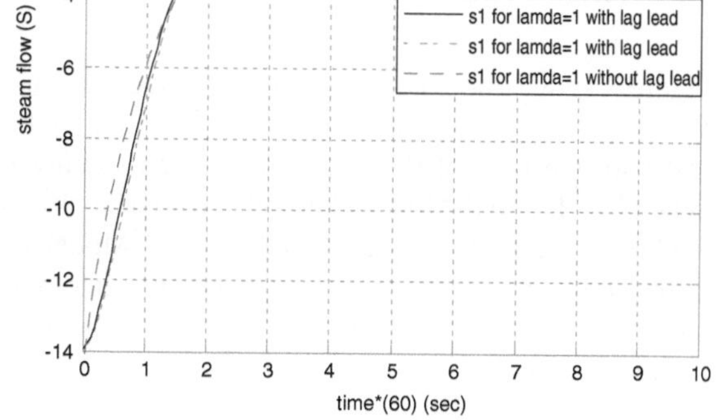

Fig. 4 Manipulated variable response

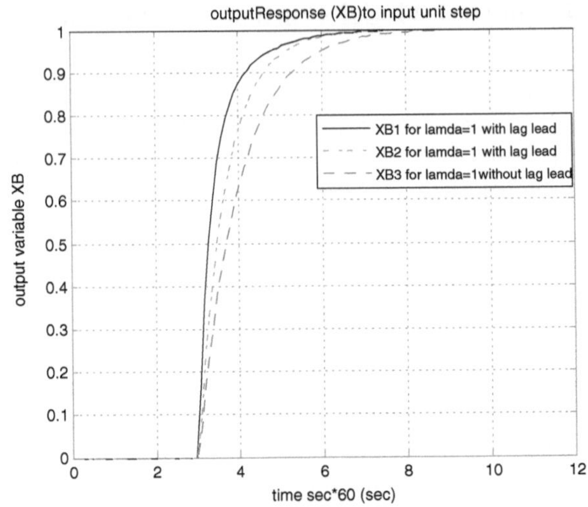

Fig. 5 Controlled variable response

4.2 When Process is Perfect and with Distrubance

When process is perfect with disturbance there is no mismatch occurs in an output response and using α, β and λ tuning parameter we remove disturbance easily. Figure 6 shows the difference between the responses of lead lag based internal model controller and generalize internal model controller. The graph shows that using lead lag based IMC for distillation column is less affected by the disturbance, here I have taken $\lambda = 1\,\text{min}$, $\alpha = 0.9\,\text{min}$, $\beta = 0.01\,\text{min}$ and $\alpha = 0.4\,\text{min}$, $\beta = 0.1\,\text{min}$ as a tuning parameter.

Figure 6 shows the difference between the responses of lead lag based internal model controller and generalize internal model controller.

4.3 When Process Only with Distrubance

Figure 7 shows the controlled variable response considering with only disturbance. We find controlled variable response using lead-lag based Internal Model Controller and with the help of this graph disturbance totally reject up to 60 min but using generalize Internal Model Controller response shows the disturbance reject up to 80–90 min. Hence lead-lag based Internal Model Control gave accurate and disturbance free response compare to generalize Internal Model Controller, here I have taken $\lambda = 1\,\text{min}$, $\alpha = 0.9\,\text{min}$, $\beta = 0.01\,\text{min}$ and $\alpha = 0.4\,\text{min}$, $\beta = 0.1\,\text{min}$ as a tuning parameter.

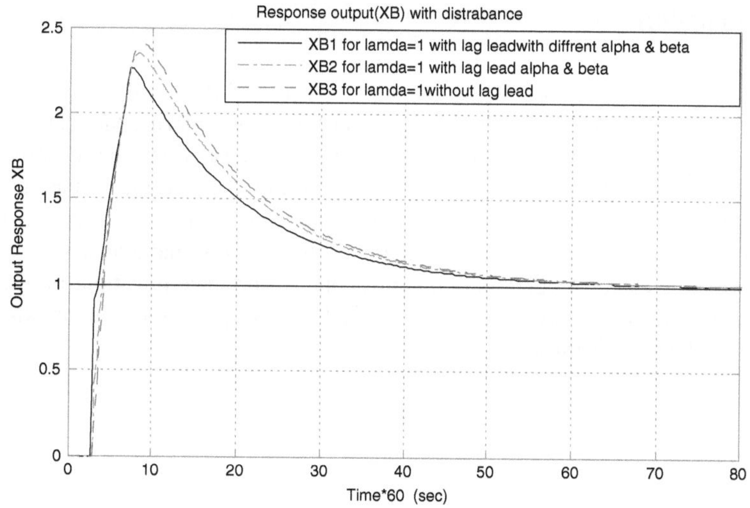

Fig. 6 Controlled variable response when model is perfect and with disturbance

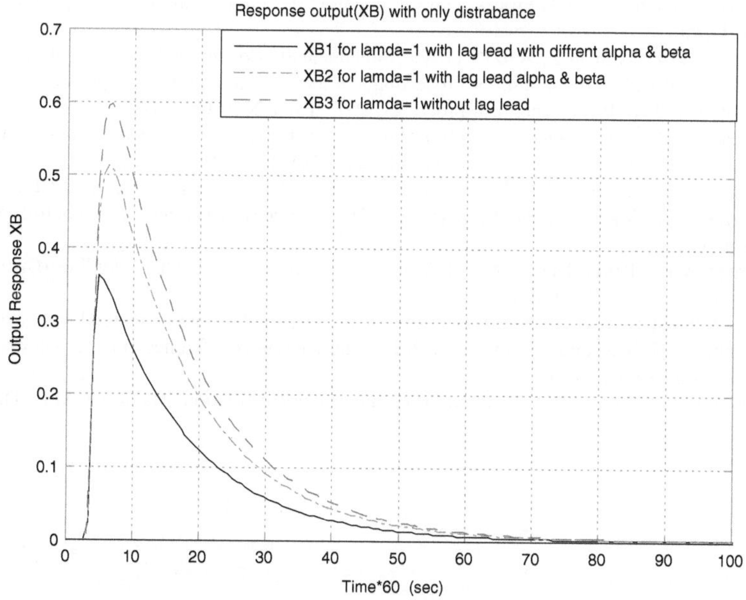

Fig. 7 Controlled variable response considering only disturbance

5 Conclusion

This paper shows a solution for composition control and disturbance rejection for a binary distillation column. We have used Lead-Lag based Internal Model Controller for composition control and disturbance rejection and comparing the response with generalize Internal Model Controller response. In this paper we have taking single input single output (SISO) process by considering only bottom product (X_B). The lead-lag based internal model control there are three tuning parameter such as α, β and λ which is used for composition control and disturbance rejection. Further this form is used for MIMO system (2 × 2, 3 × 3, 4 × 4) for good set point tracking and early disturbance rejection.

References

1. Mikles J, Fikar M (2000) Process modeling, identification and control, vol I. STU Press, Bratislava, p 22
2. Minh VT, Mansor W, Muhamed W (2010) Model predictive control of a condensate distillation column. IJSC 1:4–12
3. Wayne Bequette B (2003) Process control modeling design and simulation. PHI Publication, New Delhi
4. Muhammad D, Ahmad Z, Aziz N (2010) Implementation of internal model control (imc) in continuous distillation column. In: Proceedings of the 5th international symposium on design, operation and control of chemical processes, PSE ASIA 2010 Organizers
5. Cirtoaje V, Francu S, Gutu A (2002) Valente noi ale reglarii cu model intern. Buletinul Universit tii Petrol-Gaze Ploiesti (Seria Tehnica) LIV(2):1–6. ISSN 1221-9371
6. Morari M, Zafiriou E (1989) Robust process control. Prentice Hall, Englewood Cliffs
7. Seborge DE, Edgar TF, Mellichamp DA (2004) Process dynamics and control. Wiley, Singapore
8. Luyben WL (1987) Derivation of Transfer Functions for Highly NonlinearDistillation Columns. Ind Eng Chem Res 26:2490–2495
9. Alina-Simona B, Nicolae P (2011) Using an internal model control method for distillation column. In: Proceedings of the 2011 IEEE international conference on mechatornics and automation, August 2011
10. Băieşu Alina-Simona (2011) Modeling a nonlinear binary distillation column. CEAI 13(1):49–53

Clinical Approach Towards Electromyography (EMG) Signal Capturing Phenomenon Introducing Instrumental Activity

Bipasha Chakrabarti, Shilpi Pal Bhowmik, Swarup Maity and Biswarup Neogi

Abstract The research on electromyography (EMG) signals analysis is allied with clinical/biomedical applications, evolvable hardware chip (EHW) development, and modern human computer interaction era. EMG signals acquired from muscles require advanced methods for detection, decomposition, processing, and classification. The purpose of this paper is to illustrate the capturing phenomenon of EMG Signal introducing instrumental activity. It further point up some ideas about the block diagram of the EMG Signal recording instrument and the procedural approach towards the EMG recording techniques, provide efficient and effective ways of understanding the signal and its nature. The clinical real-time activity of EMG recording for biceps brachii muscle is presented with flow diagram. This paper provides researchers concrete and valuable information of EMG signal and its analysis procedures.

Keywords Electromyography · Motor unit action potential (MUAP) · Nervous system · Surface EMG · Invasive EMG · EMG recorder

1 Introduction

Biomedical signal means a collective electrical signal acquired from any part of our body that represents a physical variable of interest. This signal is normally a function of time and is describable in terms of its amplitude, frequency and phase.

B. Chakrabarti (✉)
Gargi Memorial Institute of Technology, Kolkata, India
e-mail: bcbipasha@gmail.com

S.P. Bhowmik · S. Maity
EIE Department, JIS College of Engineering, Kalyani, India

B. Neogi
ECE Department, JIS College of Engineering, Kalyani, India
e-mail: biswarupneogi@gmail.com

S. Gupta et al. (eds.), *Advancements of Medical Electronics*,
Lecture Notes in Bioengineering, DOI 10.1007/978-81-322-2256-9_20

The EMG signal is a biomedical signal that measures electrical currents generated in muscles during its contraction representing neuromuscular activities. The nervous system always controls the muscle activity (contraction/relaxation). Hence, the EMG signal is a complicated signal, which is controlled by the nervous system and is dependent on the anatomical and physiological properties of muscles. EMG signal acquires noise while traveling through different tissues. Moreover, the EMG detector, particularly if it is at the surface of the skin, collects signals from different motor units at a time which may generate interaction of different signals. Detection of EMG signals with powerful and advance methodologies is becoming a very important requirement in biomedical engineering. The main reason for the interest in EMG signal analysis is in clinical diagnosis and biomedical applications. Human motor system manipulates posture, strength and gestures in humans, includes central motor system and a large number of motor units (MU) [1]. Every single MU incorporates a motor neuron in spinal cord, multiple branches of its axon and muscle fibers for innervation. Human movements are possible for the relationship between central nervous system and skeletal muscle.

2 Short Informatics Study on EMG Sensor

Inherently, two types of EMG capturing techniques are broadly used: surface EMG (SEMG) and intramuscular EMG (Invasive EMG). A needle should be inserted into muscle tissue at the time of experiment of intramuscular EMG. Though intramuscular EMG is very precise in sensing muscle activity, is normally considered to be impractical for human computer interaction applications. Invasive EMG allows effective information about the state of muscle and its nerve [2]. At the time of rest, normal muscle acquires some normal electrical signal when needle is inserted into muscles but abnormal impulsive activity indicates some muscle damage. Primarily surface EMG is noisier than invasive EMG as motor unit action potentials (MUAP) must proceed through body tissues such as fat and skin before a sensor on the surface can capture them.

The raw EMG Signal consists of a series of spikes whose amplitude depends on the amount of force delivered by the muscle, the stronger the contraction of the muscle, the larger the amplitude of the EMG Signal. The frequencies of the spikes are the firing rates of the motor neurons. Since the amplitude of the EMG Signal is directly related to the force exerted by the muscle, it is used to determine the force signal sent to the exoskeleton. Fundamentally, two types of electrodes are utilized: Dry electrodes and Gelled electrodes. Dry electrodes are directly communicated with skin and are used where geometry and size of electrode does not allow gel. An electrolytic gel is required as chemical interface between skin and metallic part of electrodes in gelled electrodes. This electrode may be disposable or reusable. Disposable electrodes are very light in nature and with proper application it

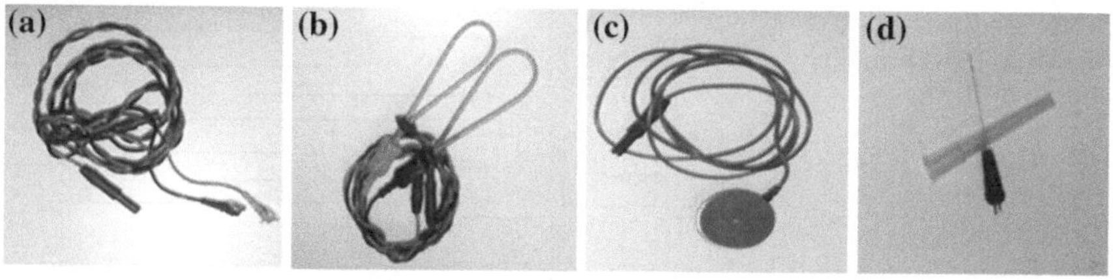

Fig. 1 **a** EMG NCV disc electrode, **b** EMG sensory ring electrode, **c** EMG GND disc electrode, **d** needle electrode

minimizes the instability of electrode displacement even during rapid movement [3]. Basically EMG sensors are essential to capture various EMG signals from different body muscles for different movements. EMG sensor can be worn around the arm or as wrist-watch. The different types of electrode are shown. Figure 1a shows that EMG NCV disc electrode, which is basically an electrode used for surface EMG signal measurement. Figure 1b shows EMG sensory ring electrode. Figure 1c gives the idea about EMG GND disc electrode. Figure 1d shows idea about needle electrode, inserted into the muscle to capture raw EMG signal.

3 Block Diagram of EMG Recorder Instrument

The schematic diagram represents the overall EMG recorder instrument and anyone can assure about their muscle condition by this recorder. Power supply feeds power to spike protector (surge protector). Surge protector is being used to protect electrical devices from voltage spikes, either obstructing or shorting ground of excess voltages beyond safe threshold. Spike protector links up two devices, personal computer besides acquisition box. Acquition of data is major goal of acquisition box. Data acquisition system adapts analog waveform (real world physical conditions) into digital numeric values.

Data acquisition incorporates sensors, signal conditioning circuitry and ADC. Physical phenomenon is being altered to equivalent electrical signals by sensor transducer. Amplification, filtering, attenuation isolation, linearization are individual parts of signal conditioning circuit. Acquisition box colligates four parts of EMG recording machine. Human muscle electrical signal is being fetched by two electrodes, plug into two channel amplifier arm, attached to CPU. Stimulus and foot switch are being employed to gather electrical signals. Headphones are being used to Bera operation. Googles is being applied for blink operation (Fig. 2).

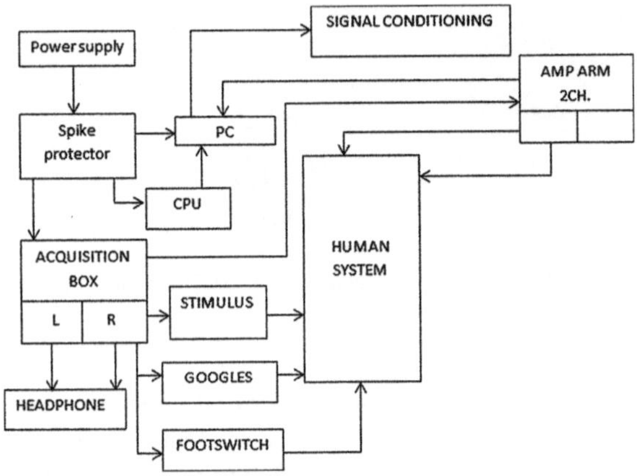

Fig. 2 Schematic block diagram of EMG recording instrument

4 Procedural Approach Towards EMG Recording Technique

The flow chart describes about the entire procedure of recording EMG data. The convenient muscle is being properly selected for descent EMG data. Reference as well as active muscles is being decently distinguished. Now, the threshold value, rise time, delay time, sweep, amplifier value are being reform in EMG window to get appropriate result. Needle/Electrode is being injected to proposed muscle (Fig. 3).

Now if the signal is not precise, each step is being scrutinized, such as muscle is being delegated accurately and ground, active muscle, reference muscle should be differentiated properly. If the signal is precise, true signal is being gathered through amplifier arm and then feed to personal computer and is being saved as patient information for future work.

5 Clinical Activity on EMG Captured and Reporting

The capability of muscle to produce tension is termed as muscle strength [5]. Muscles are innervated by nerves. Nerve simulation is the requisite condition to produce muscular force that generates mechanical work. Many motor-units (MU) incorporate a muscle and they consist of respective muscle fibers that are excited by branches of muscle fibers of a single motor nerve cell. Resting potential is similar across of both membrane and nerve fiber [6]. Depending on degree of constraint and posture desired by muscle, one MU may contain 3–2,000 muscle fibers. A low strength muscle have few muscle fibers in one MU, but the larger muscle having high strength that controls larger movement may bear 100–1,000 fibers per MU [7]. At the time of contraction, an action potential establishes muscle action that proceeds to axon and is transmitting across the motor end-plates.

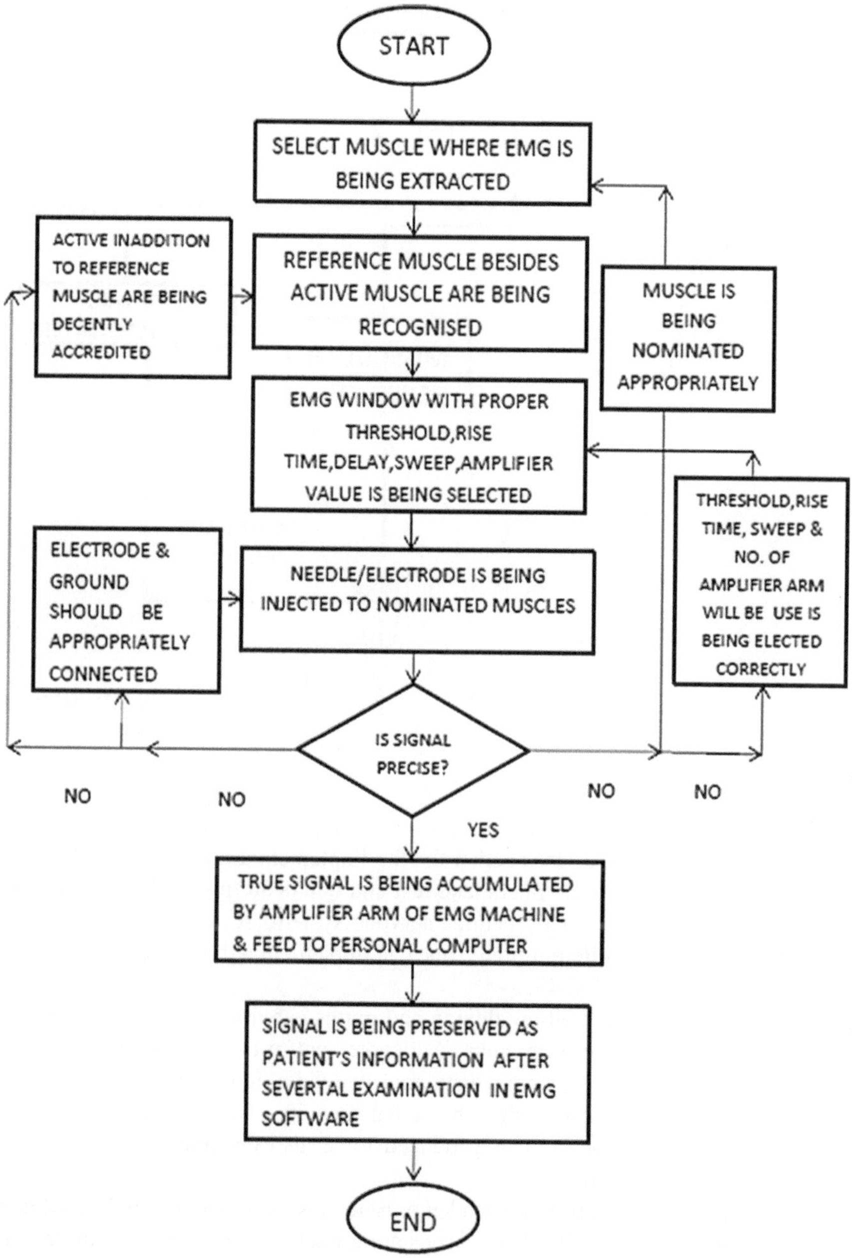

Fig. 3 Basic flow chart of EMG recording technique

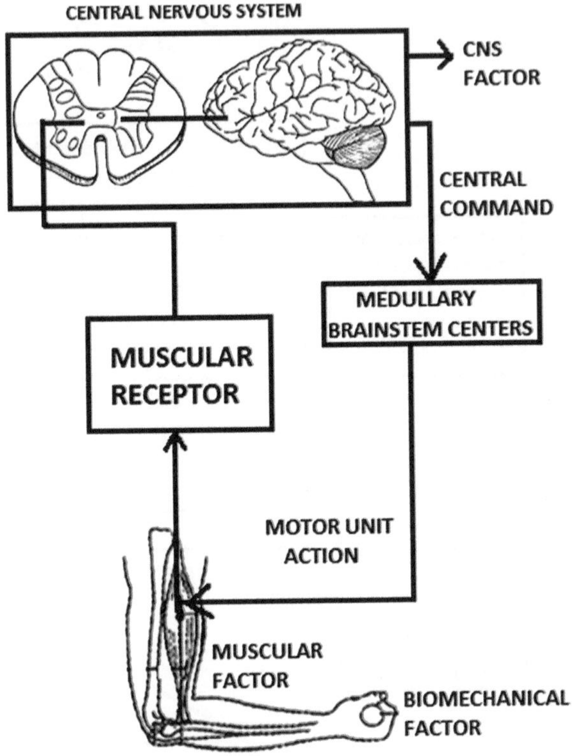

Fig. 4 Block diagram representation of motor control mechanism and different factors associated to generate muscle force

All the muscle fibers in one MU are not contracted at same time, only those fibers receive impulse that are needed for the implementation of specific function. So, at a particular time some fibers take rest and others are inspired to contract and all programs are monitored by central nervous system. As CNS handles all programs that are held by different fibers, one can implement a thought control arm that is controlled by artificial CNS.

The Fig. 4 explains the elbow flexion and in this experiment only biceps brachii is the active muscle. The term 'muscle' denotes entire muscle-tendon unit and it signifies a group that covers from a distal muscle-tendon junction to a proximal one. Muscle's basic property establishes the actual capacity of muscle to produce force or shorten at given velocity. This produced force also depends on muscle architecture and joint conception.

Biceps brachii muscle is a two headed muscle, lies on upper arm in the middle of shoulder and elbow. Both heads of biceps are joined to form a single muscle belly and is attached to upper fore arm. When it crosses shoulder and elbow, its principle function is to flex the fore arm at the elbow and supinate the forearm [8]. It is tri-effective, means it works across three junctions and its main purpose is to supinate the forearm and flex the elbow (Fig. 5).

Fig. 5 Diagram of elbow flexion, extension with null value

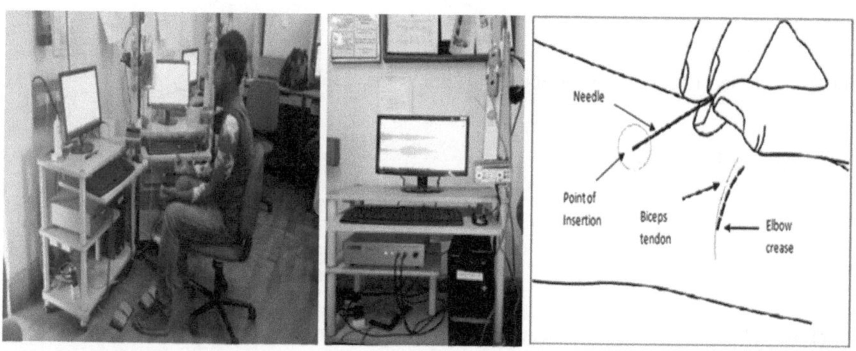

Fig. 6 EMG data collection procedure with EMG signal collecting instrument

The procedure of EMG data collection is shown in Fig. 6 and another picture is total EMG recording machine. Then collected EMG data is shown in Fig. 7 where two condition has indicated. One is action condition and another is resting condition. In action condition, EMG signal amplitude is increased. The amplitude of surface EMG signal is affected by several things. Such as thickness, conductivity and subcutaneous layer of skin, muscle fibre diameter, interval between action muscle fiber, innervations zone, tendons of active motor unit site and electrodes filtering properties. This can be calculated in terms of average rectified value (ARV) and root mean square value (RMS) [9].

(ARV) is defined as a time windowed mean of the absolute value of the signal. ARV is one of the various processing methods used to construct derived signals from raw EMG data that can be useful for further analysis.

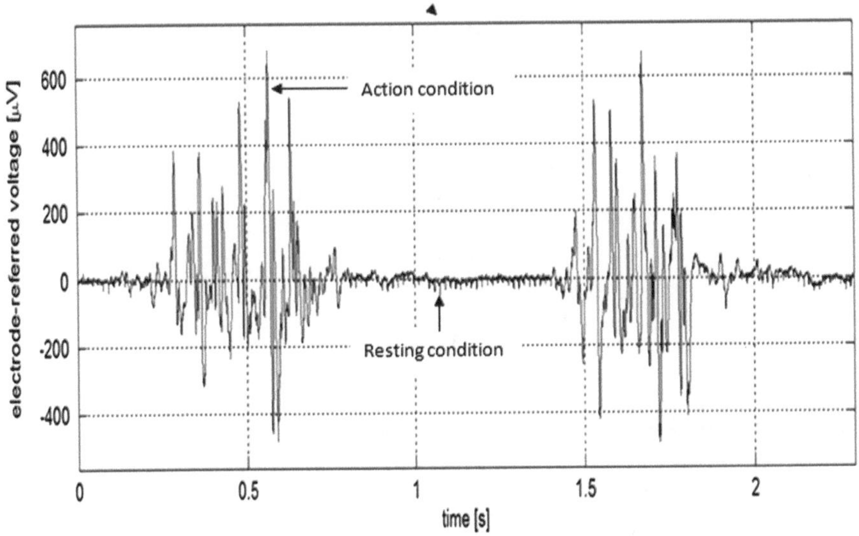

Fig. 7 Collected EMG data when biceps brachii is the active muscle

ARV can be calculated

$$\mathrm{ARV} = \frac{1}{N_e}\sum_{n=1}^{N_e}|x_n| \tag{1}$$

Here, x_n is EMG signal value at time index n and N_e is the epoch length.

Root Mean Square EMG (RMS EMG) is defined as the time windowed RMS value of the raw EMG. RMS is one of a number of methods used to produce waveforms that are more easily analyzable than the noisy raw EMG.

RMS can be calculated

$$\mathrm{RMS} = \sqrt{\frac{1}{N_e}\sum_{n=1}^{N_e}|x_n|} \tag{2}$$

6 Conclusion

EMG signal carries valuable information regarding the nervous system. So the aim of this paper is to present the detailed information about the EMG Signal recording techniques with the procedural approach of the instrument RMS EMG EP MARK-II Kit. Without acquiring the EMG signal it unable analyze. In this concern the detailed information about the instrumental activity is very necessary. This study clearly points up the method of EMG signal analysis techniques so that right methods can be applied during any clinical diagnosis, biomedical research, hardware implementations and end user applications.

Acknowledgements We would like to express our gratitude and appreciation to All India Council of Technical Education (AICTE) The Govt. of India, for encouraging financially through Research Promotion Scheme. Additionally, we acknowledge JIS College of Engineering, Kalyani and Gargi Memorial Institute of Technology, Kolkata for the support towards this research paper.

References

1. Rissanen S (2012) Feature extraction methods for surface electromyography and kinematic measurements in quantifying motor symptoms of Parkinson's Disease. Eastern Finland, Kuopio, 24 Feb 2012
2. Akumalla SC (2012) Evaluating appropriateness of emg and flex sensors for classifying hand gestures. University of North Texas, Oct 2012
3. Day SJ (1997) The Properties of Electromyogram and Force in Experimental and Computer Simulations of Isometric Muscle Contractions. Data from an Acute Cat Preparation. Dissertation, University of Calgary, Calgary
4. Finni T (2001) Muscle mechanics during human movement revealed by in vivo measurements of tendon force and muscle length. Neuromuscular Research Center, Department of Biology of Physical Activity, University of Jyväskylä, Jyväskylä
5. Acierno SP, Baratta RV, Solomonow M (1995). A practical guide to electromyography for biomechanists. Bioengineering Laboratory/LSUMC Department of Orthopaedies, Louisiana
6. Englehart K, Hudgin B, Parker PA (2001) A wavelet-based continuous classification scheme for multifunction myoelectric control. IEEE Trans Biomed Eng 48:302–311
7. Kuriki HU, de Azevedo FM, Ota Takahashi LS The relationship between electromyography and muscle force, University de São Paulo, Brazil
8. Lippert Lynn S (2006) Clinical kinesiology and anatomy, 4th edn. F. A. Davis Company, Philadelphia
9. Rissanen S (2012) Feature extraction methods for surface electromyography and kinematic measurements in Quantifying motor symptoms of Parkinson's Disease. Eastern Finland, Kuopio, 24 Feb 2012
10. Henneberg K, Plonsey R (1993) Boundary element analysis of the directional sensitivity of the concentric EMG electrode. IEEE Trans Biomed Eng 40:621
11. Light CM, Chappell PH (2000) Development of a lightweight and adaptable multiple-axis hand prosthesis. Med Eng Phys 22:679–684
12. Vinet R, Lozach N, Beaudry N, Drouin G (1995) Design methodology for a multifunctional hand prosthesis. J Rehabil Res Dev 32:316–324

Brain Machine Interface Automation System: Simulation Approach

Prachi Kewate and Pranali Suryawanshi

Abstract Brain Machine Interface (BMI) till now is generally preferred only for repairing damaged hearing, sight and movements with the help of neuroprosthetics application. These applications consist of some external unit which gathers some information in the form of signals from the brain and processes it so as to transfer them to the implanted unit. In this way these applications had helped the people to bring back their ability from various neuromuscular disabilities. Similarly, the BMI can be very useful for automation system. It will help in reducing accidents which had contributed to high mortality rate. A brain actuated automation system will also help motor disabled person to move independently. Signals from brain will be acquired with the help of dry electrodes and those signals will be processed in the system processor. The signal after processing will be then applied to the system depending on the instructions given by the person sitting on it.

Keywords Brain Machine Interface (BMI) · Neural network

1 Introduction

A Brain Machine Interface (BMI) is a correspondence framework that does not rely on upon the brains typical yield pathways of peripheral nerves and muscle. It is another approach to impart between a working human brain and the computerization framework. These are implanted interfaces with the mind, which can possibly transmit and get receive replies from the brain. This interface changes mental choices and responses into control motions by investigating the bioelectrical signals.

P. Kewate (✉) · P. Suryawanshi
Department of Computer Science & Engineering, G.H. Raisoni Institute of Engineering and Technology for Women's, Nagpur, India
e-mail: prachikewate@gmail.com

P. Suryawanshi
e-mail: pranali.suryawanshi@gmail.com

S. Gupta et al. (eds.), *Advancements of Medical Electronics*,
Lecture Notes in Bioengineering, DOI 10.1007/978-81-322-2256-9_21

It was once recognized as a science fiction to connect the machines straightforwardly with the brain, however now-a-days with huge improvement in the neuroscience it is conceivable to do so. Also still the examination is going on. Brain is a complex sensory system whose operation we will need to investigate it in point of interest. By utilizing new innovation and hi-tech technology, incapacitated individual will have the capacity to paint with the assistance of automated arm, to fly a plane and to drive vehicle.

Miscellaneous physiological activities of human will generate feeble amount of electricity; the brain is composed of neurons. Therefore, the brain signal is biological signal associated with brain, i.e. brain waves. Brain is generally ruled with its certain laws related to brain's consciousness. The frequency of brainwave will be significantly different at different states of joy, uneasiness, fear, sad, trust, etc. BMI gives new yield pathways to the mind by deciphering estimations of cerebrum movement into inputs for an outside gadget. These yield pathways normally work in one of two separate ways: Process Control and Goal selection. Example for process control is moving cursor left or right and for Goal Selection is moving cursor directly to the selected goal such as at the end of the page [1].

The fundamental thought of BMI is to decipher the pattern of brain activity into corresponding brain commands. A normal BMI is made up of signal acquisition, signal analysis and automation system [2]. The key concept of BMI is to change over the EEG information of the client into a control command. The very important part of BMI research is to adjust the mutual adaptation relationship between the human brain and the BMI system, i.e. to find a suitable signal processing, making nerve signals convert into the command or an operation signal that can be recognized by the computer in real time, quickly and accurately by the means of BMI system.

BMI gives immediate pathway between human and physical gadgets which is possible with the assistance of biological signals. There are various biological signals such as EEG (electro-encephalograms), MEG (Magneto-encephalograms), BOLD (blood oxygen level dependent). But due to its low cost and feasible to use EEG signals are mostly used for BMI systems.

Neural Network has played a vital role in the system. Neural Network takes a different approach than a regular computer. Conventional computer solve the problem in the algorithmic approach i.e. the rules or program for the particular task are already set in the computer. Thus these rules are used while solving the particular problem. The computer cannot solve the problem unless knows the steps to execute them. This restricts the problem solving capability of the computer. Thus it would be very easy if computer would solve the problems which we exactly don't know. Whereas, neural network solve the problem in the same way as the brain works. There are many tasks getting processed in the parallel way. They learn from examples. They cannot be programmed to do a specific task.

The BMI which is presenting in the paper is an automation system working from the signals acquired from the brain with the help of dry electrodes. The EEG signals acquired from the electrodes with the help of signal acquisition and then those signals are processed in the processor using Neural Network. After the signals are processed they are then applied to the automation system. Thus according to the particular command given by the brain the system will work.

In this paper the First part describes the introduction of the paper has been given. The second part describes the objective of the paper. The third part will give us the research methodology about the proposed system.

Lastly, the expected outcome about the proposed system concluding with the conclusion.

2 Objective

The primary objective of the proposed system depends on the three tasks which are as follows (Fig. 1):

1. Signal Acquisition: Dry electrodes placed on the cap will acquire the EEG signal. The acquired signals are in the analog format and hence need to be converted into digital format with the help of A/D convertor.
2. Signal Processing: The digitized signals are then analysed using FFT and only the required content are then extracted and is sent to the classifier.
3. Automation System: The classified signal is then given to the automation system which performs the particular command given.

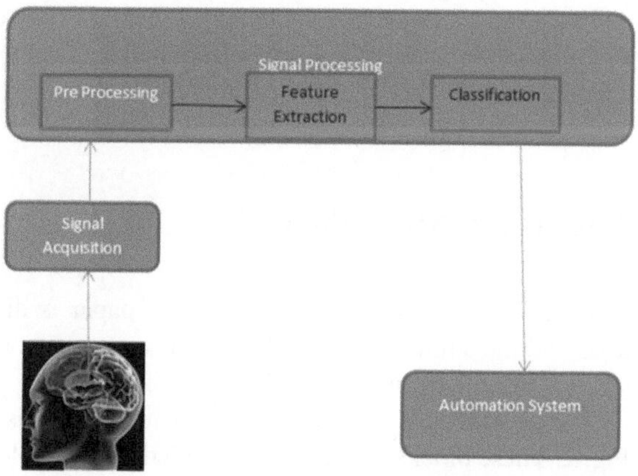

Fig. 1 Schematic view of BMI showing three main tasks of signal acquisition, signal processing and automation system

3 Literature Survey

Recently, innovative work of brain-controlled robots have accepted a lot of consideration in view of their capability to bring again to individuals with wrecking neuromuscular issue and enhance the personal satisfaction and self-reliance of these users. It incorporates different provisions, for example, a cursor on the screen [1], selecting letters from virtual console [3], searching web [4], and playing games [5].

Akce et al. [6] proposed an interface for navigating a mobile robot which moves at a fixed speed in a planar workspace [6] with noisy binary inputs. It uses hybrid strategy where measurements of brain activity as evidence to reduce the uncertainty about the entire path all at once.

Citi et al. [7] proposed BMI Mouse using the P300 waves. The system uses two approaches: first it completely dispenses with the problem of detecting P300s by logically behaving as analogue device and second approach is the single trial approach where the mouse performs an action after every trial.

A Portable BMI for expressing inner ideas by voice is designed and fabricated in Yan et al. [2]. EEG signal is recorded by dry electrode and is processed successively by the ASIC.

The same idea of BMI as illustrated in these papers above is going to be utilized as a part of the proposed work. BMI could be utilized to create such a robotization framework where engine incapacitated individual will have the capacity to move starting with one spot then onto the next. The individual who have engine handicaps and are incapacitated are not ready to move any of his appendages so in such a condition this proposed framework will be a shelter to them. They need to just think about certain course where they need to move and along these lines the framework will take them to that specific heading. A brain controlled wheelchair may serve as a requisition to them. The subtle element clarification of the proposed framework is demonstrated further in the Research system area.

4 Research Methodology

Research methodology for the proposed paper is divided according the hardware used and the simulation base. The following is according the hardware used in the system and for simulation will be explained later.

The system will consist of three main blocks as shown in the diagram above Fig. 2. These blocks are again subdivided into sub blocks which are as shown in the proposed architecture Fig. 3.

1. Input Block: The input block will perform the signal acquisition task. Signals will be acquired from the electrodes which are mounted on the cap. The cap will be placed on the scalp of the person sitting on the system. This signals acquired will be amplified with the help of op-amp. As brain is made up of numerous neurons, it will collect the large amount of data. This data will filter using notch

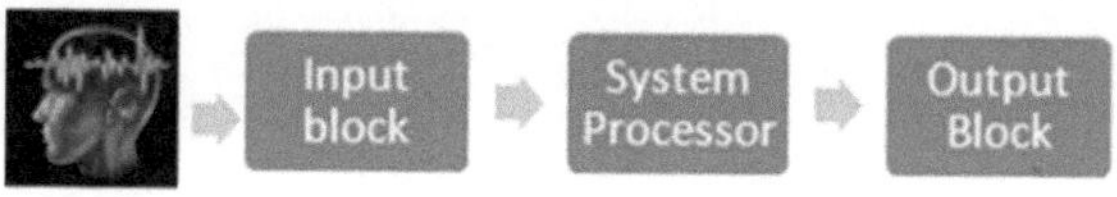

Fig. 2 Basic block diagram

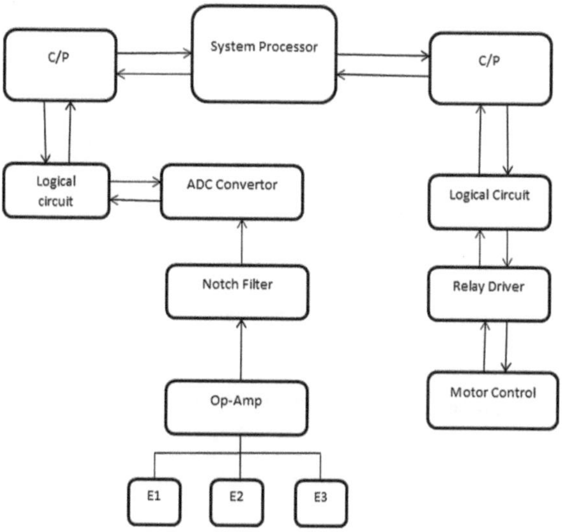

Fig. 3 Proposed architecture

filter and the required amount of data will be further sent to the ADC convertor. Then these digitized signals will be sent to the logical circuit where it will compare the real time data with the standard files stored in the system processor.

2. System Processor: The signals from the logical circuit are fed to the System Processor via Communication Protocol. Co-learning of the signals will take place in this block. After training these signals, signals will be extracted and will be classified according to mining. Single trial analysis will occur so as to consider only one signal at a time.
3. Output Block: Communication Protocol transfers the signal from system processor to the logical circuit of the output block, which are then fed to the motor control of the automation system via relay driver.

Now, the Methodology used for the simulation part of the proposed work is as follows (Fig. 4).

The first module will be of training the acquired signal. The signals which are acquired from the electrodes are stored in the processor as data set. The file is browsed and is selected for the training purpose with the help of neural network. The trained signals are then saved.

The saved file is browsed in the second module to plot them as graph as displayed in the module. The first graph is for the error rate and the second graph is for the success rate (Fig. 5).

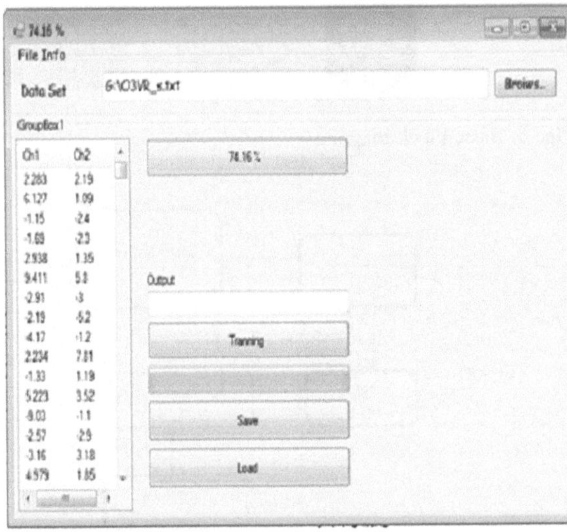

Fig. 4 Training of signals

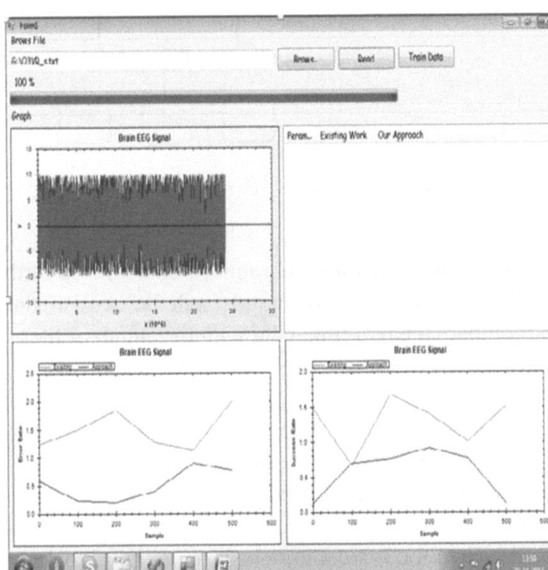

Fig. 5 Plotting the signals on graph

5 Performance Evaluation

In this section we will compare the performance achieved by the proposed system with the existing system. The parameters considered for the comparison are the error rate and the success rate.

Error rate: the number of times the system not work efficiently.

Success rate: the number of times the system has success after giving a particular command.

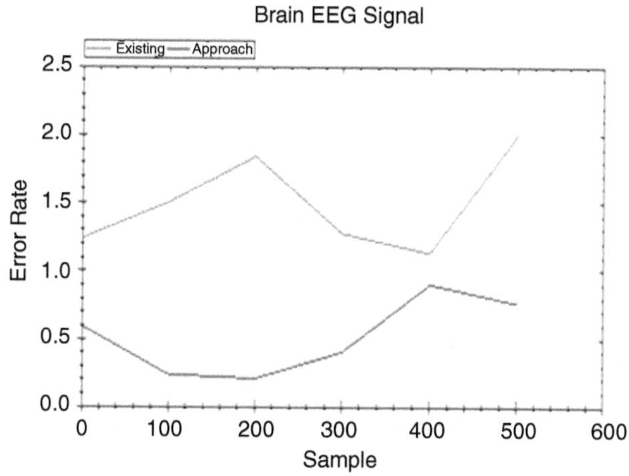

Fig. 6 Error rate for existing and our approach

The Error rate for the existing and our approach is plotted in the graph Fig. 6. On X axis we have taken the symbol length and on the Y axis percentage error rate. Thus we observe that we are getting low error rate than the error rate observed in the existing approach from the graph. Thus our approach has proved to be better than the existing approach.

Similarly, the success rate for the proposed system is compared with the existing approach. On X axis we have taken the symbol length and percentage of success rate on the Y-axis. But as can be seen from the graph it hadn't proved to be that much efficient. The graph for the success rate for the existing and for our approach is as shown in the Fig. 7.

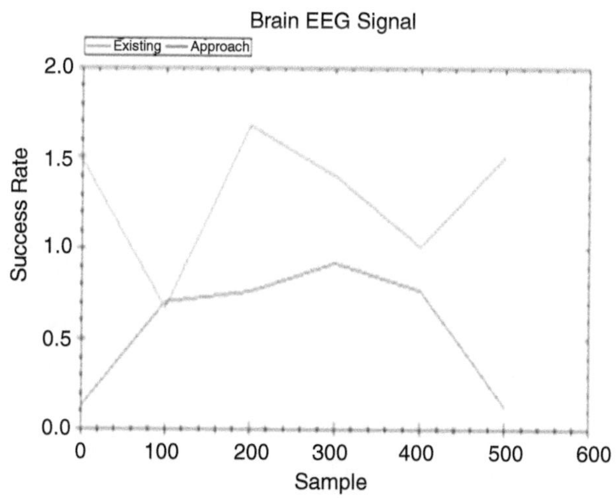

Fig. 7 Success rate for existing and our approach

6 Conclusion

Thus our proposed system has proved to be efficient in reducing the error rate than the existing system. In this way our system can be much helpful for the persons having motor disorders and can improve the quality of life. Still, there is lot of research to be made in this field. Thus BMI can be implemented on the automation system.

References

1. Wolpaw JR, McFarland DJ, Neat GW, Forneris CA (2008) An EEG-based brain-computer interface for cursor control. IEEE Intell Syst 23(3):72–79
2. Yan Y, Mu N, Duan D, Dong L, Tang X, Yan T (2013) A dry electrode based headband voice brain-computer interface device. In: International conference on complex medical engineering, May 2013
3. Donchin E, Spencer KM, Wijesinghe R (2000) The mental prosthesis: assessing the speed of P300-based brain computer interface. IEEE Trans Neural Syst Rehabil Eng 8(2):174–179
4. Karim AA, Hinterberger T, Richter J (2006) Neural Internet: web surfing with brain potentials for the completely paralyzed. Neurorehabil Neural Repair 20(2):508–515
5. Krepki R, Blankertz B, Curio G, Muller KR (2007) The Berlin brain computer interface (BBCI): towards a new communication channel for online control in gaming applications. J Multimedia Tools Appl 33(1):73–90
6. Akce A, Johnson M, Dantsker O, Bretl T (2013) A brain–machine interface to navigate a mobile robot in a planar workspace: enabling humans to fly simulated aircraft with EEG. IEEE Trans Neural Syst Rehabil Eng 21(2):306–318
7. Citi L, Poli R, Cinel C, Sepulveda F (2008) P300-based BCI mouse with genetically-optimized analogue control. IEEE Trans Neural Syst Rehabil Eng 16(1):51–61

Part III
DSP and Clinical Applications

Cognitive Activity Classification from EEG Signals with an Interval Type-2 Fuzzy System

Shreyasi Datta, Anwesha Khasnobish, Amit Konar and D.N. Tibarewala

Abstract The present work attempts to classify Electroencephalogram (EEG) signals corresponding to three different cognitive activities using an Interval Type 2 Fuzzy System (IT2FS) classifier. This approach is used in order to account for the fact that EEG signals from the same person for the same stimulus have variations in different observations in the same as well as over different days of experiments. Adaptive Autoregressive Parameters, Hjorth Parameters, Hurst Exponents and Approximate Entropy features are extracted from the acquired EEG signals to obtain the maximum possible discrimination between different classes of activities. A maximum classification accuracy of 85.33 % on an average over all subjects and classes is obtained using IT2FS classifier on a combined feature space. The study is validated using Friedman's statistical test with respect to four other classification algorithms.

Keywords Cognitive activity recognition · Electroencephalogram · Interval type-2 fuzzy systems

S. Datta (✉) · A. Konar
Department of Electronics and Telecommunication Engineering,
Jadavpur University, Kolkata, India
e-mail: shreyasidatta@gmail.com

A. Konar
e-mail: konaramit@yahoo.co.in

A. Khasnobish · D.N. Tibarewala
School of Bioscience and Engineering, Jadavpur University, Kolkata, India
e-mail: anweshakhasno@gmail.com

D.N. Tibarewala
e-mail: biomed.ju@gmail.com

S. Gupta et al. (eds.), *Advancements of Medical Electronics*,
Lecture Notes in Bioengineering, DOI 10.1007/978-81-322-2256-9_22

1 Introduction

Brain Computer Interface (BCI) technology [1] finds applications in rehabilitation, assistance and treatment of neuro-motor, sensory-motor and cognitive disabilities. An interesting avenue in BCI research is brain signal based systems with the ability to recognize the cognitive state of human beings. Such systems are important in the development of context aware pervasive computing systems [2, 3]. Human brain activities convey direct information regarding their cognitive state and location. There have been various research works that study the cognitive state of the brain through Electroencephalogram (EEG), Magnetoencephalogram (MEG) or functional Magnetic Resonance Imaging (f-MRI) [4–6]. EEG has been popular in BCI applications because of its superior temporal resolution, easy acquisition, simple processing, portability, cost-effectiveness and freedom from exposure of the subject to intense magnetic fields. There are several instances of EEG based BCI research in literature, that include decoding mental states and cognitive activities [7–9], prosthetic and robotic control through motor imagination [10], computer cursor control [11], emotion recognition [12], object shape recognition from visual and tactile exploration [13] to mention a few.

In [14, 15] various tools and techniques used commonly in EEG classification related to BCI research have been stated, that include Support Vector Machines, Linear Discriminant Analysis, Nearest Neighbour Classifiers, Neural Networks, Bayesian methods, Hidden Markov Models among others. However standard classifiers often fail to perform successfully in EEG analysis because of the nature of EEG signals. These signals are non-stationary, non-periodic [16] and have variations among different observations of the same class.

Fuzzy logic [17] can represent uncertainty associated with any system. Fuzzy logic based classifiers [18] have been successful in handling variations in signals. Type 1 Fuzzy Systems (T1FS) can be used for classification by representing the variations in the signal values at different instances using a single membership function [17, 19]. EEG responses to the same stimulus taken at different times are variable for a particular person; these responses also have variations in between persons. Therefore there exists an inherent difficulty in classification of these signals to identify the different states of the brain. A T1FS is unable to capture the variations in the memberships for different trials of the same experiment over a period of days. General Type 2 Fuzzy Systems (GT2FS) [20, 21] can overcome this difficulty by assigning a secondary membership function to model the uncertainty associated with the primary membership function. However, developing the secondary membership is challenging and the process is also computation intensive. Interval Type 2 Fuzzy Systems (IT2FS) [22, 23] employ uniform and constant secondary membership values and provide a simple solution to the problem.

In the present work, EEG signals corresponding to three activities, namely, reading, watching a video with audio and relaxing have been classified using the concept of IT2FS. As non-linear features are important in EEG classification problems [24], EEG signals have been represented through Adaptive Autoregressive Parameters and Hjorth

Parameters, as well as two non-linear features, namely Hurst Exponents and Approximate Entropy, and a combination of all the features in this study. The performance of the IT2FS classifier has been compared with T1FS classifier considering Gaussian membership function, as well as standard pattern classifiers like Support Vector Machine, Neural Network Classifier and Naïve Bayes Classifier.

The rest of the paper is structured as follows. In Sect. 2 the principles used in this work have been explained. Section 3 illustrates the experiments and the results. Finally in Sect. 4 the conclusions are drawn and future scopes of work are stated.

2 Principles and Methods

This section provides the descriptions of the tools and techniques used in the present work.

2.1 Feature Extraction Techniques

A linear approach, namely, Adaptive Autoregressive Parameters, a quasi linear approach, namely, Hjorth Parameters and two nonlinear signal features, namely, Hurst Exponents and Approximate Entropy have been used to represent EEG signals in this work.

2.1.1 Adaptive Autoregressive Parameters

Autoregressive Parameters and Adaptive Autoregressive Parameters (AAR) are time-domain features for EEG analysis [25, 26]. The AR approach models a stochastic time series assuming the data to be stationary. For estimating the non-stationary EEG signals computationally expensive methods like windowing can be used with the AR technique. Another approach is to use AAR model, where the AR parameters for representing EEG signals are estimated in a time-varying manner, as explained by (1) and (2), where the index k is an integer to denote discrete, equi-distant time points, $y(k)$ is the kth instance of the signal, p is the order of the AAR model, $y(k - i)$ with $i = 1$ to p are the p previous sample values, $a_{i,k}$ are the time-varying AR model parameters, and $x(k)$ is a zero-mean-Gaussian-noise process with time varying variance $\sigma_x^2(k)$.

$$y(k) = a_{1,k}y(k-1) + \cdots + a_{p,k}y(k-p) + x(k) \tag{1}$$

$$x(k) = N\{0, \sigma_x^2(k)\} \tag{2}$$

There are various algorithms to estimate the parameters of AAR such as, least mean square (LMS), Kalman filtering, recursive AR or recursive least square (RLS).

2.1.2 Hjorth Parameters

Activity, Mobility and Complexity, collectively known as the Hjorth Parameters [27] are another set of time domain features. For an input signal $y(k)$, Activity $A(y)$, Mobility $M(y)$ and Complexity $C(y)$ are defined by (3), (4) and (5) respectively, where $var(y)$ and y' denote the variance and first derivative of the signal $y(k)$ respectively.

$$A(y) = \text{var}(y) \tag{3}$$

$$M(y) = \sqrt{\frac{A(y')}{A(y)}} \tag{4}$$

$$C(y) = \frac{M(y')}{M(y)} \tag{5}$$

2.1.3 Hurst Exponents

Non-linear models are efficient for representations of complicated physiological processes exhibiting chaotic behavior [28]. EEG responses at different excitations may have random variations and can be described to be chaotic. Hence, Hurst Exponent (H) [24, 28], a non-linear parameter is used as an EEG feature. The Hurst Exponent is obtained using the Rescaled Range Analysis (R/S Analysis) [28] by statistical methods. It estimates the occurrence of long-range dependence and its degree in a time series by the evaluating the probability of an event to be followed by a similar event. It is described by (6), where T denotes the sample duration and R/S denotes the corresponding rescaled range value.

$$H = \log(R/S)/\log(T) \tag{6}$$

$H = 0.5$ indicates that the time-series is similar to an independent random process, $0 \leq H < 0.5$ indicates a present decreasing trend in the process implies a future increasing trend and vice versa whereas $0.5 < H \leq 1$ indicates that a present increasing/decreasing trend in the process implies a future increasing/decreasing trend.

2.1.4 Approximate Entropy

Approximate Entropy (AE) [24, 28], is another non-linear feature that provides a measure of regularity of a signal. A deterministic signal with high regularity has a very small AE value while a random signal with a low regularity has a high AE value. Two parameters, the embedding dimension m and tolerance of comparison value r are necessary for computation of AE. For a time series of length N, AE is obtained as (7) and (8), where $C_m(r)$ is the correlation integral as defined in [24].

$$AE(m, r) = \Phi^m(r) - \Phi^{m+1}(r) \tag{7}$$

$$\Phi^m(r) = (N - (m-1))^{-1} \sum_{i=1}^{N-(m-1)} \ln[C_i^m(r)] \tag{8}$$

AE can be observed as the likelihood in logarithmic terms that certain patterns in the data will be followed by similar patterns.

2.2 Interval Type 2 Fuzzy System for Classification

In a normal or Type 1 Fuzzy System (T1FS) [18, 19] every variable has a membership value in the close interval [0, 1] according to a predefined membership function. Commonly used membership functions are triangular, trapezoidal, Gaussian or sigmoidal [29]. Given a set of variables, the Gaussian membership function of a variable x can be easily determined from their mean μ and the standard deviation σ according to (9), and can be implemented in classification problems using Fuzzy logic.

$$m(x) = \exp\left(-\frac{(x-\mu)^2}{2\sigma^2}\right) \tag{9}$$

For classification using T1FS, the membership values of the test sample for each class is computed using the parameters of the membership function (for example the mean and the standard deviation of the observations in case of Gaussian membership function) for that particular class and the sample belongs to the class having the maximum value of membership. However incase the observations have varying memberships (say, in case of Gaussian membership, the mean and standard deviation are variable over a larger number of observations) as with signals taken at different days in case of EEG signals, T1FS fail to provide accurate classification results. This uncertainty in the primary membership function $m(x)$ is overcome by assigning a secondary membership function $\tilde{m}(x, m(x))$ in T2FS [20, 21]. The union of all the primary memberships for similar set of observations forms a region called the Footprint of Uncertainty (FOU) bounded by the secondary membership function and the minimum and maximum values of the primary memberships for the observations, termed as the Lower Membership Function (LMF) and the Upper Membership Function (UMF) respectively. In Interval Type-2 Fuzzy Systems (IT2FS) [22, 23], the secondary membership function is uniform and assumes a constant value of 1 for all values of $m(x)$ in between the LMF and UMF and zero otherwise.

For using IT2FS in signal classification, let us consider P sets of similar observations of each of K classes where the signal is represented by a feature space FS of $M \times N$ dimensions, for M observations and N features in each set. For each feature F_i $(1 \le i \le N)$ in FS, considering the minimum and maximum values of F_i

over the observations over P sets, the primary memberships, and consequently the LMF, UMF and FOU are constructed. Say, a feature vector f corresponding to an unknown instance of the signal has to be classified.

Each component f_i of f $(1 \leq i \leq N)$ is projected on the corresponding FOU to find the intersections with the LMF and UMF of that component to obtain LMF_i and UMF_i. Prior to projection, if it is found that f_i falls outside the range of F_i extrapolation is used.

For a particular class k the fuzzy t-norm of all LMF_i and UMF_i for $1 \leq i \leq N$ are computed to obtain the $LMF_{T,k}$ and $UMF_{T,k}$ respectively. Using these values the strength S_k of the class k $(1 \leq k \leq K)$ is determined by the computation of the centroid given by (10).

$$S_k = \frac{UMF_{T,k} + LMF_{T,k}}{2} \tag{10}$$

Computing the strengths of all the classes, the class having the maximum strength is determined to be the class of the test sample. The entire process for class k is illustrated in Fig. 1.

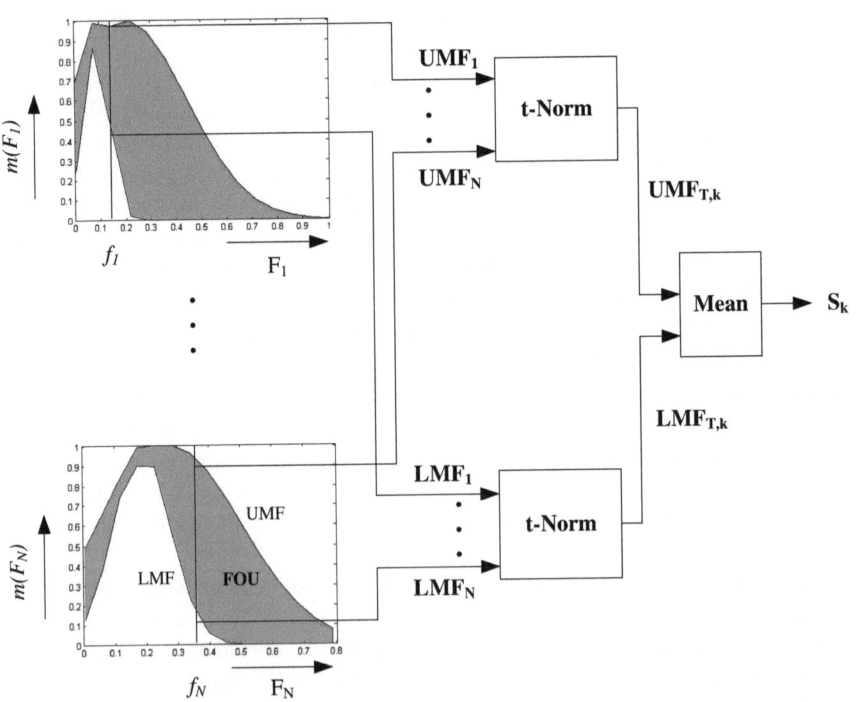

Fig. 1 Procedure for calculation of centroid or strength of a class k in the IT2FS approach, all symbols having the same meanings as explained in Sect. 2.2

3 Experiments and Results

The steps of EEG signal processing in the present work has been illustrated in Fig. 2.

3.1 EEG Acquisition

In the present work, we are interested in finding excitations corresponding to three activities, namely, reading, watching a video with audio and relaxing. Accordingly, EEG from the pre-frontal, frontal, occipital, parietal and temporal regions are continuously recorded through the 10 electrodes FP1, FP2; F3, F4; O1, O2; P3, P4 and T3, T4 respectively of a 21 channel Neurowin system according to the 10–20 international standard of electrode placement [30] as shown in Fig. 3a, at a sampling rate of 250 Hz.

EEG data is acquired from 4 subjects, 2 male and 2 female in the age group of 25 ± 5 years, for a period of 3 consecutive days. The pattern of stimulus used for data acquisition is shown in Fig. 3b, where 'STIMULUS' corresponds to the visual stimulus for activity A1 (reading the English sentences displayed) or the Audio-Video stimulus for activity A2 (watching a video clip with sound) or a black screen with instructions to relax with closed eyes for activity A3 (relaxing with closed eyes). In all cases the respective stimuli are displayed on a screen using a projector in front of the comfortably seated subject. The sequence of stimulus is repeated 5 times for each activity in random order for each day of experiment.

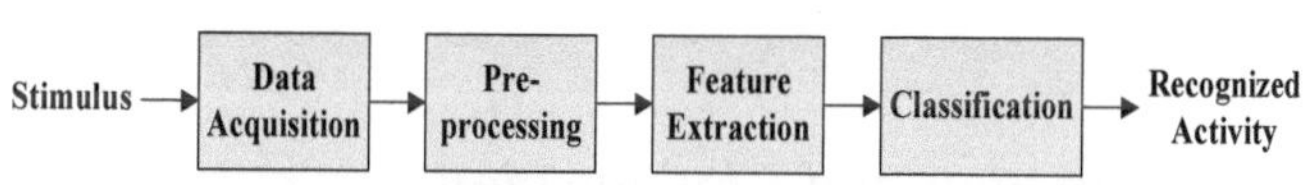

Fig. 2 Steps of EEG signal processing

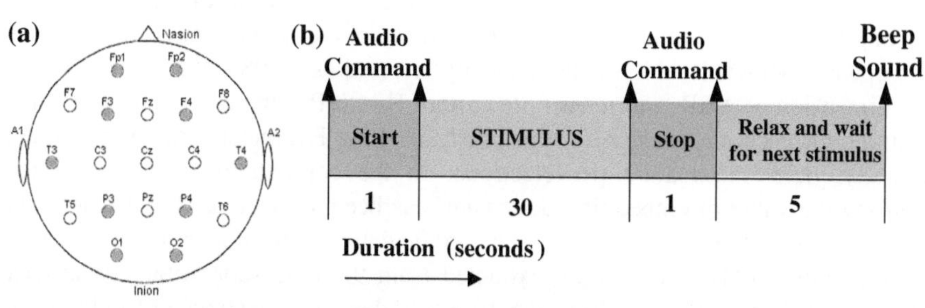

Fig. 3 **a** Electrode placement showing the selected electrodes in green, A1 and A2 are the reference electrodes. **b** An instance of presented stimulus

3.2 EEG Preprocessing

3.2.1 Filtering

The normal EEG bandwidth ranges between 0.5–70 Hz and is made up of the delta (0.5–4 Hz), theta (4–8 Hz), alpha (8–13 Hz), beta (13–30 Hz) and gamma (above 30 Hz) bands [16]. It was experimentally found that the significant changes in the EEG spectrum for decoding the three activities is limited to 4–30 Hz, hence we have considered EEG signals in the theta, alpha and beta bands for our work. To extract the EEG signals in the desired frequency range, and thereby eliminate the other frequencies, the acquired EEG is filtered using an Elliptical Band pass filter of order 6 with 1 dB passband ripple and 50 dB stopband ripple and bandwidth 4–30 Hz.

3.2.2 Common Average Referencing

Common average referencing has been performed on the acquired EEG signals for spatial filtering to remove the effect of interference between the signals of adjacent channels. For each EEG channel, all the channels equally weighted are subtracted to eliminate the commonality of that channel with the rest and preserve its specific temporal features. Let the signal at the primary and the 10 channels be $x_i(t)$ and $x_j(t)$, for $i, j = 1$–10. Then with equal weights for $x_1(t)$ through $x_{10}(t)$, we obtain (11).

$$x_i(t) \leftarrow x_i(t) - \frac{1}{10}\sum_{j=1}^{10} x_j(t) \tag{11}$$

3.3 Feature Extraction and Classification

AAR parameters have been computed using Kalman filtering as the estimation algorithm and the order of 6 has been selected after trials with different orders and finding the best performance. The AAR parameters are adapted at a rate given by the update coefficient which is heuristically selected as 0.0085. For each electrode the dimension of AAR features is 6. In case of Hjorth Parameters, the dimension of features for each electrode is 3 while each of Hurst Exponents and Approximate Entropy yield one feature per electrode. In the computation of Approximate Entropy the value of embedding dimension has been selected as 2 and that of the tolerance has been selected as 0.2× (standard deviation of the concerned time series) from literature [24]. Features are extracted from the time series obtained at each electrode and these features from the 10 electrodes are concatenated to obtain the respective feature spaces. When feature spaces are combined, each feature space is normalized with respect to its maximum value.

For classification, data from the 3 days of experiments have been utilized. 5 s of data is taken as a single instance. For each day, 5 (times) × 30 s/5 s (instances) × (250 × 5) (samples/instance) i.e. 30 total instances each of 1,250 samples are obtained from each class which is subjected to feature extraction. The resulting data is cross validated to obtain the testing and training instances. Classification is carried out in a One versus One approach i.e. binary classification considering the instances of two classes at a time. Classification has been carried out using the IT2FS approach as explained in the previous section. The performance is compared that of a T1FS classifier [17, 19], a Support Vector Machine (SVM) [14], a Neural Network (NN) classifier [14, 31] and a Naïve Bayes (NB) classifier [14]. In T1FS and IT2FS classifiers Gaussian membership functions have been used for the simplicity in determining the memberships using the mean and the standard deviations of the data and also it is appropriate because of the nature of EEG features. In IT2FS each set of observation comprise of the observations taken in a particular day. The Fuzzy t-norm has been implemented by taking the minimum of the input values. For SVM an RBF kernel is used with the width of the Gaussian kernel taken as 1. The NN classifier is implemented with back-propagation learning using gradient descent search [29] with 3 hidden layers. The NB classifier is used with the assumption that the features have a normal distribution whose mean and covariance are learned during the process of training.

3.4 Performance Analysis

The classification results are produced from the respective confusion matrices in terms of Classification Accuracy (CA), Sensitivity (ST) and Specificity (SP) given by (12), (13) and (14) respectively [32]. Here TP, TN, FP and FN denote the number of samples classified as true positive, true negative, false positive and false negative respectively. The ideal values of CA, ST and SP should each be very close to 1.

$$CA = \frac{TP + TN}{TP + TN + FP + FN} \tag{12}$$

$$ST = \frac{TP}{TP + FN} \tag{13}$$

$$SP = \frac{TN}{TN + FP} \tag{14}$$

Table 1 Classification performance of IT2FS

Class	S	Features								
		AAR + Hjorth			Hurst + AE			AAR + Hjorth + Hurst + AE		
		CA	ST	SP	CA	ST	SP	CA	ST	SP
A1 versus A2	1	0.80	0.87	0.73	0.72	0.70	0.73	0.78	0.77	0.80
	2	0.85	0.90	0.80	0.88	0.83	0.93	0.88	0.90	0.86
	3	0.74	0.77	0.72	0.68	0.73	0.63	0.78	0.83	0.73
	4	0.83	0.83	0.83	0.80	0.87	0.73	0.82	0.83	0.80
A1 versus A3	1	0.89	0.88	0.90	0.87	0.85	0.90	0.92	0.90	0.93
	2	0.86	0.83	0.90	0.89	0.85	0.93	0.89	0.94	0.83
	3	0.78	0.87	0.70	0.80	0.75	0.85	0.88	0.90	0.87
	4	0.88	0.83	0.93	0.78	0.77	0.80	0.83	0.80	0.85
A2 versus A3	1	0.77	0.85	0.70	0.72	0.80	0.63	0.92	0.93	0.90
	2	0.85	0.83	0.87	0.83	0.80	0.87	0.86	0.83	0.90
	3	0.70	0.67	0.73	0.68	0.76	0.60	0.77	0.80	0.73
	4	0.85	0.81	0.88	0.78	0.80	0.77	0.91	0.95	0.87

Table 2 Comparison of classifier performance

Parameter (mean over all subjects)	Classifiers (using AAR + Hjorth + Hurst + AE features)				
	IT2FS	T1FS	SVM-RBF	NB	NN
Classification accuracy (%)	85.33	71.65	83.50	81.27	75.45
Rank	1.25	4.75	1.75	3.75	3.5

3.5 Experimental Results and Discussions

The results of classification with IT2FS is illustrated in Table 1, where *S* denotes the Subject ID, *A1*, *A2* and *A3* denote the classes of activities as mentioned in Sect. 3.1. It is observed that using a combination of all features a maximum of 85.33 % classification accuracy is obtained on an average over all subjects and OVO classifications. The minimum classification accuracy does not go below 68 % in any of the cases.

For comparison between the different classification algorithms, Friedman Test [33], has been carried out for $N = 4$ databases of the 4 subjects and for $k = 5$ algorithms taking the mean classifier accuracy for each classification algorithm used, as shown in Table 2. The null hypothesis states that all algorithms are equivalent and hence their ranks R_j should be equal. The Friedman statistic is given by (15).

$$\chi_F^2 = \frac{12N}{k(k+1)}\left[\sum_{j=1}^{k} R_j^2 - \frac{k(k+1)^2}{4}\right] \tag{15}$$

It is distributed according to χ_{F}^2 with $k - 1$ degrees of freedom. Here k is the number of algorithms used and N is the number of datasets which are 5 and 4 in our case respectively. The null hypothesis is rejected if evaluated $\chi_{\mathrm{F}}^2 > \chi_{4,0.95}^2 = 9.49$, which indicates, for 4° of freedom, the null hypothesis is correct to an extent of only 5 %. In this case χ_{F}^2 is found to be 13.6 hence the null hypothesis is rejected and the classifier performance is evaluated by its rank.

4 Conclusion and Future Directions

In this paper cognitive activities have been classified from EEG signals using different features and classifiers, with an aim at the development of a cognitive state aware BCI device. A combination of four signal features, namely, AAR parameters, Hjorth Parameters, Hurst Exponents and Approximate Entropy yield the best results of 85.33 % classification accuracy employing an IT2FS classifier which is able to perform better due to its inherent capability to deal with the uncertainties and variations in EEG signals. The future scopes of work include real time implementation of the present scheme.

Acknowledgments This study has been supported by University Grants Commission, India, University of Potential Excellence Program (UGC-UPE) (Phase II) in Cognitive Science, Jadavpur University and Council of Scientific and Industrial Research (CSIR), India.

References

1. Vallabhaneni A, Wang T, He B (2005) Brain—computer interface. In: He B (ed) Neural Engineering. Kluwer/Plenum, Springer, New York, pp. 85–121. doi:10.1007/0-306-48610-5_3
2. Riva G, Vatalaro F, Davide M, Alcañiz M (2005) Ambient intelligence
3. Henricksen K, Indulska J, Rakotonirainy A (2002) Modeling context information in pervasive computing systems. In: Pervasive computing. Springer, New York, pp 167–180
4. Haynes JD, Rees G (2006) Decoding mental states from brain activity in humans. Nat Rev Neurosci 7(7):523–534
5. Mitchell TM, Hutchinson R, Niculescu RS, Pereira F, Wang X, Just M, Newman S (2004) Learning to decode cognitive states from brain images. Mach Learn 57(1–2):145–175
6. Freeman WJ, Ahlfors SP, Menon V (2009) Combining fMRI with EEG and MEG in order to relate patterns of brain activity to cognition. Int J Psychophysiol 73(1):43–52
7. Vuckovic A, Radivojevic V, Chen AC, Popovic D (2002) Automatic recognition of alertness and drowsiness from EEG by an artificial neural network. Med Eng Phys 24(5):349–360
8. Gevins A, Smith ME (1999) Detecting transient cognitive impairment with EEG pattern recognition methods. Aviat Space Environ Med 70(10):1018–1024
9. Wilson GF, Fisher F (1995) Cognitive task classification based upon topographic EEG data. Biol Psychol 40(1):239–250
10. Pfurtscheller G, Neuper C (2001) Motor imagery and direct brain-computer communication. Proc IEEE 89(7):1123–1134
11. Fabiani GE, McFarland DJ, Wolpaw JR, Pfurtscheller G (2004) Conversion of EEG activity into cursor movement by a brain-computer interface (BCI). IEEE Trans Neural Syst Rehabil Eng 12(3):331–338
12. Schaaff K, Schultz T (2009) Towards an EEG-based emotion recognizer for humanoid robots. In: The 18th IEEE international symposium on robot and human interactive communication, RO-MAN, September 2009, pp 792–796
13. Khasnobish A, Konar A, Tibarewala DN, Bhattacharyya S, Janarthanan R (2013) Object shape recognition from EEG signals during tactile and visual exploration. Pattern Recognit Mach Intell 459–464. Springer Berlin, Heidelberg
14. Lotte F, Congedo M, Lécuyer A, Lamarche F, Arnaldi B (2007) A review of classification algorithms for EEG-based brain–computer interfaces. J Neural Eng 4
15. Besserve M, Jerbi K, Laurent F, Baillet S, Martinerie J, Garnero L (2007) Classification methods for ongoing EEG and MEG signals. Biol Res 40(4):415–437
16. Teplan M (2002) Fundamentals of EEG measurement. J Measur Sci Rev 2(2):1–11
17. Zadeh LA (1988) Fuzzy logic. Computer 21(4):83–93
18. Mitra S, Pal SK (2005) Fuzzy sets in pattern recognition and machine intelligence. Fuzzy Sets Syst 156(3):381–386
19. Chakraborty A, Konar A, Chakraborty U K, Chatterjee A (2009) Emotion recognition from facial expressions and its control using fuzzy logic. IEEE Trans Syst Man Cybern Syst Hum 39(4):726–743
20. Karnik NN, Mendel JM, Liang Q (1999) Type-2 fuzzy logic systems. IEEE Trans Fuzzy Syst 7(6):643–658

21. Herman P, Prasad G, McGinnity T (2006) Investigation of the type-2 fuzzy logic approach to classification in an EEG-based brain-computer interface, In: 27th Annual international conference of the IEEE engineering in medicine and biology society. pp 5354–5357
22. Liang Q, Mendel JM (2000) Interval type-2 fuzzy logic systems: theory and design. IEEE Trans Fuzzy Syst 8(5):535–550
23. Konar A, Chakraborty A, Halder A, Mandal R, Janarthanan R (2012) Interval type-2 fuzzy model for emotion recognition from facial expression. Perception and machine intelligence. Springer, New York, pp 114–121
24. Balli T, Palaniappan R (2010) Classification of biological signals using linear and nonlinear features. Physiol Meas 31(7):903
25. Pfurtscheller G, Neuper C, Schlogl A, Lugger K (1998) Separability of EEG signals recorded during right and left motor imagery using adaptive autoregressive parameters. IEEE Trans Rehabil Eng 6(3):316–325
26. Nai-Jenand H, Palaniappan R (2004) Classification of mental tasks using fixed and adaptive autoregressive models of EEG signals. In: 26th Annual international conference of the IEEE engineering in medicine and biology society, IEMBS'04, Sept 2004 vol 1. pp 507–510
27. Vidaurre C, Krämer N, Blankertz B, Schlögl A (2009) Time domain parameters as a feature for EEG-based brain–computer interfaces. Neural Netw 22(9):1313–1319
28. Acharya UR, Faust O, Kannathal N, Chua T, Laxminarayan S (2005) Non-linear analysis of EEG signals at various sleep stages. Comput Methods Programs Biomed 80(1):37–45
29. Konar A (2005) Computational intelligence principles, techniques and applications. Springer, New York
30. Dornhege G (2007) Towards brain-computer interfacing. MIT Press, Cambridge
31. Mitchell TM (1997) Machine learning. McGraw Hill, Burr Ridge, p 45
32. Fielding AH, Bell JF (1997) A review of methods for the assessment of prediction errors in conservation presence/absence models. Environ Conserv 24(1):38–49
33. Conover WJ, Iman RL (1981) Rank transformations as a bridge between parametric and nonparametric statistics. Am Stat 35(3):124–129

Performance Analysis of Feature Extractors for Object Recognition from EEG Signals

Anwesha Khasnobish, Saugat Bhattacharyya, Amit Konar and D.N. Tibarewala

Abstract Recognition of objects from EEG signals requires selection of appropriate feature extraction and classification techniques with best efficiency in terms of highest classification accuracy with lowest run time for its applications in real time. The objective of this paper is to analyze the performance of various feature extraction techniques and to choose that particular method which can be implemented in real time system with best efficiency. The EEG signals are acquired from subjects while they explored the objects visually and visuo-tactually. Thus acquired EEG signals are preprocessed followed by feature extraction using adaptive auto-regressive (AAR) parameters, ensemble empirical mode decomposition (EEMD), approximate entropy (ApEn) and multi-fractal detrended fluctuation analysis (MFDFA). The performance of these features are analyzed in terms of their dimension, extraction time and also depending upon the classification results produced by three classifiers [Support Vector machine (SVM), Naïve Bayesian (NB), and Adaboost (Ada)] independently according to classification accuracy, sensitivity and classification times. The experimental results show that AAR parameter has an optimum dimension of 36 (not too large like EEMD i.e. 7,680 or too small like ApEn i.e. 6) and required minimum extraction as well as classification time of 0.59 and 0.008 s respectively. AAR also yielded highest maximum classification accuracy and sensitivity of 80.95 and 92.31 % respectively with NB classifier. Thus AAR parameters can be chosen for real time object recognition from EEG signal along with Naïve Bayesian classifier.

A. Khasnobish (✉) · D.N. Tibarewala
School of Bioscience & Engineering, Jadavpur University, Kolkata, India
e-mail: anweshakhasno@gmail.com

D.N. Tibarewala
e-mail: biomed.ju@gmail.com

S. Bhattacharyya · A. Konar
Department of Electronics & Telecommunication Engineering, Jadavpur University, Kolkata, India
e-mail: saugatbhattacharyya@gmail.com

A. Konar
e-mail: konaramit@yahoo.co.in

S. Gupta et al. (eds.), *Advancements of Medical Electronics*,
Lecture Notes in Bioengineering, DOI 10.1007/978-81-322-2256-9_23

Keywords Object recognition · Electroencephalography · Adaptive autoregressive parameter · Ensemble empirical mode decomposition · Approximate entropy · Multifractal detrended fluctuation analysis · Naïve bayesian classifier · Support vector machine · Adaboost

1 Introduction

Object recognition is a vital process in understanding our environment, which in turn enhances exploration and maneuvering. Both vision and tactile sensation along with cognitive processing is essential for proper object recognition. The visual sensation is related to the occipital lobe of human brain, whereas its tactile counterpart originates from the parietal lobe and the cognitive processing in frontal and fronto-central regions. The synchronous activity of both these sensory systems facilitates object classification in human brain [1, 2]. Thus analyzing visually and tactually stimulated electroencephalography (EEG) signals from occipital and parietal and fronto-central channels provide information on object perception. Object recognition has been performed by many researchers mainly using computer vision [3], tactile signals, and tactile images [4, 5]. But object recognition directly from EEG signals [6] will help in enhancing the controllability of the brain computer interface (BCI) [7] controlled robots.

Recognition of objects from EEG signals requires selection of appropriate feature extraction and classification techniques with best efficiency in terms of highest classification accuracy with lowest run time for its applications in real time. The objective of this paper is to analyze the performance of various existing feature extraction techniques to choose that particular method which can be implemented in real time system with best efficiency.

The EEG signals are non-stationary, non-Gaussian signals [8, 9]. Various linear and non linear features extraction methods in time as well as frequency domain are applied for EEG signals viz. power spectral density (PSD), Hjorth parameters, wavelet transforms (WT), autoregressive parameters (AR), adaptive autoregressive (AAR) parameters, Hurst coefficients, approximate entropy (ApEn), extreme energy ratio criterion (EER), empirical mode decomposition (EMD), detrended fluctuation analysis (DFA), and statistical parameters [10, 11]. Depending of the type of stimuli, the performances these features vary.

In this work we presented the subjects with visual and visuo-tactile stimuli of ten different objects and corresponding EEG signals are acquired. For visual stimulation the subjects are presented with pictures of the objects and they examine the objects only visually. For the visuo-tactile stimulation, the subjects explored the objects by visual examination as well as palpated them at the same time. During the object exploration the EEG signals are acquired from six electrode locations, namely FC5, FC6, P7, P8, O1 and O2. The electrodes are chosen so as to acquire the brain signals from frontal, parietal and occipital relating the cognitive processing, somato-sensory and visual sensations respectively. The acquired EEG signals are preprocessed and features are extracted. In this work we used four feature extraction techniques, comprising two linear methods viz. adaptive autoregressive (AAR) parameter and ensemble empirical mode decomposition (EEMD), and two non linear techniques viz. approximate entropy (ApEn) and multifractal detrended fluctuation analysis (MFDFA). The performances of these extracted features are analyzed in terms of their dimension, extraction time as well as classification accuracy, sensitivity and runtimes on classification by three classifiers namely, Support Vector Machine (SVM), Naïve Bayesian (NB) and Adaboost classifier [12–15]. It is observed that AAR parameter yielded best results with all the three classifiers in both visual and visuo-tactile stimulated signals. Moreover Adaboost classifier with SVM as base classifier produced better results with all four features, performing the best with AAR parameters, followed by NB and SVM classifiers. However, the run time of NB classifier is the least, which is very essential for online classifications.

The rest of the paper is presented in five sections. The introduction is followed by the methodology in Sect. 2. The selected EEG modalities are described in Sect. 3. Experimental paradigm and the results are depicted in Sect. 4. The concluding remarks are stated in Sect. 5.

2 Methodology

The acquired EEG signals are subjected to preprocessing followed by feature extraction for classification of corresponding objects.

2.1 *Preprocessing*

The power spectrum of the visual and visuo-tactual stimulated EEG signal is found to be in the 4–16 Hz (Fig. 1). The acquired EEG signal is thus filtered using an elliptical band pass filter of order 8 and bandwidth 4–16 Hz. The filtering removes the noises due to power line interference, head movement, and eye blinks as well as it extracts the signal of the required bandwidth.

2.2 *Feature Extraction*

Appropriate feature extraction is a pre-requisite for effective classifications. The aim of this paper is to compare the performances of various features based on classification results to ultimately select a competent technique for real time classification of object recognition from EEG signals. For this purpose we have considered two linear and two non-linear features. Among the two linear features, i.e. adaptive auto-regressive

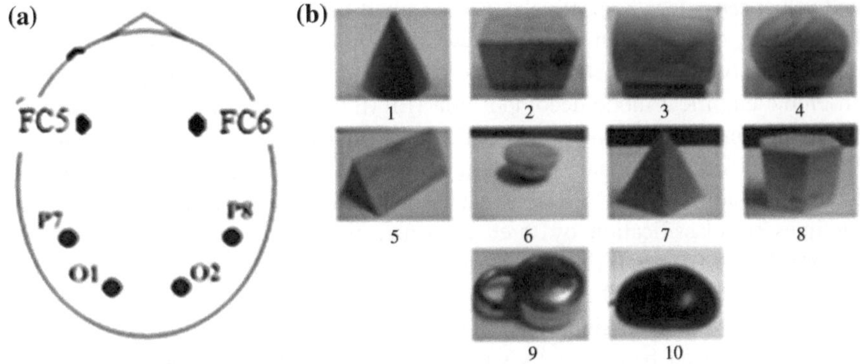

Fig. 1 **a** Electrode placement, **b** Different object shapes used in the experiments. 1 Cone, 2 Cube, 3 Cylinder, 4 Sphere, 5 Prism, 6 Hemisphere, 7 Square based pyramid, 8 Hexagonal based cylinder, 9 Lock and 10 Mouse

(AAR) parameters and ensemble empirical mode decomposition (EEMD), the first one is a time domain feature and the other is a frequency domain feature. Two non-linear features viz. approximate entropy (ApEn) and multifractal detrended fluctuation analysis (MFDFA) are also taken in account for the present work.

2.2.1 Adaptive Auto-regressive Parameter

Autoregressive parameters (AR) and adaptive autoregressive parameters (AAR) are parametric approaches for EEG analysis [16, 17]. EEG signals non-stationary random signals. AR has the capability to characterize the stochastic behavior of EEG signals without requirement of any prior information of any relevant frequencies. The AR uses various stationary processes for estimating the parameters. Since the EEG signals are non-stationary, moving window method can be applied for AR parameter estimation, but it is computationally extensive. To overcome these inherent drawbacks of AR approaches, AAR methods are implemented for EEG feature extraction since it yields time varying parameters with high temporal resolution. AAR model describes the signal $x(t)$ defined by Eq. (1).

$$x(t) = p_1(t) \times x(t-1) + \cdots + p_k(t) \times y(t-k) + s(t) \tag{1}$$

where, $p_k(t)$ are the parameters, $s(t)$ is white noise with zero mean and k is the order of the model.

There are various algorithms to estimate the parameters of AAR such as, least mean square (LMS), Kalman filter, recursive AR, recursive least square (RLS) and as like.

In this work, AAR model with Kalman filter method for parameter estimation is considered. The order and update coefficient (UC) are experimentally determined to be 6 and 0.0085 respectively.

2.2.2 Ensemble Empirical Mode Decomposition

Empirical mode decomposition (EMD) is method for adaptive analysis of non-stationary and non-linear signals [18]. This method separates a signal in fast and slow oscillations by local data-driven approach. Mode-mixing i.e. presence of very divergent oscillations in one mode or presence of similar oscillations in different modes, is a familiar problem encountered in EMD. This setback is overcome by performing EMD over an ensemble of white Gaussian noise and the signal. This progression is termed as ensemble empirical mode decomposition (EEMD) [18]. Addition of white noise takes the advantage of dyadic filter bank behavior of EMD by populating the total time-frequency space and thus eliminates the mode-mixing problem. In EMD the signal is decomposed in number of intrinsic mode functions (IMFs), also called modes. A number ensemble trials are obtained by adding different apprehensions of white noise to the original signal $x(t)$ and their corresponding IMFs computed. The ensemble trials are represented by Eq. (2).

$$x^j[n] = x[n] + g^j[n] \tag{2}$$

where $j = 1, 2, \ldots, J$, and $g^j[n]$ is Gaussian noise of various apprehensions. IMFs are obtained from completely decomposing each $x^j[n]$. EEMD calculates the mean of the obtained IMFs, termed as true IMF (IMF_{true}).

We have computed five IMFs of the preprocessed EEG signals for feature extraction.

2.2.3 Approximate Entropy

Approximate entropy (ApEn) is a statistical method proposed by Pincus [19] and [20] for quantifying the unpredictability or irregularity of stochastic signals. For non-linear time-series signals this technique measures the randomness or degree of complexity. The steps in estimation of ApEn are as follows:

I. Let a time series signal containing N data points be represented as $X = [x(1), x(2), \ldots, x(N)]$. Specify the embedding dimension or window length (l) and the tolerance (t).

II. l vectors $X(1), X(2), \ldots, X(N - l + 1)$, representing l successive values of x is defined as

$$X(i) = [x(i), x(i+1), \ldots, x(i+l-1)] \tag{3}$$

where $i = 1, 2, \ldots, N - l + 1$.

III. Compute the distance between two vectors $X(i)$ and $X(j)$ by calculating the maximum absolute difference between their corresponding scalar elements, as given by Eq. (4).

$$d[X(i), X(j)] = \max_{k=1,\ldots,l} |x(i+k-1) - x(j+k-1)| \quad (4)$$

IV. For a given $X(i)$ find the number of $j = 1, 2, \ldots, N - l + 1, j \neq i$, such that $d[X(i), X(j)] \leq t$ and denote is as $M^l(i)$. Then for $i = 1, 2, \ldots, N - l + 1$, compute

$$C_t^l(i) = \frac{M^l(i)}{N - l + 1} \quad (5)$$

This measures the frequency of patterns alike the one given by window of length l and tolerance t.

V. Compute the natural logarithm of each $C_t^l(i)$ and find the mean over all i, as given by Eq. (6).

$$\varphi_t^l(t) = \frac{1}{N - l + 1} \sum_{i=1}^{N-l+1} \ln C_t^l(i) \quad (6)$$

VI. Compute $C_i^{l+1}(i)$ and $\varphi_t^{l+1}(i)$ by increasing the dimension to $(l + 1)$ and repeating the Steps II–V. Calculate approximate entropy (ApEn) according to Eq. (7).

$$ApEn(l, t, N) = \varphi^l(t) - \varphi^{l+1}(t) \quad (7)$$

For feature extraction of the preprocessed EEG signal experimentally we have determined $l = 2$ and $t = 0.02$.

2.2.4 Multifractal Detrended Fluctuation Analysis

If a time-series signal repeats itself on the subintervals of the signal, then it posses scale invariant structures. For EEG signals, the scale invariant structures of inter-spike interval of firing of the neurons are capable of discriminating between the neural activities of brain. Alterations in scale invariant structure of bio-signals indicate adaptability of physiological processes. Detrended fluctuation analysis (DFA) quantifies the self-resemblance of a signal [21, 22]. Unlike fluctuation analysis (FA), DFA is not affected by non-stationarities of a signal which measures long range correlations of non-stationary signals. Time series with complicated time behavior necessitate different scaling exponents for different part of the series. In such case multi-fractal analysis is performed with provides multitude scaling exponents for complete description of complex scaling behavior. Regular partition function multi-fractal formalism developed for multifractal characterization and normalized stationary measure is the basis of simplest MFA. Non-stationary time series signals are affected by trends that cannot be normalized. Due to this characteristic of non-stationary signals MFA produces erroneous results, which is overcome by implementing multi-fractal detrended fluctuation analysis (DFA) [21, 22].

Let x_k is a time series of length N of compact support that $x_k = 0$ for an insignificant amount of values. The steps required to estimate MFDFA is as follows:

I. Compute the 'Profile' $P(i)$

$$P(i) = \sum_{k=1}^{i} [x_k - \bar{x}], \quad i = 1, \ldots, N \tag{8}$$

II. Divide profile in a number of non-overlapping segments denoted by $N_l = N/l$, of equal length l. The same process is repeated from end to start to the series to consider the small parts that can remain at the end of the series. Thus we obtain total $2N_l$ segments.

III. Compute the local trend for each of the obtained segments by least square fit of the series and calculate the variance, as depicted by Eq. (9).

$$F^2(l, v) = \frac{1}{l} \sum_{i=1}^{l} \{P[(v-1)l + i] - y_v(i)\}^2$$

where v is a segment such that $v = 1, \ldots, N_l$, and

$$\mathrm{F}^2(l, v) = \frac{1}{l} \sum_{i=1}^{l} \{P[N - (v - N_l)l + i] - y_v(i)\}^2 \tag{9}$$

for $v = N_l + 1, \ldots, 2N_l$, where $y_v(i)$ is the fitting polynomial in the segment v.

IV. Calculate qth order fluctuation function by averaging over all segments.

$$F_q(l) = \left\{ \frac{1}{2N_l} \sum_{v=1}^{2N_l} [F^2(l, v)]^{q/2} \right\}^{1/q} \tag{10}$$

where q can take any value other than zero.
To determine the dependency of generalized q dependent fluctuations on time scale l, repeat Steps II–IV.

V. Analyze the log-log plots of $F_q(l)$ versus l for each value of q for quantifying the scaling behavior of fluctuation function. For long-range power-law corrected series, x_i, $F_q(l)$ increases as power-law for large values of l.

$$F_q(l) \sim l^{h(q)} \tag{11}$$

The generalized Hurst exponent $h(q)$ may be q-dependent.

We have used MFDFA of 7th order fitting polynomial (i.e. $y_v(i)$) and varied q in the range −5 to 5 with 101 discreet intervals.

2.3 Classification

The extracted features are classified separately using three supervised classifiers namely support vector machine (SVM), Naïve Bayesian (NB) and Adaboost (Ada) [12–15]. These features are classified in corresponding object classes by implementing these classifiers. The one-against-all (OAA) approach is implemented for classification of ten objects.

SVM is a supervised, non-probabilistic binary classifier that separates the data points by a hyperplane, by maximizing its distance from the support vectors. For classification, it considers only the support vectors, thus the search space is reduced, and pattern recognition is performed in less time. Due the formation of the hyperplane with maximum margin, the efficiency is also high in case of SVM classifiers. On the other hand Naïve Bayesian is a probabilistic classifier, which makes conditional independence assumptions which further facilitates the reduction of high complexity in case of general Bayesian classifiers. Adaboost (adaptive) is an ensemble classifier, utilizing a number of weak learners to build a strong learner iteratively and is suitable for both binary and multi-class classification problems. In the present study, for Adaboost we have used SVM as the base classifier.

3 EEG Modalities

3.1 Frequency Band Selection

The bandwidth of EEG signals is in the range 0.5–70 Hz. As evident from the power spectrum of the visually and visuo-tactually stimulated EEG signals' activity lie in the range of 4–16 Hz. Thus we have considered 4–16 Hz bandwidth, comprising of theta, alpha and central beta rhythms.

3.2 Electrode Selection

The acquired EEG signals are either visually or visuo-tactually stimulated. Here we are concerned with decoding object recognition from EEG signals. Complete course of object recognition is related to visual and tactile sensory information and also cognitive processing. The brain regions related to vision, tactile sensation, and cognition are occipital, parietal, and fronto-central regions. Thus we have acquired the EEG signals from six electrodes, viz. O1, O2; P7, P8; and FC5, FC6, where each pair are positioned on occipital, parietal and fronto-central regions (Fig. 1a).

4 Experiments and Results

4.1 Experimental Paradigm

Ten rigid objects consisting of eight geometrical shapes and two non-geometrical objects are considered for experiments (Fig. 1b). Thirteen subjects, (8 male and 5 female) of age group 22 ± 5 years, took part in the experiments. The subjects are all right handed with normal or corrected to normal vision. The experimental procure is described to them and they signed a consent form before starting the experiments. EEG from six electrode channels (FC5, FC6, P7, P8, O1 and O2) is acquired using Emotive EEG.

The subjects are presented with audio-visual cues. At the start of the experiment there appears a blank screen for 30 s to relax the subject and detect the baseline EEG, followed by a fixation cross, indicating the subject to get ready. For only visual stimulation phase, a picture of a particular object appears on screen for 5 s. In case of visuo-tactile stimulation, followed by the fixation cross the subjects are instructed to explore the object by palpating for 5 s whose picture appears on the screen. At the end of 5 s a beep sound is indicates the end of object examination and again a blank screen appears for 10 s. In this manner every object is examined 10 times by each subject in two experimental phase i.e. visual stimulation and visuo-tactual stimulation phases.

5 Results

Four features (i.e. AAR, EEMD, ApEn, MFDFA) are extracted from the preprocessed EEG signals. Each feature space is tenfold cross validated to form corresponding training and test instances which are classified independently using three classifiers, i.e. SVM, NB and Adaboost. The efficacy of each feature is analyzed depending upon the classification results based on three metrics viz. classifier run time, classification accuracy and sensitivity. The feature dimension and extraction time are also considered for their performance estimation.

Table 1 presents the feature dimensions over all six electrodes and also their mean extraction times. It is evident that EEMD has highest dimension as well as needs longest time for extraction, thus is inapplicable in real-time scenarios. As noted from Table 1, AAR requires the minimum amount of computational time.

Table 1 Features' dimension and extraction time

Features	Feature dimension	Mean extraction time (s)
1. AAR	36	00.59
2. EEMD	7,680	43.24
3. ApEn	6	00.83
4. MFDFA	606	03.42

Table 2 Classification results

Features	Classifier	Run time (s)	Classification accuracy (%)		Sensitivity (%)	
AAR	SVM	0.015	Min.	52.38	Min.	78.57
			Max.	85.71	Max.	94.12
			Av.	72.12	Av.	85.55
	NB	0.008	Min.	52.38	Min.	83.33
			Max.	80.95	Max.	92.31
			Av.	70.06	Av.	85.93
	Ada	0.926	Min.	66.67	Min.	82.35
			Max.	95.24	Max.	94.74
			Av.	80.95	Av.	86.53
EEMD	SVM	0.055	Min.	57.14	Min.	82.35
			Max.	80.95	Max.	88.89
			Av.	65.98	Av.	83.36
	NB	0.032	Min.	51.14	Min.	75.00
			Max.	80.50	Max.	100
			Av.	69.12	Av.	84.40
	Ada	12.37	Min.	66.67	Min.	82.35
			Max.	85.71	Max.	89.47
			Av.	78.23	Av.	85.54
ApEn	SVM	0.006	Min.	47.62	Min.	76.92
			Max.	71.43	Max.	92.86
			Av.	56.46	Av.	82.39
	NB	0.004	Min.	66.67	Min.	73.73
			Max.	80.95	Max.	88.89
			Av.	64.62	Av.	82.69
	Ada	0.881	Min.	71.43	Min.	81.25
			Max.	85.71	Max.	88.89
			Av.	75.5	Av.	85.06
MFDFA	SVM	0.013	Min.	47.62	Min.	49.21
			Max.	71.43	Max.	71.40
			Av.	53.74	Av.	80.65
	NB	0.012	Min.	47.62	Min.	61.90
			Max.	76.19	Max.	76.19
			Av.	59.86	Av.	83.26
	Ada	1.172	Min.	76.21	Min.	80.23
			Max.	81.17	Max.	89.33
			Av.	80.95	Av.	85.00

The classification results are depicted in Table 2. The run time is the time in seconds taken by the classifier to classify the corresponding features. The classification accuracy [23] is presented in three terms, i.e. minimum (Min.), maximum

(Max.) and average (Av.) over all subjects and classes for both visual and visuo-tactile stimulation phases. The sensitivity [23] is the measure of detecting the positives correctly, as defined by Eq. (12).

$$Sensitivity = \frac{True\,Positive}{True\,Positive + False\,Negative} \tag{12}$$

Since classification is performed on one-against-all basis, thus we are concerned with detecting a particular class with respect to other classes taken altogether as another class. So the rate of correctly detecting the true positives, i.e. our class of concern for particular instance, is the main aim. The sensitivity is also presented as Min., Max., and Av.

It is observed from the Table 2 that the AAR parameter requires the least classification time of 0.008 s with Naïve Bayesian (NB) classifier. The mean classification rates over all three classifiers for AAR, EEMD, ApEn and MFDFA are 74.37, 71.11, 65.52 and 64.85 % respectively. Their corresponding mean sensitivities are 86, 84.43, 83.38, and 82.97 % respectively. In terms of both classification accuracy and sensitivity AAR parameter performed best followed by EEMD, ApEn and MFDFA. Among the three classifiers though Adaboost showed highest classification accuracies and sensitivities with every feature, but its run time is also the longest, making it unsuitable for online classifications. The next best classification accuracy and sensitivity is observed for Naïve Bayesian classifier for each feature.

6 Conclusions

EEG signals are acquired from thirteen subjects while they explore ten rigid objects visually and visuo-tactually. The acquired EEG signals are preprocessed followed by feature extraction using four techniques i.e. adaptive autoregressive (AAR) parameters, ensemble emipirical mode decomposition (EEMD), approximate entropy (ApEn) and multi-fractal detrended fluctuation analysis (MFDFA). The performance of these features are analyzed in terms of their dimension, extraction time and also depending upon the classification results produced by three classifiers (SVM, NB, and Adaboost) independently according to classification accuracy, sensitivity and classification times. The experimental results show that AAR parameter has an optimum dimension (not too large like EEMD or too small like ApEn) and required minimum extraction as well as classification time. The classification accuracies and sensitivities are also found to be the highest for AAR parameters. Though Adaboost performed with highest classification accuracy and sensitivity, but due to is long execution time it cannot be applied in real time recognitions. The NB classifier classified the features with classification accuracy and sensitivity better than SVM with least execution time. Thus it can be concluded that among all the features considered, AAR parameters can be chosen for real time object recognition from EEG signal along with Naïve Bayesian classifier.

In future we will implement more feature and classifiers for analysis. We are also working on implementation of feature selection techniques to improve the efficacy of object recognition from visually and visuo-tactually stimulated EEG signals.

Acknowledgments This study has been supported by University Grants Commission, India, University of Potential Excellence Program (UGC-UPE) (Phase II) in Cognitive Science, Jadavpur University and Council of Scientific and Industrial Research (CSIR), India.

References

1. Schacter DL, Gilbert DL, Wegner DM (2009) Psychology, 2nd edn. Worth Publishers, New York
2. Mishkin M, Ungerleider LG (1982) Contribution of striate inputs to the visuospatial functions of parieto-preoccipital cortex in monkeys. Behav Brain Res 6(1):57–77
3. Kuo CC, Yau HT (2006) A new combinatorial approach to surface reconstruction with sharp features. IEEE Trans Visual Comput Graphics 12(1):73–82
4. Pezzementi Z, Reyda C, Hager GD (2011) Object mapping, recognition and localization from tactile geometry. In: Proceedings of IEEE international conference robotics and automation, pp 5942–5948
5. Singh G et al (2012) Object shape recognition from tactile images using regional descriptors. In: Fourth world congress on nature and biologically inspired computing (NaBIC) 2012, pp 53–58
6. Khasnobish A, Konar A, Tibarewala DN, Bhattacharyya S, Janarthanan R (2013) Object shape recognition from EEG signals during tactile and visual exploration. In: Accepted in international conference on pattern recognition and machine intelligence (PReMI), 10–14 Dec 2013
7. Vallabhaneni A, Wang T, He B (2005) Brain–computer interface in neural engineering. Springer, Heidelberg, pp 85–121
8. Teplan M (2002) Fundamentals of EEG measurement. Meas Sci Rev 2(2):1–11
9. Sanei S, Chambers JA (2007) Brain computer interfacing. EEG Signal Process, pp 239–265
10. Yom-Tov E, Inbar GF (2002) Feature selection for the classification of movements from single movement-related potentials. IEEE Trans Neural Syst Rehabil Eng 10:170–178
11. Bhattacharyya S, Khasnobish A, Konar A, Tibarewala DN (2010) Performance analyisis of LDA, QDA and KNN algorithms in left-right limb movement classification from EEG data. In: Accepted for oral presentation in international conference on systems in medicine and biology, IIT Kharagpur, 2010
12. Tae-Ki A, Moon-Hyun K (2010) A new diverse AdaBoost classifier. In: International conference on artificial intelligence and computational intelligence (AICI) 2010, pp 359–363
13. Cunningham P (2009) Evaluation in machine learning: objectives and strategies for evaluation. In: European conference on machine learning and principles and practice of knowledge discovery in databases 2009, p 26
14. Daniel WW (2002) Biostatistics. Hypothesis testing, 7th edn. Wiley, New York, pp 204–229
15. Thulasidas M, Guan C, Wu J (2006) Robust classification of EEG signal for brain computer interface. IEEE Trans Neural Syst Rehabil Eng 14(1):24–29
16. Vickenswaran J, Samraj A, Kiong LC (2007) Motor imagery signal classification using adaptive recursive band pass filter and adaptive autoregressive models for brain machine interface designs. J Biol Life Sci 3(2):116–123
17. Schloegl A et al (1997) Using adaptive autoregressive parameters for a brain-computer-interface experiment. In: Proceedings of the 19th annual international conference of the IEEE engineering in medicine and biology society 1997, vol 4, pp 1533–1535

18. Torres ME et al (2011) A complete ensemble empirical mode decomposition with adaptive noise. In: IEEE international conference on acoustics, speech and signal processing (ICASSP) 2011, pp 4144–4147
19. Pincus SM (1991) Approximate entropy as a measure of system complexity. In: Proc Natl Acad Sci USA 88(6):2297–2301
20. Lei W et al (2007) Feature extraction of mental task in BCI based on the method of approximate entropy. In: 29th annual international conference of the IEEE engineering in medicine and biology society, EMBS 2007, pp 1941–1944
21. Kantelhardt JW et al (2002) Multifractal detrended fluctuation analysis of nonstationary time series. J Physica A 316, 82:1–14
22. Ihlen EAF (2012) Introduction to multifractal wavelet and detrended fluctuation analyses. Front Physiol: Fractal Physiology 3(141):1–18
23. Mahajan K, Rajput SM (2012) A comparative study of EEG and SVM for EEG classification. J Eng Res Technol 1(6):1–6

Rectangular Patch Antenna Array Design at 13 GHz Frequency Using HFSS 14.0

Vasujadevi Midasala, P. Siddaiah and S. Nagakishore Bhavanam

Abstract This paper presents a new element antenna array of rectangular topology microstrip patches is introduced to operate at Ku band. The antenna has been designed as an arrays of patches, where number of elements, spacing's and feeding currents has been optimized to fulfil the requirements of low side lobe level and good cross polarization. The operating frequency range of antenna array is from 12 to 18 GHz. The antenna has been designed and simulated on FR4 Substrate with dielectric constant of 4.4. This paper also presents that, the detail steps of designing and simulating the rectangular patch antenna and rectangular patch antenna Array, in Ku-band. The design is analysed by Finite Element Method (FEM) based HFSS Simulator Software 14.0 by which return loss, Impedance, 3D polar plot, Directivity and Gain of the antenna are computed. The simulated results are shows that the proposed antenna provides good performance in terms of return loss and radiation pattern for dual frequency applications.

Keywords Microstrip antenna · Rectangular patch antenna · HFSS 14.0 · Return loss · Impedance · 3D polar plot · Directivity · Gain

1 Introduction

In Modern Communication Systems the Antennas are the most important components to create a communication link. Microstrip Patch antennas are widely used in wireless communication systems because of, they are low profile, of light weight,

V. Midasala (✉)
Department of ECE, JNTUA, Ananthapuranu, India
e-mail: vasujadevi@kluniversity.in

P. Siddaiah · S.N. Bhavanam
University College of Engineering and Technology, Acharya Nagarjuna University, Nagarjuna Nagar, Guntur, Andhra Pradesh, India
e-mail: kishore.ece.anu@gmail.com

S. Gupta et al. (eds.), *Advancements of Medical Electronics*,
Lecture Notes in Bioengineering, DOI 10.1007/978-81-322-2256-9_24

low cost, conformal design, low power handling capacity and easy to integrate and fabricate. They can be designed in a variety of shapes in order to obtain enhanced gain and bandwidth. Microstrip Patch Antenna implementation is a mile stone in wireless communication systems.

The design of microstrip patch antenna, operating in Ku band is a very difficult task. Ku band is primarily used in the satellite communication. Mostly for, fixed and broadcast services and for specific applications for NASA. Ku-band is also used for satellite from Remote Location (RL) back to a television network studio for editing and broadcasting. Ku-band provides reliable high-speed connectivity between the personal organizers and other wireless digital appliances.

The proposed model is one such antenna which is a Rectangular Microstrip fed patch antenna. It can be operated at Ku Band. In addition to this, Rectangular Patch Antenna Arrays designed. Its operation in Ku band the proposed antenna is also a dual band antenna. Dual frequency antenna is mainly used in applications where transmission and reception should be done by using same antenna. Many dual band antennas have been improved to face the rising demands of a modern portable wireless Communication Devices.

2 Design of Proposed Antenna

In this paper Rectangular Microstrip Patch Antenna Array Design at 13 GHz frequency has been modelled and simulated at Ku-band. The patch (radiating part) is the dominant figure of a microstrip antenna; the other components are the substrate and ground, which are the two sides of the patch.

The above figures can represent that, the Rectangular Microstrip Patch Antenna and Rectangular Microstrip Patch Antenna Array Design. Figure 1 can explains about single element patch antenna design but from Fig. 2 we can found that, it consists an array of Rectangular Microstrip Patch Antennas by this we can get better gain.

Method used to analyze antenna:	FEM (Finite Element Method)
Excitation Technique Used:	Probe Feeding
The spacing between antenna elements:	λ/4

Fig. 1 Rectangular microstrip patch antenna design

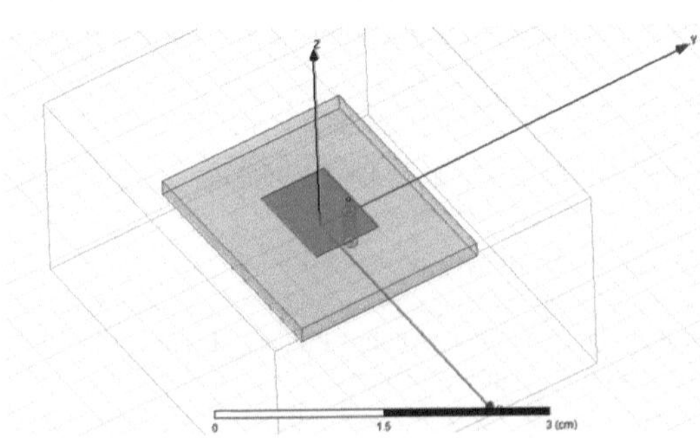

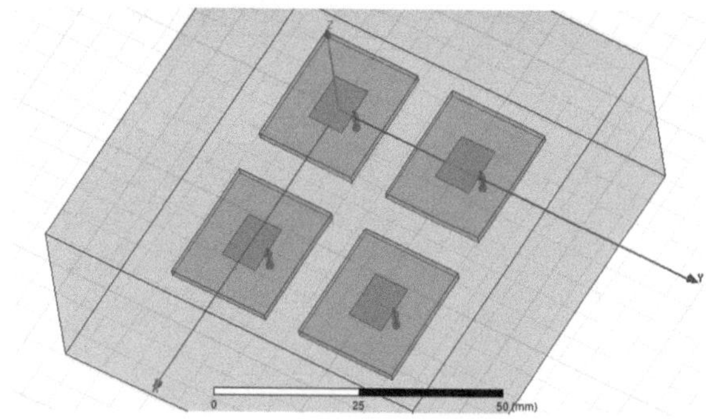

Fig. 2 Rectangular microstrip patch antenna array design

The design considerations are also shown below:

Design Considerations

Substrate material: *FR4*		
Relative permittivity 2.2, loss tangent 0.0009		
Position "−subX/2, −subY/2, 0 cm" "−1.15cm, −0.95cm, 0 cm"		
XSize	subX	2.3 cm
YSize	subY	1.9 cm
ZSize	subH	62 mil
Patch:		
Position "−patchX/2, −patchY/2, subH" "−0.455 cm, −0.335 cm, 62 mil"		
Axis	Z	
XSize	patchX	0.91 cm
YSize	patchY	0.67 cm

The Table 1 can give the corresponding values of Design Parameters.

Table 1 Design parameters and corresponding values

Design parameter	Value
Operating frequency	13.28 GHz
Dielectric constant of the substrate	4.4
height of the substrate	0.157 cm
Length of substrate	3 cm
Width of substrate	3 cm
Height of coax pin	0.157 cm
Radius of coax pin	0.2 mm

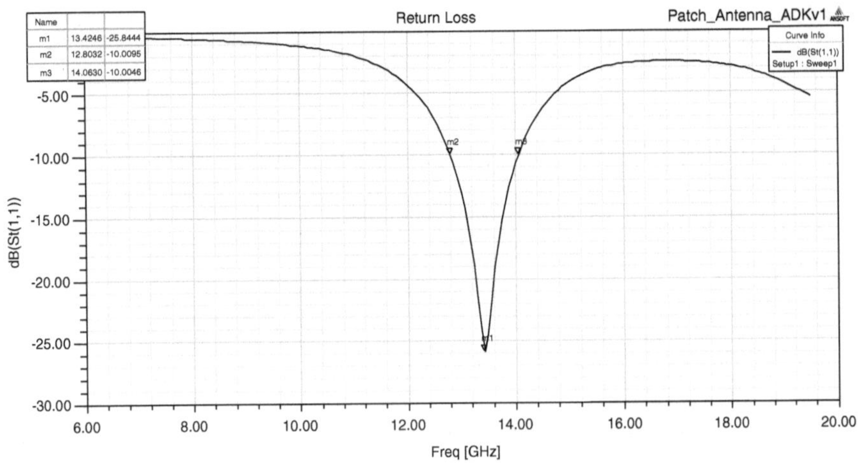

Fig. 3 Return loss for single patch antenna

3 Simulation Results Using HFSS

The design is analyzed by Finite Element Method. The return loss, Impedance, 3D polar plot, Directivity and peak gain is obtained by using HFSS 14.0. The results are shows below:

Return Loss

The Fig. 3 can represents the Return Loss of Single patch antenna. At −10 db we can get a Band Width −1.25 Ghz. And have a gain of 6.844 db.

The Fig. 4 can represents the Return Loss of patch antenna Array Design. At −10 db we can get a Band Width −1.25 Ghz. And have a gain of 8.5548 db.

Impedance Diagrams

Figures 5 and 6 can shows about Impedance diagram of Single Patch antenna and patch antenna Array Design.

3D Polar Plots

Figures 7 and 8 can show 3D Polar Plots of Single Patch antenna and patch antenna Array Design.

Directivity

For Single Patch Antenna: 6.375
For Array Antenna Design: 6.874

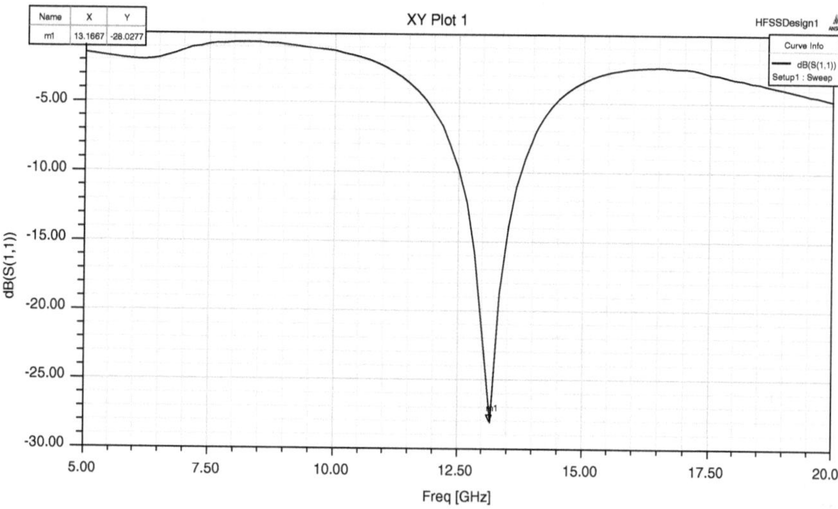

Fig. 4 Return loss for array design

Fig. 5 Impedance for single patch antenna

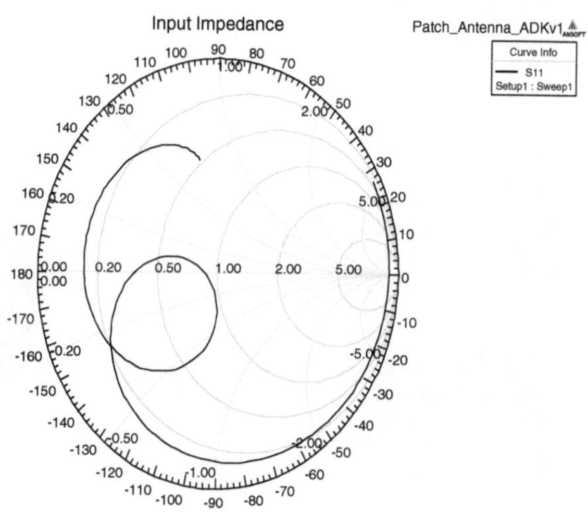

Peak Gain

For Single Patch Antenna: 6.436
For Array Antenna Design: 7.169

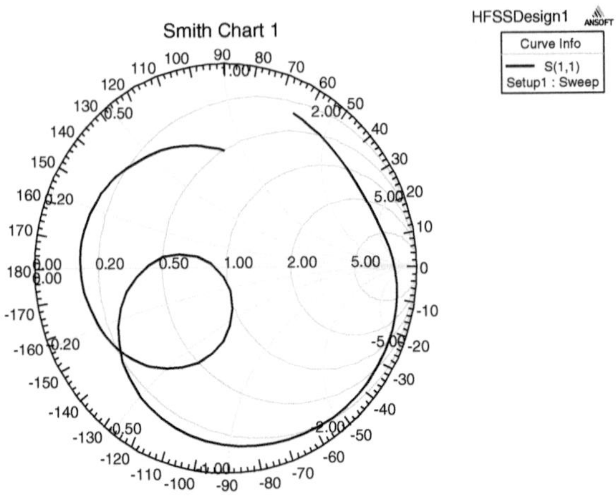

Fig. 6 Impedance for array antenna design

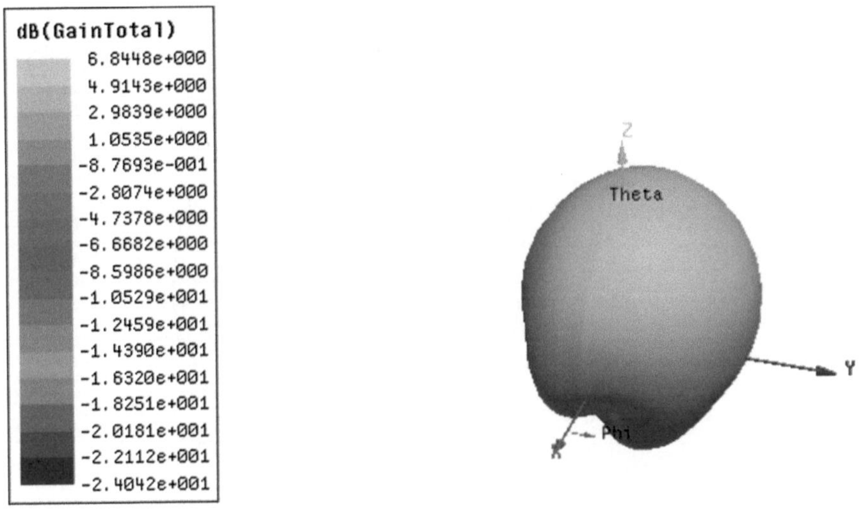

Fig. 7 3D polar plot for single patch antenna

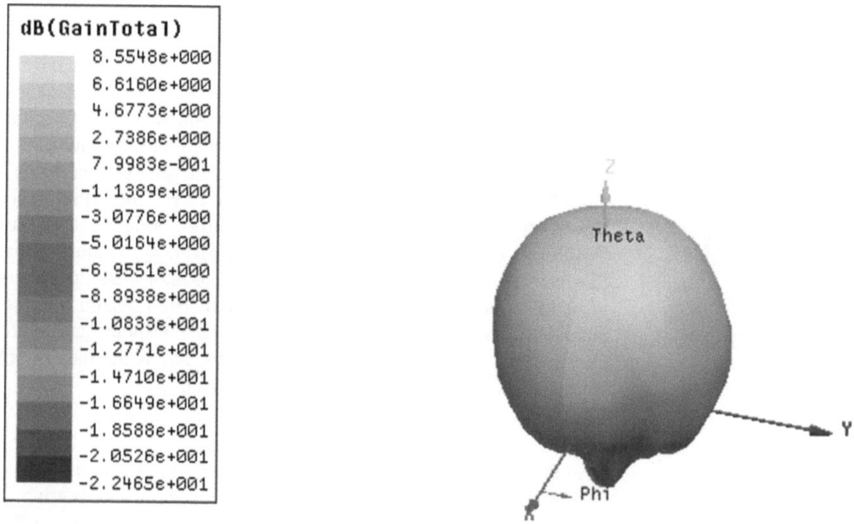

Fig. 8 3D polar plot for array antenna design

4 Conclusion

To improve the better gain of different antennas and also to nullify the side lobes we can go for Phase array antennas. This paper explains about a Rectangular patch antenna and also Rectangular Patch Antenna by Phased Arrays using HFSS 14.0. For rectangular Microstrip Patch Antenna Design, this paper got a gain of **6.844 db**, for rectangular Microstrip Patch Antenna with Array Design got a gain of **8.5548 db**. By using arrays this paper gets better gain. And also return loss, Impedance, 3D polar plot, Directivity and Gain of the antenna are computed and corresponding results are shown. The proposed antenna provides good performance in terms of return loss and radiation pattern and gain.

Acknowledgments Extending our grateful thanks to the authorities of JNTUA, Acharya Nagarjuna University and K L University for their support and encouragement to write this paper and to use R&D Labs.

Author Biographies

Smt. Vasujadevi Midasala is presently pursuing her Ph.D degree from JNTUA, Anantapuram in the area of Antennas and Communications. She received her degree in Electronics and communication Engineering from S.V.V.S.N Engineering College, Acharya Nagarjuna University, Guntur in 2008, M.Tech Degree from BSIT, JNT University, Hyderabad in 2011. She is currently working as **Assistant Professor**, Deptartment of ECE in **K L University**, Guntur, India. She has 6 years of teaching experience. She has published '26' papers in national/international journals/conferences. His research interests include Antennas, Signal Processing, Communications and VLSI Design. He has guided '04' M.Tech Projects. She is the member of IEEE and ISTE. e-mail: vasujadevi@gmail.com

Dr. P. Siddaiah obtained B.Tech degree in Electronics and communication Engineering from JNTUA college of engineering in 1988. He received his M.Tech degree from SV University Tirupathi. He did his Ph.D program in JNTU Hyderabad. He is the Chief Investigator for several outstanding Projects sponsored by Defense Organizations, AICTE, UGC and ISRO. He is currently working as Professor and **Principal**, Department of ECE in University College of Engineering and Technology, **Acharya Nagarjuna University**, Guntur, India. He has taught a wide variety of courses for UG and PG students and guided several projects. Several members successfully completed their Ph.D under his guidance. Several members pursuing their Ph.D degree. He has published '76' papers in National and International Journals and Conferences. He is the life member of FIETE, IE and MISTE.

Mr. S. NagaKishore Bhavanam is presently pursuing his **Ph.D** degree from JNTUA, Anantapuram in the area of Signal Processing and Communications. He obtained **M. Tech** Degree from Aurora's Technological and Research Institute, JNT University, Hyderabad in 2010. He obtained **B.Tech** degree in Electronics and communication Engineering from S.V.V.S.N Engineering College, Acharya Nagarjuna University, Guntur in 2008, He is currently working as **Assistant Professor**, Department of ECE in University College of Engineering and Technology, **Acharya Nagarjuna University**, Guntur, India. He has 6 years of teaching experience. He has published '44' papers in national/international journals/conferences. His research interests include Antennas, Signal Processing, Electro Magnetic Field Theory and VLSI Design. He has guided around '06' M.Tech Projects. He is the student member of IEEE. e-mail: kishore.ece.anu@gmail.com

Automated Neural Network Based Classification of HRV and ECG Signals of Smokers: A Preliminary Study

Suraj Kumar Nayak, Ipsita Panda, Biswajeet Champaty, Niraj Bagh, Kunal Pal and D.N. Tibarewala

Abstract Smoking of cigarettes has been reported to alter the cardiac electrophysiology by modulating the autonomic nervous system. A preliminary investigation of the heart rate variability (HRV) parameters suggested sympathetic predominance in smokers. An in-depth analysis of the time domain and wavelet processed ECG signals indicated that the automated neural networks (ANNs) were able to classify the signals with an accuracy of ≥85 %. This suggested that smoking not only modulates the functioning of the autonomic nervous system but is also capable of modulating the cardiac conduction pathway.

Keywords Smokers · Heart rate variability · Autonomic nervous system · Automated neural network

1 Introduction

Electrocardiogram (ECG) is the electrical potential which is associated with the functioning of the heart. The potential is initiated from the sino-atrial (SA) node. The functioning of SA node is controlled by the Autonomic Nervous System (ANS). ANS consists of two major subsystems, namely, parasympathetic and sympathetic nervous systems. The subsystems of ANS control the pacing of the SA node. ANS try to maintain the heart rate of a person by either increasing or decreasing the activities of the parasympathetic and sympathetic nervous systems. Due to this reason, there is a variation in the timing of the subsequent heart beats.

S.K. Nayak · I. Panda · B. Champaty · N. Bagh · K. Pal (✉)
Department of Biotechnology and Medical Engineering,
NIT-Rourkela, Odisha 769008, India
e-mail: pal.kunal@yahoo.com

D.N. Tibarewala
School of Bioscience and Engineering,
Jadavpur University, Kolkata 700032, India

S. Gupta et al. (eds.), *Advancements of Medical Electronics*,
Lecture Notes in Bioengineering, DOI 10.1007/978-81-322-2256-9_25

An in-depth study of this variation is regarded as Heart Rate Variability (HRV). The measure of the HRV parameters divulges information about the condition of the ANS in a non-invasive manner [1].

The functioning of the ANS has been reported to be affected by training the persons with regular exercises or by regular smoking. Regular exercise helps in improving the health of the cardiovascular system by inducing changes of the structural and the functional capability of the heart. The functional changes due to the exercises are mainly attributed to the parasympathetic dominance [2]. On the contrary, cigarette smoking has often been associated with cardiovascular diseases (e.g. coronary heart disease, aortic aneurysm, sudden death and peripheral artery disease). This can be explained by the reduced activity of the cardiac autonomic function, which in turn, results in the cardiac vulnerability of the smokers [3]. Due to the above reason, there is an increased risk of cardiac mortality in the smokers.

Keeping the above facts in mind, in this study, we have tried to understand the cardiac activity of the smokers taking athletes as the control group. A conscious effort was made for not selecting the sedentary group as the control because our current study is based on the short-term HRV studies, where the differences in most of the parameters of the cardiac activities of the smokers and the sedentary groups might be statistically insignificant [4]. Further, the statistical parameters of the ECG signals were calculated. The HRV and the ECG signal parameters were used for classifying the smokers from the athletic group using ANN.

2 Methods

Forty volunteers were invited to participate in the study. Out of the forty volunteers, twenty volunteers were the smokers who smoke at least ten cigarettes per day and twenty volunteers were the sports personnel (member of various athletic teams of NIT Rourkela) who practiced for at least 2 h/day. All the volunteers were the students of NIT Rourkela and were within the age group of 21–27 years (24.27 ± 1.82). The smokers were leading sedentary life. The invited athletes didn't smoke at all. All the volunteers were informed about the study and a written consent form was obtained from the volunteers before the start of the study. The ECG of the volunteers was recorded for 5, 90 min after dinner. The HRV parameters, time domain parameters and wavelet processed parameters of the ECG signal were calculated using the trial version of the software Biomedical Workbench (National Instruments, USA) and tabulated in the statistica spreadsheet. The important parameters were determined using t-test, classification and regression tree (CART), boosted tree (BT) and random forest (RF) analysis. The important parameters were used in various combinations as input for ANN based classification.

3 Results and Discussions

Thirty five HRV features were obtained from the study. To determine the important parameters, various statistical methods (t-test, CART, BT and RF) were employed. The t-test is a linear importance predictor whereas CART, BT and RF are non-linear classifiers for predicting important predictors.

A predictor importance of ≥95 % was considered acceptable. The analysis of the HRV features using t-test suggested that VLF % (FFT), VLF Power (AR) and VLF % (AR) were the important predictors. All the important parameters were higher in smokers as compared to the control group (athletes). This can be explained by the sympathetic dominance in the smokers [5].

The importance prediction using CART analysis showed that the important predictor was SD1. SD1 has been reported to be a marker of short-term variability of the heart rate [6]. Since the HRV parameters were calculated using 5 min of ECG signal, the occurrence of SD1 as the major predictor is justified. The SD1 values were higher in smokers as compared to the athletes. As per the reported literatures, a higher value of SD1 is associated with the sympathetic dominance of the autonomic control of the heart. The boosted tree classification showed that the VLF Power (FFT) and the VLF % (AR) were the main important predictors. Both the values were higher in smokers as compared to the athletes indicating a sympathetic dominance in smokers [7]. Similar to the previous linear and non-linear classification, Random Forest also suggested that the VLF Power (AR) was the major important predictor indicating the sympathetic dominance in smokers (Table 1).

The above important parameters were used in various permutations and combinations as the input variables for automated neural network (ANN) based classification. Best classification efficiency was obtained using the features VLF power (AR) and VLF % (FFT) with both MLP and RBF algorithms. The classification efficiency using MLP algorithms was found to be 82.5 % (Table 2) whereas the classification efficiency was found to be 95 % (Table 3) when RBF algorithm was used. The architecture properties of the MLP and RBF networks have been tabulated in Table 4.

Similar to the HRV parameters, the important time domain ECG signal parameters were calculated using t-test, CART, BT and RF. t-test and BT suggested that kurtosis and skewness were the important predictors, whereas CART and RF suggested that only skewness was the important predictor during classification (Table 5). Both the kurtosis and the skewness were found to be higher in the control group. Since only two important predictors were obtained, both the predictors were used as input for probable classification in ANN. The MLP algorithm showed a classification efficiency of 77.5 % (Table 6) whereas the RBF algorithm has showed a classification efficiency of 90 % (Table 7). The parameters of the network architecture have been provided in Table 8.

The ECG signals were decomposed using db6 wavelet up to the level 8. The reconstruction of the signal was achieved using d7 + d8 levels. The analysis of the parameters using t-test didn't show any important predictor. CART analysis showed

Table 1 Important predictors from HRV features

Classifiers	HRV features	Mean A	Mean SM	SD A	SD SM	p	Predictor importance
t-test	VLF (%) (FFT)	7.995	14.125	10.47776	8.348433	0.047692	–
	VLF Power (AR)	94.3575	252.451	123.4777	223.7172	0.008693	–
	VLF (%) (AR)	6.768	13.2665	6.294338	6.537892	0.00555	–
CART	SD1 (ms)	29.7365	36.828	22.0389	19.65057	–	1
BT	VLF power (FFT)	177.4445	315.0765	459.3199	289.5487	–	0.96826
	VLF (%) (AR)	6.768	13.2665	6.294338	6.537892	–	1
RF	VLF power (AR)	94.3575	252.451	123.4777	223.7172	–	1

Table 2 Confusion matrix of MLP network

	Category-A	Category-SM	Total
Total	20	20	40
Correct	16	17	33
Incorrect	4	3	7
Correct (%)	80	85	82.5
Incorrect (%)	20	15	17.5

Table 3 Confusion matrix of RBF network

	Category-A	Category-SM	Total
Total	20	20	40
Correct	18	20	38
Incorrect	2	0	2
Correct (%)	90	100	95
Incorrect (%)	10	0	5

Table 4 Parameters of the RBF and MLP networks

Networks	Features used	C.E.	Algorithm	E.F.	Hidden act	Output act
RBF 2-25-2	VLF power and VLF %	95.00	RBFT	CE	Gaussian	Softmax
MLP 2-16-2	VLF power and VLF %	82.50	BFGS 40	CE	Logistic	Softmax

C.E. classification efficiency, *E.F.* error function

Table 5 Important predictors from time domain features

Classifiers	Time domain features	Mean A	Mean SM	SD A	SD SM	p	Predictor importance
t-test	Kurtosis	5.074	3.819	1.879	1.437	0.022	–
	Skewness	1.341	1.122	0.299	0.264	0.019	–
CART	Skewness	1.341	1.122	0.299	0.264	–	1
BT	Kurtosis	5.074	3.819	1.879	1.437	–	1
RF	Skewness	1.341	1.122	0.299	0.264	–	1

Table 6 Confusion matrix of MLP network

	Category-A	Category-SM	Total
Total	20	20	40
Correct	14	17	31
Incorrect	6	3	9
Correct (%)	70	85	77.5
Incorrect (%)	30	15	22.5

Table 7 Confusion matrix of RBF network

	Category-A	Category-SM	Total
Total	20	20	40
Correct	17	19	36
Incorrect	3	1	4
Correct (%)	85	95	90
Incorrect (%)	15	5	10

Table 8 Parameters of the RBF and MLP networks

Networks	Features used	C.E	Algorithm	E.F	Hidden act	Output act
RBF 2-25-2	Kurtosis skewness	90	RBFT	CE	Gaussian	Softmax
MLP 2-4-2	Kurtosis skewness	77.5	BFGS 5	CE	tanh	Softmax

C.E. classification efficiency, *E.F.* error function

Table 9 Important predictors from time domain wavelet features

Classifiers	Wavelet features	Mean A	Mean SM	SD A	SD SM	p	Predictor importance
CART	AM	0.00	0.000	0.002	0.001	–	1
	Skewness	−0.291	3.184	12.140	6.127	–	1
BT	RMS	0.124	0.111	0.043	0.040	–	0.95
	Std. dev	0.124	0.111	0.043	0.040	–	0.95
	Variance	0.017	0.014	0.011	0.009	–	0.95
	Median	0.002	0.002	0.007	0.011	–	1
	Mode	0.002	0.002	0.007	0.011	–	1
RF	Kurtosis	3.743	3.329	0.814804	0.574	–	1

that the arithmetic mean (AM) and skewness were the important predictors. BT analysis indicated that root mean square (RMS), standard deviation (Std. Dev.), variance, median and mode were the important parameters. RF analysis indicated that kurtosis was the only important predictor (Table 9). Like HRV and time domain signal parameters, the important parameters of the wavelet processed signals were used in various combinations and permutations as input for ANN classification. The best classification efficiency using MLP algorithm was obtained when AM and skewness were used as the input parameters. An efficiency of 72.5 % was achieved (Table 10). On the contrary, an efficiency of 85 % was achieved when RMS and standard deviation were used as the input parameters for the ANN classification using RBF algorithm (Table 11). The architecture of the best networks have been provided in Table 12.

As a matter of fact, a classification efficiency of ≥85 % has been considered as a good efficiency while classifying ECG signals using automated neural networks [6]. During classification using HRV parameters a classification efficiency of 95 % was

Table 10 Confusion matrix of MLP network

	Category-A	Category-SM	Total
Total	20	20	40
Correct	14	15	29
Incorrect	6	5	11
Correct (%)	70	75	72.5
Incorrect (%)	30	25	27.5

Table 11 Confusion matrix of RBF network

	Category-A	Category-SM	Total
Total	20	20	40
Correct	18	16	34
Incorrect	2	4	6
Correct (%)	90	80	85
Incorrect (%)	10	20	15

Table 12 Parameters of the RBF and MLP networks

Networks	Features used	C.E	Algorithm	E.F	Hidden act	Output act
RBF 2-16-2	RMS std. dev	85.00	RBFT	CE	Gaussian	Softmax
MLP 2-30-2	AM skewness	72.50	BFGS 22	CE	tanh	Softmax

C.E. classification efficiency, *E.F.* error function

achieved using RBF algorithm. Similarly, classification efficiency of 90 and 85 % were achieved when time domain and wavelet processed signals, respectively, were used for classification using RBF algorithm. From the above results, it can be observed that the classification efficiency of the HRV, time domain and wavelet processed signals were ≥85 %, which is within the accepted limit of the recommended classification efficiency for the ECG signals. In our study, we have seen that the RBF algorithms showed better results compared to the MLP algorithms. The observation may be explained by the architecture of the ANN networks. MLP uses an inner product of the input and the weight vectors whereas RBF uses the Euclidean distance between the input and the weight vectors [7]. The MLP algorithm consists of multilayer hidden layers. Each layer is connected to the other layers using a linear combination function. In other words, the connection between each layer with the next layer is given by a linear mathematical equation. On the contrary, RBF algorithms usually contain only one hidden layer and the inputs are not weighted before connecting the hidden layers. Due to this reason, the development of RBF model requires less time to obtain an optimum model parameter. This can be explained by the fact that there is no need for repetition during the training period [8]. Additionally, the RBF type ANNs have been reported to allow universal approximations and learning without having any local minimum. This, in turn, allows quick convergence of the model parameters to provide results. Because of the above mentioned facts it has been reported that the RBF model provides better accuracy in prediction than the MLP models.

4 Conclusion

In the current study, the effect of smoking on the cardiac activity of the smokers was analyzed and compared with cardiac activities of athletes (control group). The athletic group was taken as a control. This was done, because the athletic activities have been reported to improve the cardiac activity. Hence, a pronounced difference in the cardiac activity between the groups was expected. The HRV analysis of the ECG signals suggested sympathetic predominance in the smokers as compared to the athletes. Even though the sympathetic activity was found to be prominent from the VLF % (FFT), VLF Power (AR), VLF % (AR), SD1 and VLF Power (FFT) features, the differences in the heart rate of the smokers and athletes was not statistically significant. This may be explained by the participation of a less number of volunteers. Apart from the linear classifier (t-test), the important predictors from the non-linear classifiers (CART, BT, RF) also suggested sympathetic dominance. RBF ANN network was able to efficiently classify the smokers from the athletes with a classification efficiency of 95 %. The ECG signals provide information about the cardiac electrophysiology. In our study, we found that though the linear classifier was not able to classify the ECG signals (time domain and wavelet processed), the non-linear classifiers were able to figure-out the important predictors. When the important predictors were used in probable classification using ANN, a classification efficiency of ≥85 % was achieved [8]. This indicated that the smoking of cigarettes have a marked effect on the cardiac electrophysiology. The changes in the cardiac electrophysiology can't be detected using linear classifiers.

References

1. Champaty B et al (2013) Artificial intelligence based classification of menstrual phases in amenorrheic young females from ECG signals. In: India conference (INDICON), 2013 annual IEEE, pp 1–6
2. Saboul D et al (2013) The impact of breathing on HRV measurements: implications for the longitudinal follow-up of athletes. Eur J Sport Sci 13:534–542
3. Harte CB et al (2013) Association between smoking and heart rate variability among individuals with depression. Ann Behav Med 46:73–80
4. Pal K et al (2013) Heart rate variability and wavelet-based studies on ECG signals from smokers and non-smokers. J Inst Eng (India) Ser B 94:275–283
5. Harte CB, Meston CM (2014) Effects of smoking cessation on heart rate variability among long-term male smokers. Int J Behav Med 21:302–309

6. Ciszewski P et al (2013) Lower preoperative fluctuation of heart rate variability is an independent risk factor for postoperative atrial fibrillation in patients undergoing major pulmonary resection. Interact Cardiovasc Thorac Surg 17:680–686
7. Saperova E, Dimitriev D (2014) Effects of smoking on heart rate variability in students (545.6). FASEB J 28:545.6
8. Baxt WG (1990) Use of an artificial neural network for data analysis in clinical decision-making: the diagnosis of acute coronary occlusion. Neural Comput 2:480–489

Reliable, Real-Time, Low Cost Cardiac Health Monitoring System for Affordable Patient Care

Meghamala Dutta, Sourav Dutta, Swati Sikdar, Deepneha Dutta, Gayatri Sharma and Ashika Sharma

Abstract It is increasingly being made aware by various organizations including the WHO (World Health Organization) that Cardiovascular Disease (CVD) is the 'Silent Killer' and is the perhaps the most important cause for death globally in the present times. The prime reason being the fact that CVD often goes unnoticed at the early stages and clinical intervention is soughted only during emergency. A thorough understanding of the nature of the disease reveals the association of physiological abnormalities and subdued physiological disorders that might have raised due to stress, lifestyle related disorders, pathological conditions etc. Studies have revealed that all these problems if addressed to at an early stage can be controlled and the potential risks and hazards lessened to a considerable extent. Early detection is possible in developed countries with expensive and advanced machineries, which can keep the data recorded in a user-friendly device affordable by the patient. But in a developing country like ours a low cost affordable device is yet to be available to the rural population especially addressing the problem. The aim of this study is to design an affordable and reliable Cardiac Health Monitoring platform for identification, data storage and availability to the physician even after a considerable time interval for diagnosis of the condition and post diagnosis continued therapy. The Cardiac Health Monitoring Systems (CHMS) was developed to capture and pre-process physiological parameters (ECG, Heart Sound, Heart Rate in Normal, Stressed and Physiologically abnormal conditions) in real-time and transmit the same using wired as well as wireless communication technology. The CHMS has been so designed is designed to capture the patient's/user's physiological parameters in electrical form e.g. the ECG signal, apply preliminary adaptive filtering to shape the requisite signals and forward the processed data. This data

Presented at International Conference on Advancements of Medical Electronics ICAME 2015, 29–30th January 2015.

M. Dutta (✉) · S. Sikdar · D. Dutta · G. Sharma · A. Sharma
Department of Biomedical Engineering, JIS College of Engineering, Kalyani, India
e-mail: dr.m.dutta@ieee.org

S. Dutta (✉)
IBM, New Delhi, India

S. Gupta et al. (eds.), *Advancements of Medical Electronics*,
Lecture Notes in Bioengineering, DOI 10.1007/978-81-322-2256-9_26

is acquired by Smart hand held mobile device, and the excess processing power of the device is utilized for further processing as well as storage of medical records and transmission of data to the hospital management system or for isolated diagnosis and assessment. Critical readings crossing the set threshold (per WHO standards) would set off an alarm [locally as well as over the mobile network] and thus alert any imminent life-threatening situation.

Keywords ECG · Heart sound · CHMS · LPF · FFT · Smartphone

1 Introduction

It has been proved by scientific research and years of collected data that Heart rate, ECG, Heart Sound signals can be considered as monitoring parameters related to Cardiovascular Functioning. Timely monitoring and analysis can predict subdued cardiovascular disorders [1]. The Gold standard for Heart Beat sets the normal Heart Beat for adults at 60–100 beats/min and any deviation is considered to be abnormal or irregular indicating subtle or pronounced malfunctioning of the heart, the effects needless to say might be fatal [2]. Though till date the Electrocardiogram (ECG) is the most widely adopted clinical tool to diagnose and assess the risk of CVD, ECG is clinically effective when the disturbances of the Cardiovascular system are reflected in the recording [1]. Often at an early stage of CVD, recordings in an ECG are near normal and the physician fails to diagnose and treat CVD [1]. Hence other physiological signals like Heart Sounds and Heart Rate should be seriously considered to be an important clinical marker for diagnosis and treatment of CVD in addition to ECG [3]. Monitoring mechanisms, let alone the continuous ones, are often absent in majorities of townships and rural areas in a developing country, like ours and it becomes difficult and expensive for the patient to visit the nearest hospital or diagnostic center for a sensation of increased heart beat (Tachycardia), or feelings of breathlessness. This early warning mechanism by which the body indicates its problems goes unnoticed or are neglected. There has been increasing demand for an easy to use and low cost device suitable for the patient as well as the physician, which would be able to forecast at an early stage any future possibilities of CVD and related disease.

Studies have clearly demonstrated that cigarette smoking, physical inactivity, and increased body mass index (BMI) are associated with increased risks of premature cardiovascular disease (CVD) and death especially in the age group 35–60 years [4]. Though the urban population gets the benefit of state-of-the-art hospitals with CVD monitoring devices the rural population especially in the economically backward countries is ignorant about the devastating effects of CVD. Hence a large population of men and women continue with subdued or untreated CVD with fatal results. Monitoring of abnormal Heart Sound signals in real time is an effective way to indicate the probable onset of CVD and CVD related deaths especially in this age group [3]. The pathogenesis of CVD is poorly understood. The untimely occurring CVD leading to heart attacks and deaths can be largely prevented by the timely diagnosis of CVD and its arrest with the available medicines [5]. The widely used ECG studies is limited to detection of abnormalities rising due to disturbances in the conduction of electrical impulse across the heart musculature while Heart sound studies e.g. the Phonocardiograph or Echo cardiograph deal with the mechanical defects of the heart. However, these diagnostic tools have their own limitations, as they require skilled medical personnel for interpretation of result, laboratory or hospital infrastructure for carrying out the tests thus making the tests too expensive and are condition specific. Hence cases of mild or subdued cardiac abnormalities often go unnoticed and results in under treatment.

The ECG signals along with the heart sound signals, being a direct expression of the mechanical activity of the cardiovascular system, is potentially an unique source of information for identifying significant events in the cardiac cycle and detecting irregular heart activity [6]. A simple recording of ECG along with Heart Sound signals from audible sounds at the heart apex, in real time can be a useful marker for CVD, which would enable early diagnosis and treatment. Conversion of these Biosignals into electrical form using a microphone can enable us to visualize, interpret and analyze the sound information [7].

Considering the fact that these abnormalities are unpredictable and may occur at anytime and anywhere especially with those with hypertension and high serum cholesterol levels, a strong need has been felt for the development of a unique device working on the Cardiac Health Monitoring Platform combining an unique indigenously developed device and the mobile phone which will be non-invasive, cost effective, simple to operate and can be used by anybody for prolonged home monitoring application. Our approach was to couple existing technology with a user-friendly device like the smart phone, which is within the affordable reach of the large populace and at the same time has the extra capacity to process the signal as well as store it for later reference and use. The prototype designed by us could be used for recording and reading the both the ECG signals and Heart Sound signals in real time and is aware of any abnormality with these parameters. If the intensity of Heart Sound increased to very high or dropped to very low, a signal would be given suggesting urgent attention required. Similarly for the ECG signals,which is of single electrode as well as multi-electrode type, the device can filter out the noise and accurately show the signal on the smart phone's screen.

2 Previous Work

Earlier works attempted to develop a mobile Biosignal recording device for signals originating from the Heart and adjoining areas. Prior research activities were mainly focused on developing:

A. Electronic ECG Sensing Device [8]
B. Recording and data storage of the ECG signals in the mobile [7]

Most of the mobile ECG research used either a PDA or a cell phone only as a data display or a transmission device, which sends the collected ECG data to a remote PC for later review and diagnosis by a physician [7, 9]. But this prototype has the capability to apply advanced signal processing techniques to filter out the noises (which is very common while acquiring heart sound) and electrical interferences in the smart phone itself. Although many devices are being worked upon on similar areas, no single device can be considered as the ultimate one; each has its own deficiencies and limitation.

Our work was mainly focused on developing an indigenously made low cost, reliable portable ECG system capable of transmitting signals via wired as well as wireless mode to the users' regular smart phone. The device has the capability of acquiring signals through various input sources—single electrode, multiple electrode and through microphone. The unused and unutilized storage space in the mobile phone would collect, record, display, and transmit the signals originating from the heart data in real time, but is also analyze the acquired Signal data, match them with existing available data and predict in advance possible CVD conditions.

3 Procedure

The real challenge faced by the team was keeping the signals noise free and noise filtering. Interference with the ambient sound and sound arising due to other physiological activities like blood flow, surrounding electrical signals arising due to intrinsic muscle activities. The signals were picked by the microphone needed to be filtered and processed by using several filtering techniques to isolate the meaningful electrical and acoustic signals generated due to the functioning of the heart and its surrounding tissues. Similarly for the single electrode and multiple electrode ECG signal acquisition, commercial adhesive conductor strips as well as clamps can be used to get commercial grade signal.

Our indigenous CHMS combines the simplicity of acoustic stethoscope with advanced electronics and information technology to facilitate better performance, recording of the heart and lung sounds and analysis of the recorded signals. The same CHMS has the capability of acquiring ECG signals through various leads, which need to connected appropriately. It also sends the data to a computer either by wireless or with a wired connection. The primary limitation of the analog or the conventional stethoscope is that it is unable to capture several physiological sound signals as they are below the threshold of hearing. This problem can be easily overcome using the microphone aided digital stethoscope although the amplified sound signal of interest contains enough noise, which interferes with the signal generation process [10].

Digital stethoscopes available are intended for observation and recording of heart sounds and murmurs as well as lung and airway sounds. There are other types of digital stethoscopes, which can be used, for monitoring heart and lung sounds during anesthesia and some even combine with synchronous ECG monitoring. But the CHMS was even more cost effective with same functional capability as that of the Digital Stethoscope. The CHMS is a good tool for medical education also. Heart sounds recorded with CHMS can be visually observed and listened to [10]. These recording can be used to build multimedia tools to improve the quality of physical exam and education. By providing connectivity solutions either wired or wireless, a digital stethoscope can be used in telemedicine applications facilitating remote diagnosis by specialists [11]. The recorded signals can further be spectrally analyzed and used for automated cardiac auscultation and interpretation. Our CHMS coupled to a portable computer/tablet with suitable software and connected to internet for automated or remote diagnosis by a specialist, will be the future trend.

4 Results

4.1 Design of the Prototype

In order to acquire real time Heart Sound signals, it is first essential to filter the associated noise due environment and other physiologic activities like talking, breathing, etc. which might be mixed with the heart sound. The design of the prototype was done in two phases.

Phase 1: The Hardware system of CHMS with Projected Mobile Application (Fig. 1)

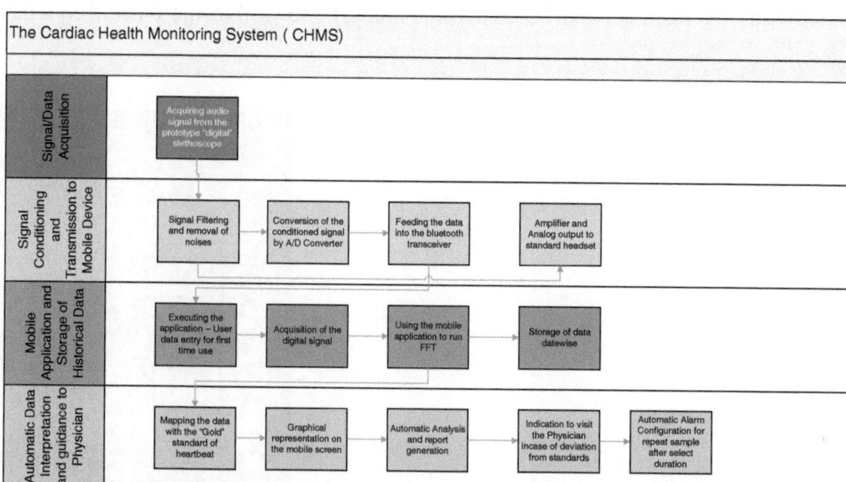

Fig. 1 The process for the heart sound capture

Phase 2: Further Work involves signal processing and signal analysis using DSP techniques and FFT for conversion of Electrical and Acoustic signals into the frequency domain for further analysis.

4.2 Prototype of the CHMS Used to Capture the ECG Signals in Real Time

The CHMS circuit uses a very sensitive common off-the-shelf Quad Op–Amp in a Low Pass Filter [LPF] setup with differential voltage amplifier configured for human interfacing with high impedance and high-noise environment (Fig. 2).

4.3 Frequency Response Analysis and Development of Algorithm of the Signal Generated by CHMS

In this paper we would concentrate on acquisition of the various signals (ECG, Heart Sound) through the custom designed front end analog hardware (Fig. 3) based on the concept that appropriate digital and analog filter combination should allow selective hearing of heart sounds (low frequency) and ECG signals in the frequency spectrum (Fig. 4) and the development of the algorithm of the various signal processing portions for removal of certain types of noises using the digital signal processing techniques.

The frequency spectrum illustrated in Fig. 4 shows that the Frequency Response of the normal heart beat sound signal captured by the CHMS. The frequency content varying from around 50–250 Hz. Figure 4 shows that the basic frequency

Fig. 2 Prototype for Cardiac Health Monitoring System

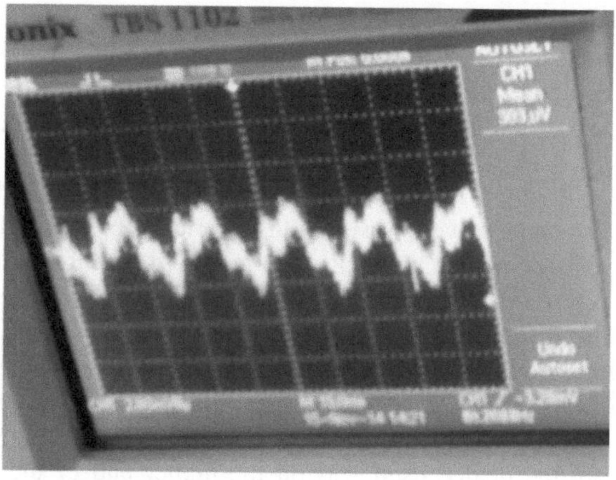

Fig. 3 Acquired ECG Signals on our CHMS

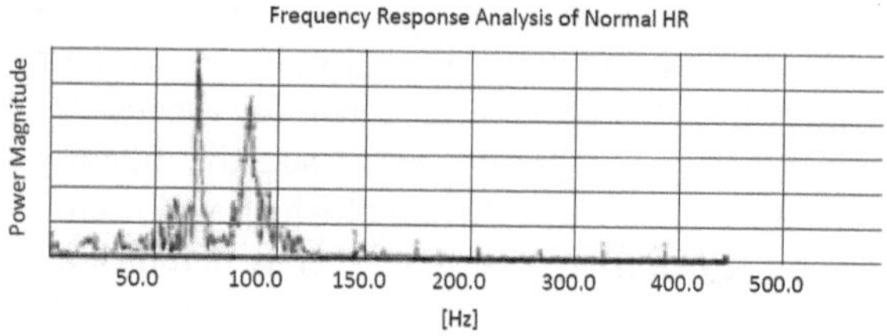

Fig. 4 Frequency response

components are obviously detected by the Fourier transform but not the time delay between these components. Possibility lies that the interference of surrounding noise as a result of other physiological activities may be one of the factors. Hence removal of this noise using adaptive filtering and low pass filters prior to recording of the said sound signal becomes very important. The two components (due to the closure of the aortic valve) and (due to the closure of the pulmonary valve) are obvious in Fig. 3. However, the FFT analysis cannot tell which of two, precedes the other or the value of the time delay between them. These parameters (position and time delay) are very important to detect some pathological cases.

5 Discussion

The impact of capturing and analyzing cardiac signals is realized when it is integrated within a clinical information system. Probabilistic modeling is a requirement to predict automated patient diagnosis and always requires accompanying clinical history to optimize discrimination and calibration. The device stores the data in a particular format and can be seamlessly uploaded to the hospital's database using the internet. If the mobile device does not have the connectivity for transmitted data over the network, it can be paired with the hospital Bluetooth hotspot for easy transmission. This requires customization for each hospital care system. The crucial component in the success of these techniques is to get the universal model in place and enable in providing sufficient signal quality to perform an accurate analysis of the data.

It may be noted that while using various mobile devices to record heart signals and sounds, we observed a high variance in quality between hardware, with some units being completely unable to record useful data because of the low frequency response characteristics of the Bluetooth audio section. This necessitated the incorporation of the converting the analogue signal into a digitally sampled one using an A/D converter. (This however had a big impact in the total cost of ownership.) While studying the characteristics of some readily available branded phones, we found that the Apple's devices had very good low frequency response, but the device itself was on the higher end of the cost spectrum with the ECG attachment costing more than the smart phone itself.

An important factor in using digital auscultation techniques is the problem of ambient noise and movements. Efforts have been made to remove such additional signals from the actual heart sound for accurate analysis [10, 12]. There has been use of LECG [12] for the use of adaptive filtering but has the problem of proper placement in the chest, failure of which will result in the actual heart sound getting cancelled out. The band pass filter is helpful to a certain extent after which the differentiation of heart, lung and ambient noise becomes quite difficult for automated study and analysis. Since the noise overlaps in the time and frequency domain, filtering is extremely difficult, [7] used a second off-body microphone to record ambient noise to provide information for an adaptive filter.

Finally, we once again point out the fact that it is possible to reuse the existing technology and hardware in countries where health care is expensive as well as not easily accessible to provide low-cost reliable diagnostics. We also note that these standalone systems would form a complimentary diagnostic framework that can be leveraged by the higher health care mechanisms in providing timely and cost effective treatment. We also note that the diagnostic capability of the stand-alone advanced mobile device is not confined to heart rate analysis, but can be extended to HRV, heart valve issues, lung function, infection, sleep structure and even depression using various transducers.

6 Conclusion

There have been many attempts to bring advanced health monitoring devices at affordable cost for the rural and masses below the poverty line. With the advancement in microprocessor technologies, more and more computing power is being provided at a comparatively lesser-cost month on month. This has led to splurge in sophisticated mobile devices, which have become ubiquitous. The digital auscultation procedure has been there for some while but has its own challenges of cost, accuracy etc. We have tried to address the existing problems and develop a low-cost solution for the mass. So far the results have been satisfactory under controlled conditions. Further work would include removal of extraneous noise, optimum placement of electrical activity probe in the chest, improvement in getting frequency response below 100 Hz and packaging of the device for ease of handling and use.

Acknowledgments This work has been funded by AICTE under AICTE Research Proposal Scheme.

References

1. Goldberger AL, Goldberger E (2006) Clinical electrocardiography: a simplified approach, 7th edn. Elsevier/Mosby Inc, St Louis
2. American Heart Association (2008) Heart disease and stroke statistics-2008 update. AHA, Dallas, Texas
3. WHO (2007) Prevention of cardiovascular disease. Guidelines for assessment and management of cardiovascular risk. Geneva 47(12):1555–1569
4. Amin DSM, Fethi B-R (2008) Features for heartbeat sound signal normal and pathological. Recent Pat Comput Sci 1:1–8
5. Mackay J, Mensah G (2004) Atlas of heart disease and stroke. World Health Organization, Geneva
6. Mittra AK, Choudhari NK (2009) Development of a low cost fetal heart sound monitoring system for home care application. JBiSE 2(6):380–389
7. He B, Cohen RJ (1992) Body surface laplacian ECG mapping. IEEE Trans Biomed Eng 39 (11):1179–1191
8. Oresko J, Cheng J, Jin Z, Duschl H, Cheng A (2009) Detecting cardiovascular diseases via real-time electrocardiogram processing on a smart phone. In: Proceedings of the workshop on biomedicine in computing: systems, architectures, and circuits (bic), June 2009, Austin, TX, pp 13–16
9. Bhatikar SR, DeGroff C, Mahajan RL (2005) A classifier based on the artificial neural network approach for cardiologic auscultation in pediatrics. Artif Intell Med 33(3):251–260
10. Zhang GQ, Zhang W (2009) Ageing Res Rev 8(1):52–60. doi:10.1016/j.artmed.2004.07.00
11. Fox K et al (2007) Resting heart rate in cardiovascular disease. J Am Coll Cardiol 50:823–830
12. Tan A, Masek M (2009) TT fetal heart rate and activity monitoring via smart phones. health summit 2009. Washington DC. http://research.microsoft.com/en-us/events/mhealth2009/tan-masek-presentation.pdf. Accessed 29–30 Oct 2009

Part IV
Embedded Systems and Its Applications in Healthcare

An Ultra-Wideband Microstrip Antenna with Dual Band-Filtering for Biomedical Applications

Subhashis Bhattacharyya, Amrita Bhattacharya and Indranath Sarkar

Abstract Anultra-wideband (UWB) microstrip antenna with dual band filtering to prevent interference from coexisting WLAN and downlink of X-band satellite communication system is proposed for use in biomedical applications. Dual notch bands are achieved by etching two nearly half wavelength inverted U shaped slots on the radiating patch. Impedance bandwidth has been improved by asymmetric optimization of patch width with respect to the feed line. The antenna achieves −10 dB return loss bandwidth of about 10.45 GHz (2.35–12.796 GHz) while rejecting 4.9–5.98 and 7.15–7.99 GHz bands. The antenna has wide bandwidth, similar and stable radiation performance at the passband frequencies.

Keywords Half wavelength slot · Microstrip antenna · Notchband · Ultra-wideband (UWB)

1 Introduction

The approval and allocation of the frequency band 3.1–10.6 GHz for ultra-wideband (UWB) systems by Federal Communications Commission (FCC) [1], has motivated researchers to design UWB antennas for short-range high data rate wireless communication, high accuracy radar and imaging systems with low power consumption. UWB indoor wireless communication devices can be utilized in modern electronic-healthcare system and Wireless Body Area Network (WBAN).

S. Bhattacharyya (✉) · A. Bhattacharya · I. Sarkar
Department of ECE, JIS College of Engineering (Autonomous Institution Under WBUT), P III, A5, Kalyani, Nadia 741235, West Bengal, India
e-mail: subhashis.ece.jis@gmail.com

A. Bhattacharya
e-mail: amrita.ece.jis@gmail.com

I. Sarkar
e-mail: indranath_sarkar07@rediffmail.com

S. Gupta et al. (eds.), *Advancements of Medical Electronics*, Lecture Notes in Bioengineering, DOI 10.1007/978-81-322-2256-9_27

The printed planar monopole antennas [2, 3] have attractive features like wide bandwidth, compact size, simple structure, low cost, ease of fabrication and omnidirectional radiation pattern [4]. Owing to coexistence of the UWB system with other wireless standards such as the WLAN systems (5.1–5.825 GHz) and downlink of X-band satellite communication system (7.25–7.75 GHz), UWB antennas with filtering property are required. To reject the two above mentioned bands, two band stop filters are necessary which will increase the cost and complexity of the system. A simpler way to solve this problem is to design an UWB antenna with double band-rejection characteristics. UWB antennas with band-notched function have been reported, mostly with one notched band [5–7] for WLAN (5.15–5.825 GHz). Recently several antennas with dual notched bands [8–10] were presented. Etching slots on the patch or on the ground plane and adding parasitic elements are two widely used methods to obtain notched bands [11–13].

In this paper, a single layer microstripline-fed monopole UWB antenna with dual notched bands for the rejection of WLAN and downlink of X-band satellite communication systems is proposed. Dual notched bands have been achieved by introducing two inverted U shaped slots of different lengths on the radiating patch. Parametric study has been carried out through simulation and analyzed thoroughly. Simulated return loss (S_{11}) characteristics, surface current distribution and radiation patterns have been illustrated and discussed. Surface current distributions analyze the creation of notches.

2 Antenna Design

Figure 1a shows the configuration of the proposed antenna. Dimensions of the antenna have been optimized using IE3D software [13] to achieve the desired result. This antenna with dimensions of $40 \times 34 \times 1.6$ mm^3, has been fabricated on a PTFE substrate with relative permittivity (ϵ_r) of 2.4. A rectangular beveled patch has been created on the front side of the substrate while a ground plane with a rectangle slot of dimension 30×16 mm^2 etched on it has been designed on the back side of the substrate. The width of the microstrip feed line is kept uniform of about 3.9 mm to achieve 50-Ω characteristic impedance. The bevelled patch is asymmetrical with respect to the microstrip feed line. Two bevelling angles have been created with $\alpha_1 = 10.56°$ and $\alpha_2 = 26.19°$ on either side of the patch respectively.

The design of the slots on the patch involves guided wavelength $\lambda_g = (\lambda_0/\sqrt{\epsilon_{eff}})$ [14] where 'λ_0' given as $\lambda_0 = c/f_c$ is the free space wavelength corresponding to the notch centre frequency f_c, and where c is the speed of electromagnetic wave in free space. $\epsilon_{eff} = (\epsilon_r + 1)/2$ [14] is the effective dielectric constant averaged over two mediums, the substrate and air. The longer inverted U slot (slot-1) has an optimized length of 21.5 mm and uniform slot width of about 0.625 mm. This length of slot-1 is about $0.51\lambda_g$ with respect to the simulated notch centre frequency of 5.44 GHz. The total length of the shorter inverted U slot (Slot-2) is 15.5 mm with slot width t = 0.25 mm. The length of this slot-2 is about 0.51 λ_g at the simulated notch centre

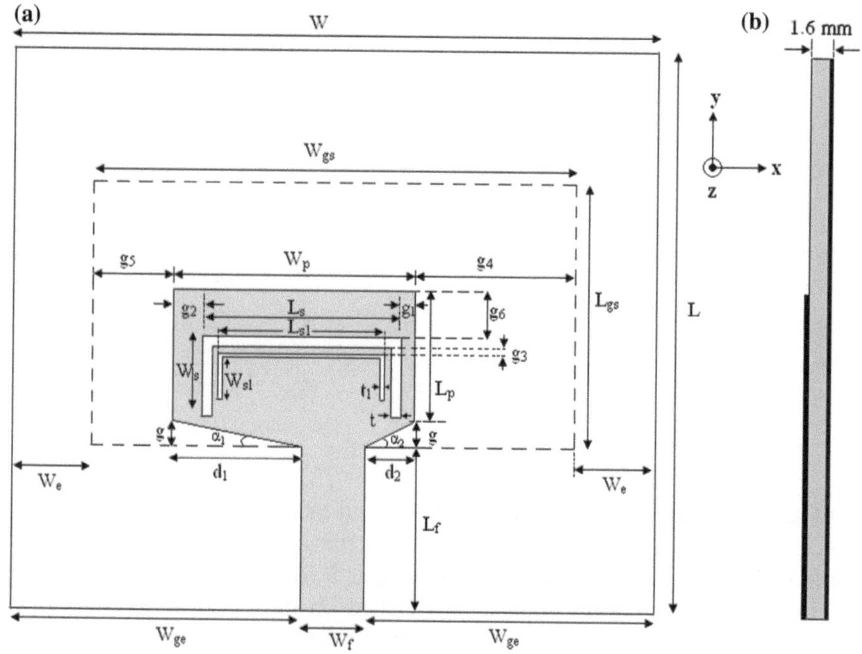

Fig. 1 **a** Configuration of the proposed antenna. **b** Side view

frequency of 7.6 GHz approximately. Therefore both the slots become resonant at those frequencies where their length is approximately 0.51 of the guided wavelength. Slot-2 is etched 0.375 mm below slot-1.

The optimized parameters of the antenna are given as follows:

$W = 40$ mm, $L = 34$ mm, $W_p = 15$ mm, $L_p = 8$ mm, $g = 1.5$ mm, $d_1 = 8.05$ mm, $d_2 = 3.05$ mm, $\alpha_1 = 10.56°$, $\alpha_2 = 26.19°$, $L_f = 10$ mm, $W_f = 3.9$ mm, $L_s = 12$ mm, $W_s = 4.75$ mm, $t = 0.625$ mm, $L_{s1} = 10$ mm, $W_{s1} = 2.75$ mm, $t_1 = 0.25$ mm, $W_{gs} = 30$ mm, $L_{gs} = 16$ mm, $W_{ge} = 18.05$ mm, $W_e = 5$ mm, $g_1 = 1$ mm, $g_2 = 2$ mm, $g_3 = 0.375$ mm, $g_4 = 10$ mm, $g_5 = 5$ mm, $g_6 = 3$ mm.

3 Results and Discussion

3.1 Parametric Study

Several antenna parameters are varied with a range of values as part of the optimization process.

Figure 2 shows that when the patch width increases asymmetrically due to d_1 increase, the return loss improves throughout the UWB range and resonances get prominence at lower frequencies also.

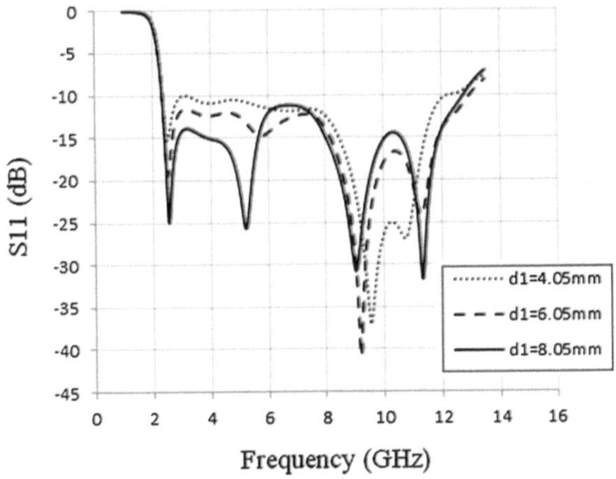

Fig. 2 Simulated S_{11} of the antenna for different values of d_1

When only one slot is etched on the patch with total slot length of 21.5 mm and slot width of 0.5 mm then only one notched band is achieved from 5.012 to 6.104 GHz with notch band centre frequency at 5.56 GHz. When another slot of length 15.5 mm and width 0.25 mm is etched on the patch in addition to the comparatively longer slot of length 21.5 mm as stated previously in this paragraph and width changed to 0.625 mm, a second notched band is achieved from 7.15 to 7.99 GHz with notch band centre frequency of 7.57 GHz along with the first notched band ranging from 4.9 to 5.98 GHz with centre frequency 5.44 GHz as inferred from Fig. 3. So it can be concluded that etching of slots plays a vital role in generation of notched bands. Each slot creates a notch with a centre frequency which is related to the length of the respective slot.

Figure 4 shows the effect of variation of W_{s1} on the second notch band centre frequency and its notch bandwidth. As slot-2 dimension (W_{s1}) is increased, the total slot length increases which implicitly corresponds to higher guided wavelength and thus a corresponding lower notch centre frequency. From Table 1 it is observed that with increasing W_{s1}, the notch centre frequency attains lower values with gradually

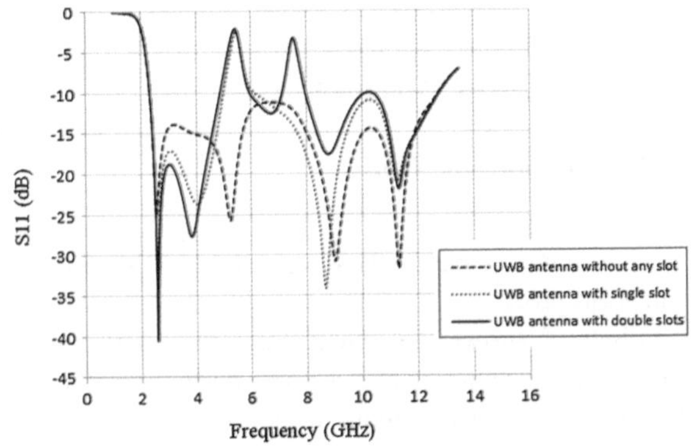

Fig. 3 Simulated S_{11} of the antenna with and without slot on patch

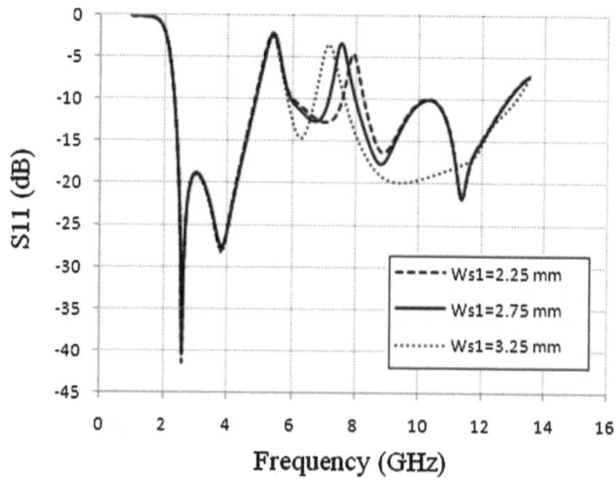

Fig. 4 Simulated S_{11} of the antenna for various W_{s1}

Table 1 Effect of W_{s1} on the notch-2 characteristics

W_{s1} (mm)	Notch-2 bandwidth (%)	Notch-2 centre frequency (GHz)
2.25	8.85	7.91
2.75	11.096	7.57
3.25	12.81	7.18

increasing notch bandwidth. So it can be concluded that the notch band centre frequency is controllable by varying the length of the slot.

Figure 5 shows the effect of variation of the vertical gap g_3 between slot-1 and slot-2. As the slot-2 position is moved on the patch vertically away from or towards slot-1, the gap g_3 varies. The details of this variation are listed in Table 2. It has been observed that if any slot is shifted closer to the feed line (closer to the bottom

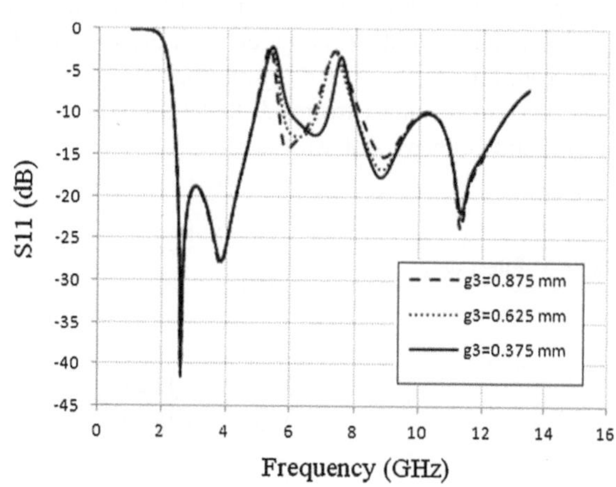

Fig. 5 Simulated S_{11} of the antenna for different g_3

Table 2 Effect of G3

g3 (mm)	Notch-2 bandwidth (%)	Notch-2 centre frequency (GHz)
0.875	20.12	7.405
0.625	16.4	7.375
0.375	11.096	7.570

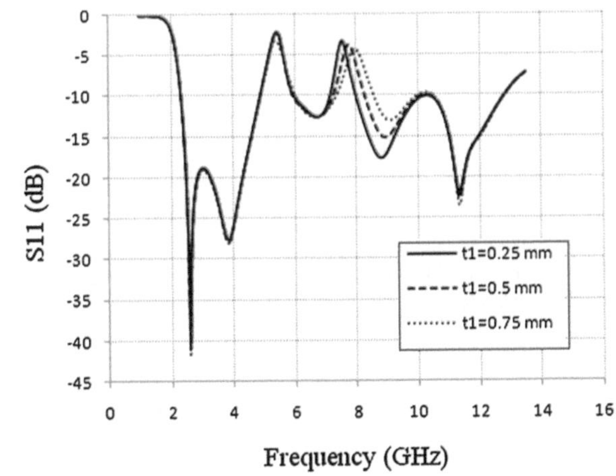

Fig. 6 Simulated S_{11} of the antenna for different slot-2 width t_1

edge of the patch), notch bandwidth, created by that slot, increases as well as the notch peak level (in terms of Return Loss or S_{11} in dB) increases along with slight decrease in notch centre frequency.

Figure 6 shows the effect of different widths of slot-2 on the S_{11} characteristics of the antenna. With the increase in the slot-2 width 't_1', the second notch bandwidth increases gradually as displayed in Table 3. Also the notch band centre frequency increases by small amounts. So it is an important generic conclusion that the notch bandwidth can be controlled by varying the slot width.

When slot-1 width 't' is varied then the first notch higher cut off is changed as seen from Fig. 7. For t = 0.525 mm the 1st notch ranges from 4.87 to 5.8 GHz and for t = 0.625 it ranges from 4.9 to 5.98 GHz. So it again validates the control of notch bandwidth by slot width.

Table 3 Effect of T_1

t_1 (mm)	Notch-2 bandwidth (%)	Notch-2 centre frequency (GHz)
0.25	11.096	7.570
0.5	13.26	7.765
0.75	15.19	7.965

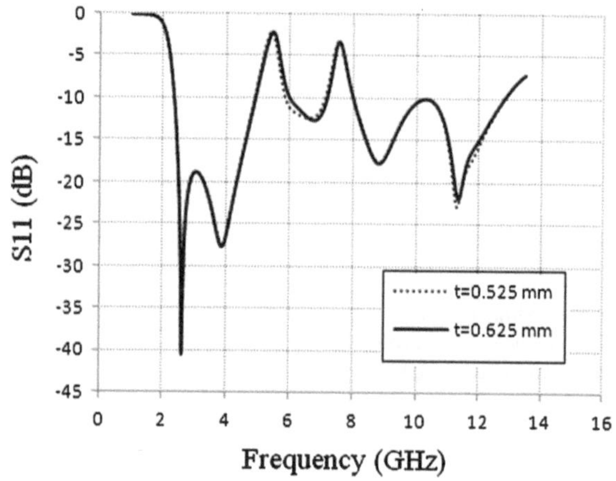

Fig. 7 Simulated S_{11} of the antenna for different slot-1 width t

From the above slot width variations it has been observed that the higher cut off frequency is affected more than the lower cut off value of the notched bands.

3.2 Simulation Results

The proposed antenna has been studied using IE3D software.

3.2.1 S_{11} Characteristics

The simulated impedance bandwidth of the antenna ranges from 2.35 GHz to about 12.79 GHz with dual notched bands, one ranging from 4.9 GHz to about 5.98 GHz there by rejecting the WLAN band (5.15–5.825 GHz) completely and another ranging from 7.15 to 7.99 GHz thereby completely rejecting the 7.25–7.75 GHz band reserved for downlink of X-band satellite communication systems. This makes it usable for WBAN applications within the UWB range and also making it interference free from the two rejection bands. The details of the passbands are provided in Table 4. It is observed from Fig. 8 that there are five prominent resonances within the $S_{11} < -10$ dB range. A sharp resonance at 2.6 GHz (−40 dB) and others at 3.85 GHz (−27.62 dB), 6.75 GHz (−12.52 dB), 8.82 GHz (−17.6 dB) and 11.34 GHz (−21.8 dB) are observed.

3.2.2 Surface Current Distributions

The excited surface current distributions, simulated using Zeland's IE3D simulation software [13], for the proposed antenna at five different frequencies are presented in

Table 4 S_{11} performance of the proposed antenna

Lower cut off (GHz)	Higher cut off (GHz)	% Bandwidth	Centre frequency (GHz)
2.35	4.88	70	3.6
5.99	7.14	17.517	6.565
8	12.796	46.12	10.4

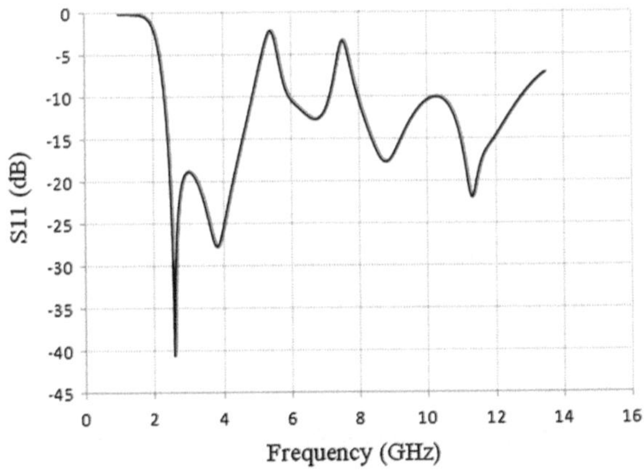

Fig. 8 Simulated S_{11} performance of the proposed dual band-notched UWB antenna

Fig. 9. At the passband frequencies of 2.6, 6.76, 8.8 GHz the distribution of surface current is uniform over the patch as well as the ground as shown in Fig. 9a, c and e respectively. The current is spread over a greater area in the ground planet 2.6 GHz compared to other passband frequencies, as observed from Fig. 9a which may attribute to better radiation. So it may be the cause of the sharp resonance of −40 dB, achieved at 2.6 GHz.

However when the antenna is operating at 5.44 and 7.6 GHz, which are the centre frequencies of the two notched bands, respectively, it is observed that stronger current is concentrated around the edges of the longer inverted U slot (slot-1) only at the vicinity of 5.44 GHz as shown in Fig. 9b whereas stronger current forms around the edges of the shorter inverted U slot (slot-2) only at the vicinity of 7.6 GHz as shown in Fig. 9d. This leads to impedance mismatch between the feed line and the patch at these notch frequencies and thus radiation from the antenna is much less in the two notched band regions. The current around the edges of these slots is oppositely directed to the current emerging from the microstrip feed line at each notch band. So the radiation fields generated by them are neutralised due to destructive interference taking place. This leads to the desired high attenuation around the notch centre frequency. So it can be derived that the longer etched slot is responsible for creation of the notched band with lower notch centre frequency of 5.44 GHz and the shorter slot is responsible for the notched band with higher notch centre frequency of 7.6 GHz.

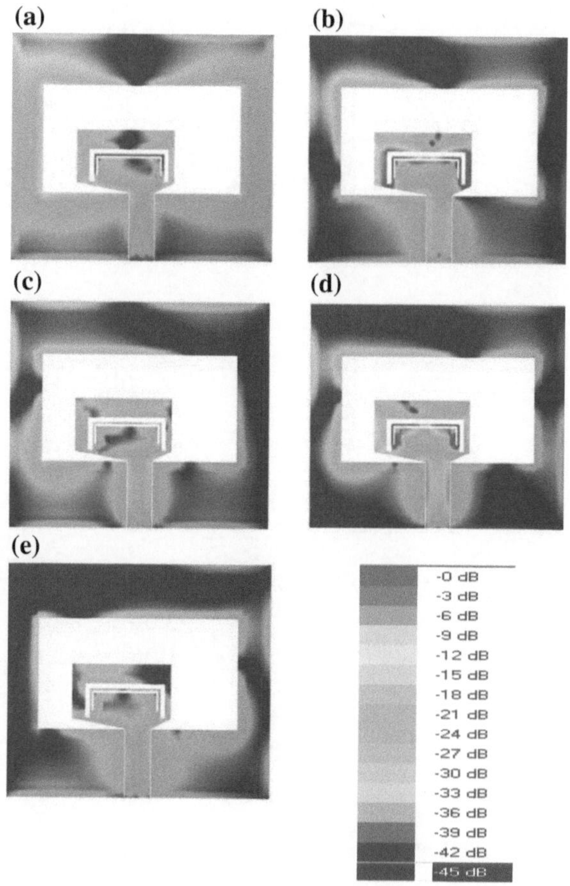

Fig. 9 Simulated surface current distributions of the proposed antenna at **a** 2.6 GHz, **b** 5.44 GHz, **c** 6.76 GHz, **d** 7.6 GHz, **e** 8.8 GHz

3.2.3 Radiation Pattern

The simulated radiation patterns in the elevation direction at different frequencies are illustrated in Fig. 10. For $\Phi = 0°$, E_{Φ} represents the co polarization (co-pol) and E_{θ} represents the cross polarization (x-pol) whereas for $\Phi = 90°$, E_{θ} represents co polarization and E_{Φ} represents cross polarization.

At 2.6 GHz pass band frequency, the co-pol is dominating the cross-pol highly in both $\Phi = 0°$ and $\Phi = 90°$ as shown in Fig. 10a, b respectively. We see similar scenario at 8.8 GHz observed from the Fig. 10e, f. At the notch band centre frequencies, the co-pol levels are usually lower than the x-pol levels. The patterns for 5.44 GHz notch band frequency is illustrated in Fig. 10c, d.

The simulated patterns at the pass band frequencies are similar which is expected from a wideband antenna. So it can be a general conclusion that in the notch band, useful radiation that is the co-pol is less compared to the co-pols at all other pass

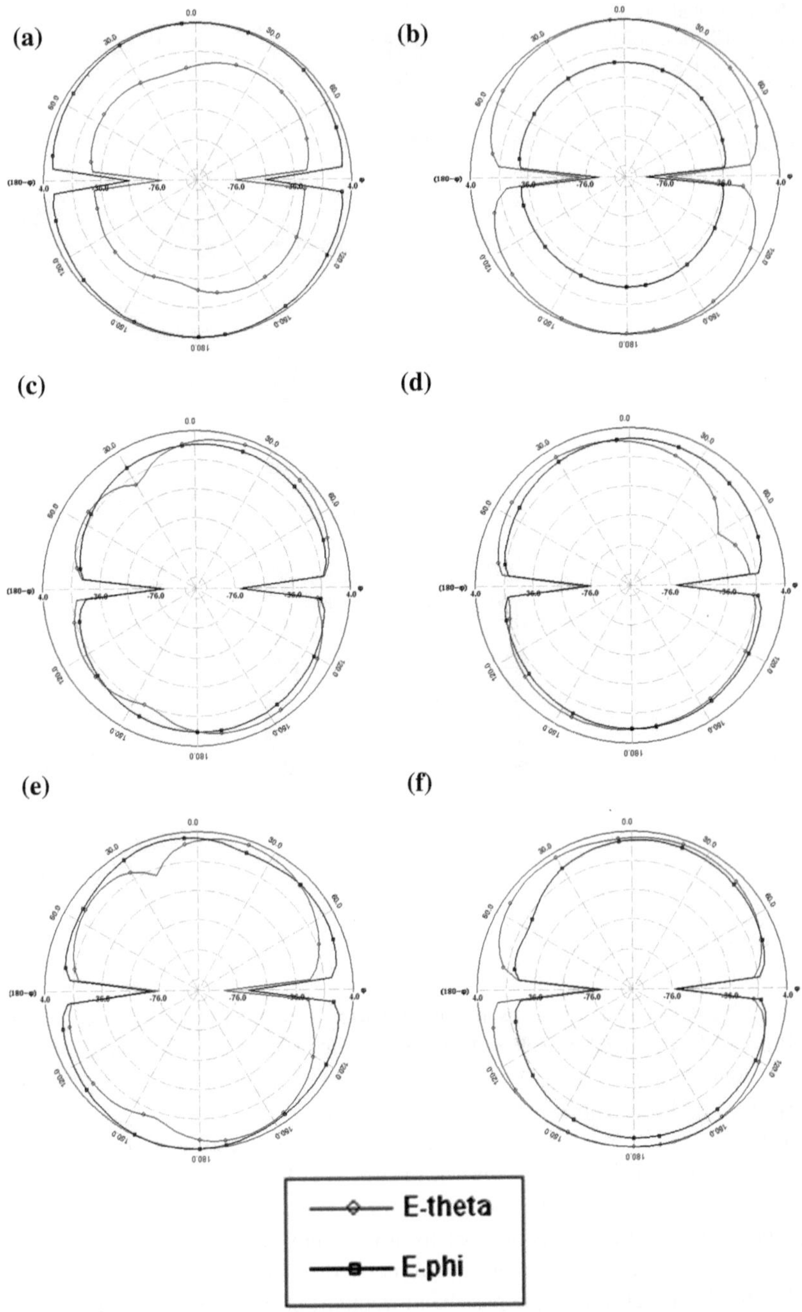

Fig. 10 Simulated radiation patterns of the proposed antenna in the elevation direction at **a** $\Phi = 0°$, **b** $\Phi = 90°$ at 2.6 GHz; **c** $\Phi = 0°$, **d** $\Phi = 90°$ at 5.44 GHz; **e** $\Phi = 0°$, **f** $\Phi = 90°$ at 8.8 GHz

band frequencies. This may provide tolerable link loss at the pass band frequencies for a satisfactory on-body antenna required for biomedical applications.

3.3 Applications

3.3.1 Bio-medical Imaging Systems

In the UWB microwave imaging systems, a very narrow pulse is transmitted from a UWB antenna to penetrate the body. As the pulse propagates through the various tissues, reflections and scattering occur at the interfaces. A particular interest is in the scattered signal from a small size denser tissue representing a tumour or other abnormalities. The reflected and scattered signals can be received using an UWB antenna, or array of antennas, and used to map different layers of the body. Microwave Imaging is used in breast cancer detection which is a common diagnosis in women. A typical detection system is illustrated in Fig. 11a. Normal breast tissue is largely transparent to microwave radiation, whereas malignant tissues, which contain more water and blood, cause microwave signal backscattering. This scattered signal can be picked by an array of microwave antennas and analysed using a computer [15]. Imaging result on an artificial breast model is shown in Fig. 11b.

3.3.2 Microwave Hyperthermia

Hyperthermia means repeated heating of the tumour to just over 40 °C. This treatment is toxic for the tumour itself and has shown doubled cure rates when combining hyperthermia with traditional cancer treatment. The microwave technique enables heat treatment of tumors which are deep-seated and/or relatively hard to access in the body. The microwaves are transmitted from a number of antennas

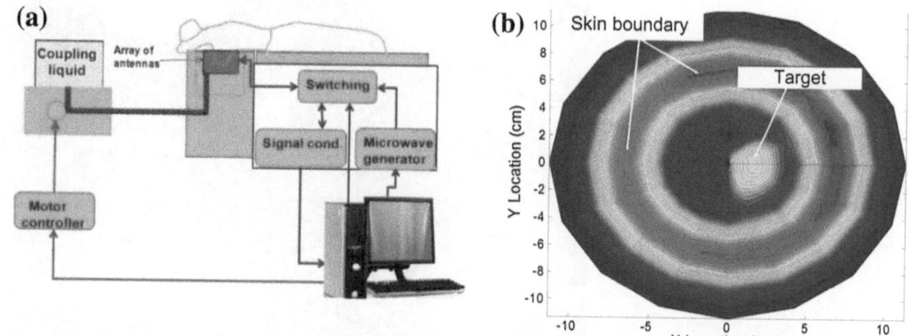

Fig. 11 **a** Configuration of the breast cancer detection UWB radar system, **b** imaging result on an artificial breast model

enclosing the relevant body part. The heat effect is developed in the tumor by the transmission of microwaves which are adjusted in time, frequency and strength in order to work together to form a focus in the desired location. This requires high precision of the system, in order for the heating to be concentrated on just the tumor, without heating surrounding healthy tissues.

Figure 12 shows scan reports of a 24-year-old man with grade III recurrent glioma at three various stages as (A, B), (C, D), (E, F). In the hyperthermia group, primary cases received hyperthermia treatment, and patients with recurrent tumors were treated with hyperthermia in combination with radiotherapy and chemotherapy. Electrodes were inserted into the tumor with the aid of a CT-guided stereotactic apparatus and heat was applied for 1 h. During 3 months after hyperthermia, patients were evaluated with head CT or MRI every month. Gliomas in the hyperthermia group exhibited growth retardation or growth termination. Necrosis was evident in 80 % of the heated tumor tissue and there was a decrease in tumour diameter [16].

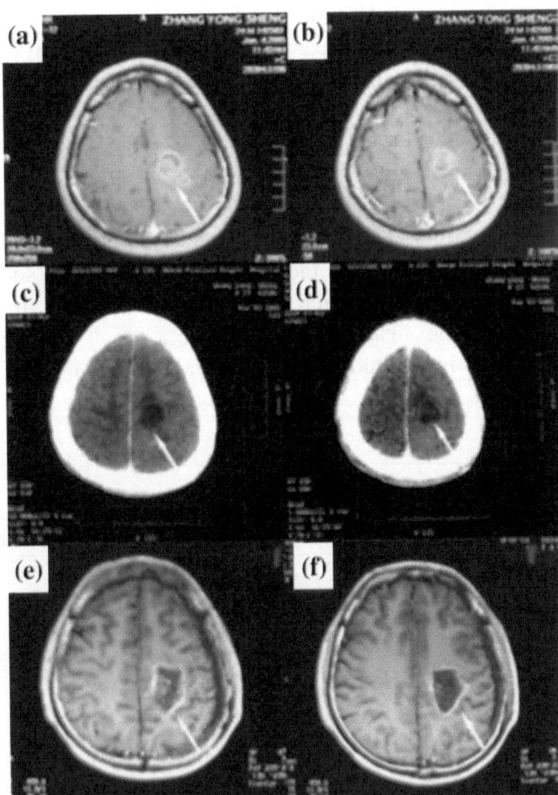

Fig. 12 **a**, **b** Brain tumor showing an irregular enhancing signal shadow (*arrows*) on CT before treatment; **c**, **d** CT scan showing internal liquefaction and tumor necrosis (*arrows*) in low-density images at 7 days after hyperthermia (hyperthermia + radiotherapy + chemotherapy); MRI scan showing increased tumor necrosis (*arrows*) and surrounding tissue with no obvious enhancement at 12 months after hyperthermia

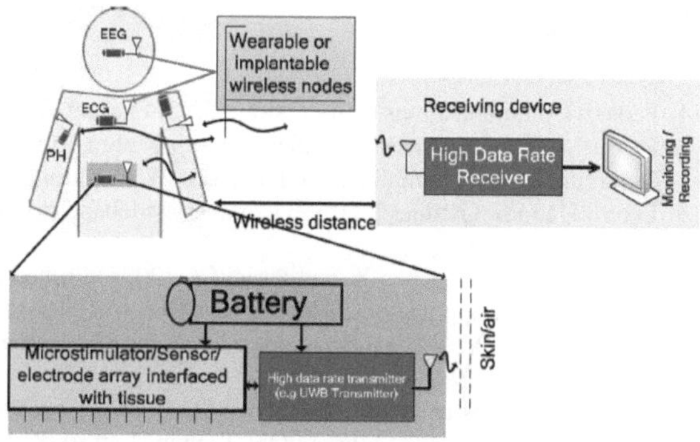

Fig. 13 A modern medical monitoring and detection system

3.3.3 Medical Telemetry

Miniaturized sensors can be worn on the body or implanted inside the body or kept at a distance to monitor a person's physiological state continuously in his free living conditions.

UWB telemetry systems are suitable for high data rate transmission such as wireless endoscope and multi-channel continuous biological signal monitoring such as Electroencephalography (EEG), Electrocardiogram (ECG) and Electromyography (EMG). A telemetry system is shown in Fig. 13. Their benefits are low power UWB transmitter to increase battery life, high data rate to increase the resolution and performance, and less interference from other wireless systems. Generally a UWB receiver consumes more power than narrow band systems. So an off shelf receiver is placed at 0.5–10 m away to detect the transmitted signal from body. For a UWB transmitter the FCC regulation requires the signal output to be −41 dBm/Hz and lower [17]. However the power levels should not reach above the regulated in-body tissue absorption levels and it is better to use microstrip antennas as external nodes which are kept away from the body. This reduces long term exposure to radiation.

4 Conclusions

An Ultra-wideband microstrip antenna with filtering property has been investigated for use in important biomedical applications like microwave imaging, breast cancer diagnosis, microwave hyperthermia, telemetry and many WBAN applications with little or no interference from WLAN and downlink of X band satellite communication systems. Dual band notch characteristic has been achieved by embedding two inverted U shaped slots on the patch. The designed antenna exhibits broad bandwidth, good and similar radiation performance at Yucethe pass band frequencies.

References

1. Federal Communications Commission (2002) First report and order in the matter of revision of Part 15 of the commission's rules regarding ultra-wideband transmission systems. Federal Communications Commission, ET-Docket, Washington, pp 98–153
2. Lee E, Hall PS, Gardner P (1999) Compact wideband planar monopole antenna. Electron Lett 35(25):2157–2158
3. Liang J, Chiau CC, Chen X, and Parini CG (2004) Printed circular disc monopole antenna for ultra-wideband applications. Electron Lett 40(20):1246–1247
4. Chen ZN (2007) UWB antennas: from hype, promise to reality in IEEE Antennas Propagation Conference. pp 19–22
5. Qu X, Zhong SS, and Wang W (2006) Study of the band-notch function for a UWB circular disc monopole antenna. Microw Opt Technol Lett 48(8):1677–1670
6. Cho YJ, Kim KH, Choi DH, Lee SS, Park SO (2006) A miniature UWB planar monopole antenna with 5-GHz band-rejection filter and the time-domain characteristics. IEEE Trans Antennas Propag 54(5):1453–1460
7. Hong CY, Ling CW, Tarn IY, Chung SJ (2007) Design of a planar ultra wideband antenna with a new band-notch structure. IEEE Trans Antennas Propag 55(12):3391–3397
8. Chu QX, Yang YY (2008) 3.5/5.5 GHz dual band-notch ultra wideband antenna. Electron Lett 44(3):172–174
9. Yin K, Xu JP (2008) Compact ultra wideband antenna with dual bandstop characteristic. Electron Lett 44(7):453–454
10. Lee WS, Kim DZ, Kim KJ, Yu JW (2006) Wideband planar monopole antennas with dual band-notched characteristics. IEEE Trans Microw Theory Tech 54(6):2800–2806
11. Chung Kyungho, Kim Jaemoung, Choi Jaehoon (2005) Wideband microstrip-fed monopole antenna having frequency band-notch function. IEEE Microwave Wirel Compon Lett 15 (11):766–768
12. Trang ND, Lee DH, Park HC (2011) Design and analysis of compact printed triple band-notched UWB antenna. IEEE Antennas and Wirel Propag Lett 10:403–406
13. Kelly James R, Hall Peter S, Gardner Peter (2011) Band-Notched UWB antenna incorporating a microstrip open-loop resonator. IEEE Trans Antennas Propag 59(8):3045–3048
14. Zeland Software Inc. IE3D: MoM-Based EM Simulator. http://www.zeland.com/
15. Cohn SB (1969) Slot line on a dielectric substrate. IEEE Trans Microw Theory Tech 17 (10):768–778
16. Younis M. Abbosh (2014) Breast cancer diagnosis using microwave and hybrid imaging methods. Int J Comp Sci Eng Surv 5(3):41
17. Sun J, Guo M, Pang H, Qi J, Zhang J, Ge Y (2013) Treatment of malignant glioma using hyperthermia. Neural Regen Res 8(29):2775–2782

Design of Cryoprobe Tip for Pulmonary Vein Isolation

B. Sailalitha, M. Venkateswara Rao and M. Malini

Abstract Pulmonary vein isolation is a method used for the treatment of arrhythmias originating within pulmonary veins. Cryoablation is a technique which employs extreme freezing to treat diseased or abnormal tissue. Cryoablation has gained importance in the treatment of pulmonary vein isolation for the patients suffering from atrial fibrillation. The proposed ellipsoidal ring shaped cryoprobe tip has eliminated the need of continuous point to point lesion ablation by ablating the entire circumferential area at a time. The aim of this paper is to discuss the design and advantages of the proposed tip. The desired time for the entire tip to reach a uniform temperature is ideally less than 2 s. But with the proposed tip the equilibrium condition was obtained in much lesser time of 0.5 s.

Keywords Atrial fibrillation · Cryoablation · Pulmonary vein isolation

1 Introduction

The cardiovascular system involves heart and blood vessels which carry blood to and away from the heart [1]. One of the commonly observed cardiovascular diseases is cardiac arrhythmia. Fibrillation is one type of cardiac arrhythmia. On an average the normal heart rate of a healthy person is 72 beats/min. When the electrical activity or heart rate is irregular i.e., faster or slower than normal, then the person is said to be suffering from cardiac arrhythmia (http://en.wikipedia.org/wiki/Electrical_activity). Arrhythmias can occur in any part of the heart i.e., in atria or ventricles. They can affect a person at any age and can cause a sudden cardiac arrest if they are not treated. When fibrillation occurs in atria it is classified as atrial fibrillation. In atrial fibrillation, the electrical impulses are not only produced from SA node; instead many impulses commence and spread chaotically through the atria. As a result, the

B. Sailalitha (✉) · M. Venkateswara Rao · M. Malini
Department of Biomedical Engineering, Osmania University, Hyderabad, India
e-mail: sailalithabme@gmail.com

S. Gupta et al. (eds.), *Advancements of Medical Electronics*,
Lecture Notes in Bioengineering, DOI 10.1007/978-81-322-2256-9_28

heartbeat is aberrant and there is a chance of formation of clots in the atria because the blood received by the atria is not fully pumped into the ventricles. One of the most commonly observed atrial fibrillations is arrhythmia originating within the pulmonary veins. Doctors have discovered that there is a narrow band of muscle tissue around each of the pulmonary veins near the opening of the left atrium that may trigger these extra electrical signals as shown in Fig. 1.

In a human heart there are four pulmonary veins that carry oxygenated blood from the lungs back to the left atrium. When the patient is diagnosed with atrial fibrillation originating within the pulmonary veins the doctor suggests pulmonary vein isolation as a mode of treatment [2–5]. Pulmonary vein isolation is an ablation technique where the surgeon uses cryoablation method to destroy this small area of abnormal tissue which results in a scar/lesion. These "cryo lesions" are introduced in a pattern around each pair of pulmonary veins on each side (Fig. 2). As a result the scar tissue blocks the fast and irregular electrical impulses produced by pulmonary veins reaching the left atrium thus preventing atrial fibrillation.

To perform this ablation technique a cryoprobe is used to ablate the tissue. The probe tip is made up of thermally conductive material to get cooled when the cryogenic gases are allowed to circulate along the probe [6]. The size and shape of the lesion created on the tissue depends on the size and shape of the tip. So according to the size and shape of the scar tissue required, respective tip is selected by the surgeon [7–10].

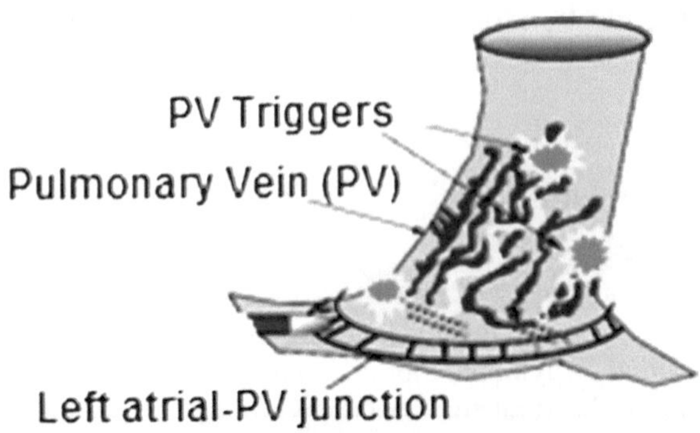

Fig. 1 Fibrillation originating within pulmonary vein

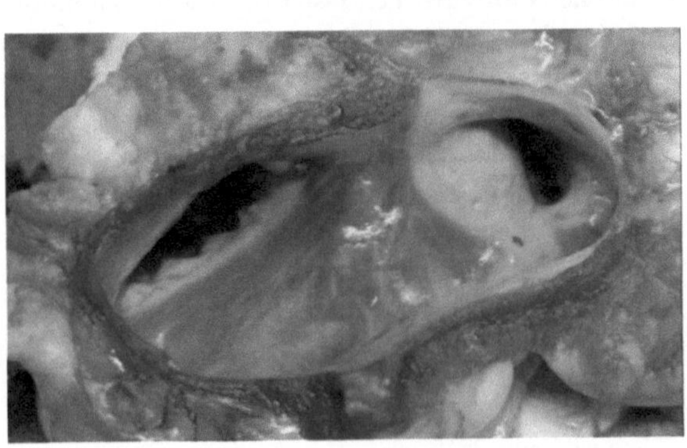

Fig. 2 The required ablating area on side of pulmonary veins

2 Methodology

The open heart surgery carried out for pulmonary vein isolation treatment lasts for around 7 h. For 3 h during the treatment time the cardiopulmonary function of the patient is maintained by a heart lung machine. When the blood is bypassed and the heart is frozen the surgeon selects the set of points that are needed to be ablated. The ablation is done along the circumferential area of the pulmonary veins using a point to point ablation method. This method requires more than 30 min because the selection of points has to be done carefully otherwise it may damage a healthy tissue.

Presently, a 1 cm nipple shaped tip (Fig. 3) is being used by the cardiac surgeons for creating the lesion. The main drawbacks of using such a shaped tip, is that it takes more than 30 min for the surgeon to complete the procedure. This is because, the surgeon has to select the location, ablate it and move to the next location and repeat the same procedure till the required area is ablated. During the entire procedure, the patient is kept on a heart lung machine. Prolonged usage of heart lung machine poses certain problems, and even the patient is prone to infection.

Considering the limitations of the presently used tip, a new ellipsoidal ring shaped tip is designed. The tip shape is designed such that it almost corresponds to the area to be ablated. So, it eliminates the need for point to point selection and subsequent ablation.

The procedure performed is same but the tip is placed on pulmonary veins and the total desired area is ablated at once. The time taken for the total ablation process with the proposed tip is less than 10 min. As the tip is designed according to the measurements of the pulmonary veins, only one time selection is required.

The major advantages of the newly proposed tip are (i) its simplifies the procedure as it requires only one time ablation (ii) reduces the time of the procedure and (iii) Reduces the patients dependence on heart lung machine.

Fig. 3 Tip presently used for pulmonary vein isolation

3 Results and Discussion

Different prototypes were designed and shown to the cardiovascular surgeon for scrutiny. Based on their feedback, the ellipsoidal ring shape is finally selected the prototype of the proposed tip (Fig. 4). By using this as a model, the evaluation and design is performed using the mathematical tool (cad software 'creo'). This is shown in Fig. 5.

3.1 Design Details of the Tip

Considering the normal measurements of pulmonary veins the dimensions of the tip were finalised as follows:

Major axis of the ellipsoid: 40 mm
Minor axis of the ellipsoid: 30 mm
Thickness of the ellipsoid: 1 mm

The time taken for the tip to reach equilibrium temperature after the application of cryo gases as required by the surgeon is <2 s. The cryogenic gas chosen was carbon dioxide which operates at −78 °C.

Fig. 4 Prototype of the tip

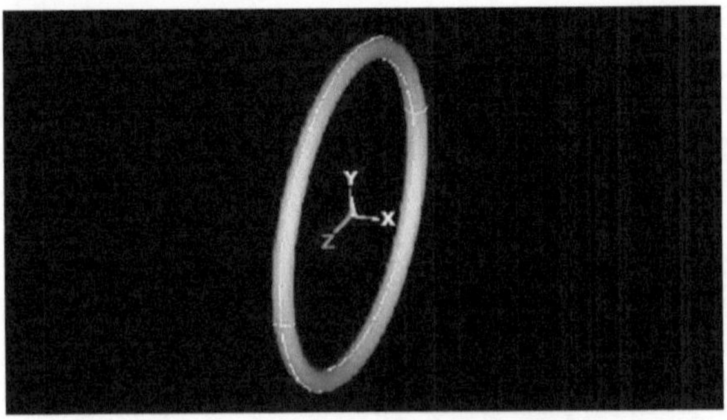

Fig. 5 The design of the tip in 'creo' (CAD software)

The austenitic stainless steel 304 (1.4301) is considered as "tough" at cryogenic temperatures. AISI 304 is called as "cryogenic steel". Therefore AISI 304 can be considered suitable for sub-zero ambient temperatures and hence is chosen to design the proposed cryoprobe tip.

Volume of ellipsoid, $v_1 = \pi(a_1 b_1)(t)$

where

a_1 = length axis radius

b_1 = width axis radius

t = diameter of the ellipsoid

$$= \pi \frac{(20)}{1000} \frac{(15)}{1000} \frac{(1)}{1000}$$
$$= 942 \times 10^{-9}\,\text{m}^3$$
$$v_2 = \pi(a_2 b_2)(t)$$

where

a_2 = length axis radius

b_2 = width axis radius

t = diameter of the ellipsoid

$$= \pi \frac{(19)}{1000} \frac{(14)}{1000} \frac{(1)}{1000}$$
$$= 835 \times 10^{-9}\,\text{m}^3$$

∴ Volume of the ellipsoid ring (V), $v_1 - v_2 = 187 \times 10^{-9}\,\text{m}^3$

(1) Mass, m = volume × density = V × d

$$d = \text{density of 304 stainless steel}$$
$$d = 8{,}030\ \text{kg/m}^3$$
$$= 187 \times 10^{-9} \times 8030$$
$$= 0.000857282\,\text{kg}$$

(2) Temperature drops from 21 to −78 °C

∴ Heat removed from the ring = m c_p (ΔT)

where

c_p = specific heat of 304

c_p = 502 k/kg*k

$$= m\ (502)\ (21 - (-78))$$
$$= 42.60\ \text{J}$$

If a solid CO_2 (Dry Ice) in gases condition,

If Mass, m = 0.0083 kg/s

Temperature at −78 °C = 195 K

∴ Heat content of CO_2 in gases from = m c_p (ΔT) = 0.0083(1.84 × 10^3) (5)

In absolute 5 °C raise in CO_2 (Dry Ice) = 76.36 J/s

∴ Time taken $= \frac{42.60}{76.36} = 0.55$ s

Through the desired time taken for the tip to reach equilibrium temperature is less than <2 s. The calculated time for the designed tip is 0.5 s.

4 Conclusion

In the present work, an ellipsoidal ring shaped tip has been designed for the cryoprobe used for the ablation of pulmonary veins. The limitations of the presently-being-used nipple shaped tip have been overcome by proposed tip. The proposed tip also provides the advantages of simplified procedure, reduced time of the procedure and reduced patient's dependence on heart lung machine. The presently designed tip is made of rigid material and the dimensions are based on average normal values. To suit other dimensions, the tip may be made with flexible material and tested.

Acknowledgments The authors would like to acknowledge the support from the UGC project universities with potential for excellence awarded to Osmania University.

References

1. Etheridge ML, Choi J, Ramadhyani S, Bischof JC Methods for characterizing convective cryoprobe heat transfer in ultrasound gel phantoms. doi:10.1115/1.4023237
2. Kuhne M, Sticherling C (2011) Cryoballoon ablation for pulmonary vein isolation of atrial fibrillation: a better way to complete the circle? J Innov Cardiac Rhythm Manage 2:264–270
3. Wellens HJJ (2000) Pulmonary vein ablation in atrial fibrillation: hype or hope? Circulation 102:2562–2564. doi:10.1161/01.CIR.102.21.2562
4. Takase et al. (2004) Pulmonary vein dimensions and variation of branching pattern in patients with paroxysmal atrial fibrillation using magnetic resonance angiography. Jpn Heart J 45:81–92
5. Fraunfelder FW (2008) Liquid nitrogen cryotherapy for surface eye disease. Trans Am Ophthalmol Soc 106:301–324
6. Rostagno C, Berioli MB, Stefàno PL Treatment of atrial fibrillation in patients undergoing mitral valve surgery. Dipartimento Area Critica Università Firenze, Cardiochirurgia AOU Careggi Firenze, Italia
7. Iarocci M et al (1997) Advances in cryogenics engineering. In: Kittle P (ed) vol 43. Plenum Press, New York, p 499
8. Kubota H, Sudo K, Takamoto S, Endo H, Tsuchiya H, Yoshimoto A, Takahashi Y, Inaba Y, Furuse A Clinical result of epicardial pulmonary vein isolation (LAVIE) by cry ablation as concomitant cardiac operation and clinical application of new ablation device(KIRC-119 infrared coagulator) to treat atrial fibrillation

9. Cabrera JA, Pizarro G, Sanchez-Quintana D (2010) Transmural ablation of all the pulmonary veins: is it the Holy Grail for cure of atrial fibrillation? Eur Heart J 31:2708–2711. doi:10.1093/eurheartj/ehq241 (online publish-ahead-of-print 6 Sept 2010)
10. Hurlich Low temperature metals, general dynamics astronautics, San Diego, California

Designing of a Multichannel Biosignals Acquisition System Using NI USB-6009

Gaurav Kulkarni, Biswajeet Champaty, Indranil Banerjee, Kunal Pal and Biswajeet Mohapatra

Abstract The current study delineates the designing of a three channel biosignals acquisition system. The acquisition system was made using NI USB-6009. A LabVIEW based graphical user interface (GUI) program was made to simultaneously acquire the biosignals. Electrocardiogram, spirogram and body surface temperature signals were used as the representative signals for testing the device. The device was tested successfully and may allow the acquisition of the signals under ambulatory conditions.

Keywords Multichannel · Biosignal acquisition · NI USB-6009 · LabVIEW · Ambulatory conditions

1 Introduction

There has been an increased effort in the development of ambulatory monitoring devices. This may be attributed to the fact that the ambulatory devices provides the healthcare providers in diagnosing the cause of the medical conditions at an early stage [1]. This is specifically important if the condition of the patient is critical. In general, the ambulatory monitoring devices should be user friendly, low cost, non-invasive and light weight. Even though the ambulatory devices are expected to be low cost, the accuracy and the sensitivity of the devices should not be compromised [2]. This is because a lower accuracy and sensitivity of a medical device may sometimes lead to wrong diagnosis of the medical condition. This, in turn, will lead to wrong treatment of the patient thereby aggravating their medical condition.

G. Kulkarni · B. Champaty · I. Banerjee · K. Pal (✉)
Department of Biotechnology and Medical Engineering,
NIT-Rourkela, Rourkela 769008, Odisha, India
e-mail: pal.kunal@yahoo.com; kpal.nitrkl@gmail.com

B. Mohapatra
Vesaj Patel Hospital, Rourkela 769008, Odisha, India

S. Gupta et al. (eds.), *Advancements of Medical Electronics*,
Lecture Notes in Bioengineering, DOI 10.1007/978-81-322-2256-9_29

Keeping this in mind, in this paper we have tried to develop an ambulatory device capable of acquiring ECG, spirogram and body surface temperature. NI USB-6009 data acquisition device was used for the acquisition of the signals into a computer. NI USB-6009 was chosen for efficient and accurate acquisition of the biosignals. An acquisition program was made in LabVIEW.

2 Materials and Methods

2.1 *Materials*

NI USB-6009 (NI, US), ECG (Vernier, US), spirometer (Vernier, US) and body surface temperature (Vernier, US) sensors were used in this study. LabVIEW 2010 (NI, US) was used for making the signal acquisition program.

2.2 *Methods*

The outputs from the ECG, spirometer and surface temperature sensors were connected to the AI0 (pin no. 2), AI1 (pin no. 5) and AI2 (pin no. 8) analog input terminals, respectively, of USB 6009 [3]. The sensors were powered with the 5 V, acquired from the USB-6009. The DAQ was connected with the computer using USB port. It drew power (5 V) from the USB port of the computer. It transmitted the signal information serially to the computer using USB port. A LabVIEW program was made to acquire the signals from the USB acquisition device. A schematic diagram of the setup has been shown in Fig. 1.

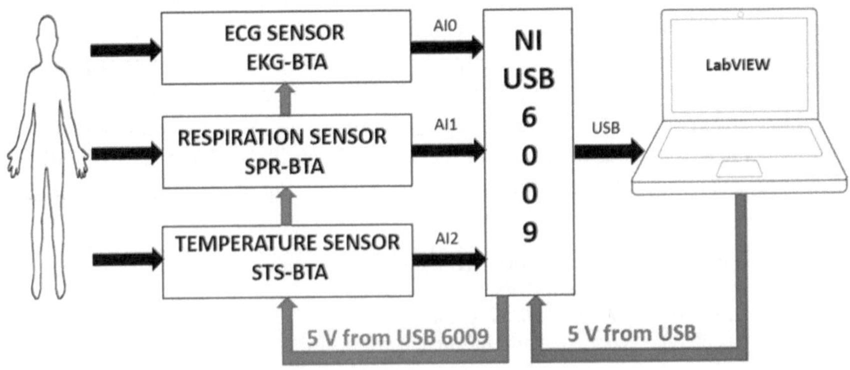

Fig. 1 Schematic diagram of the setup

3 Results and Discussions

3.1 Designing of the Multichannel Acquisition

The ECG, spirometer and surface temperature sensors were connected with the USB-6009 using prototype connector board. The pictograph of the sensor connection with the USB-6009 has been shown in Fig. 2. This device is compatible with LabVIEW and does not require any extra drivers for interfacing. A LabVIEW program for the simultaneous acquisition of the 3 biosignals was developed. The schematic representation of the LabVIEW program has been shown in Fig. 3. The program was made such that the signals from the analog input channels are acquired simultaneously.

The controls for each of input channels was put in a while loop. The DAQ assistant is configured to acquire analog input voltages from each channel. Acquisition mode was set to continuous samples. The sampling rate was set to 1,000 Hz/s and the number of samples to be displayed were chosen to be 10,000 [4]. The signals were then spilt and processed based on the type of biosignals. The output from ECG channel was processed using multiresolution wavelet analysis.

Since the ECG signal is considered as the time-varying signal, Wavelet transform is being used in ECG signal analysis. The db 06 (Daubechies) wavelet was used to decompose ECG signal into 8 levels. The signal was reconstructed using

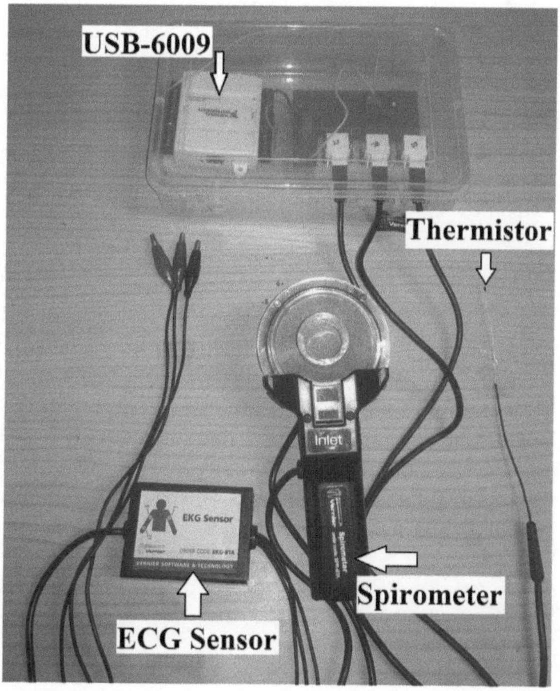

Fig. 2 Sensor connection with USB-6009

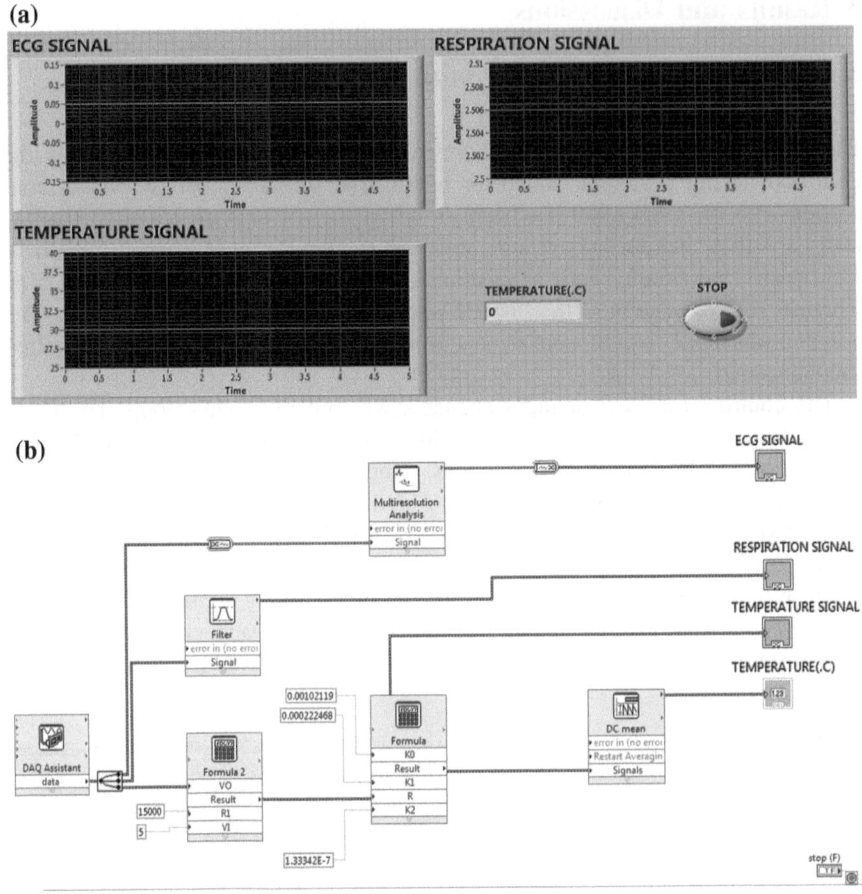

Fig. 3 LabVIEW program: **a** front panel, **b** block diagram

D6 + D7 levels as most of the energy of the ECG signal lies between 0.1 and 40 Hz [5]. The signal from spirometer was smoothened using smoothing filter using triangular moving average algorithm using 16 data points. The purpose of smoothening is to eliminate high frequency noises and power line interference. The output from body surface temperature was fed into express formula vi, which contained formula for the conversion of the voltage signal into temperature (°C) [6, 7]. The mean DC of this signal was extracted and displayed in a numeric indicator. The waveform signals from all the 3 sensors were visualized in three different waveform graphs on the front panel.

3.2 Acquisition of the Signals

The acquisition of the ECG signals was carried out in the lead—I configuration. For the purpose, red electrode was placed in left arm, green electrode in the right arm and black electrode (right leg drive) was placed near the elbow [8]. The placement of the electrode was done as per the manual provided by the manufacturer. The output of the ECG sensor was fed into AI0 channel of the DAQ. The output of the spirometer was fed into AI1 channel of USB-6009, whereas, output of the surface temperature sensor was fed into AI2 channel of USB-6009 [9]. The spirometer was placed near the mouth. The surface temperature sensor was placed on the forearm and was secured using a white plaster. The representative output from the sensors have been shown in Fig. 4. The results were found to be as reported in the literature [10].

After designing the ambulatory device, 5 volunteers were invited to test the functioning of the device. The experimental details were described to the volunteers. A written consent was taken from the volunteers, if they agreed to participate for the study. An ethical clearance was previously taken from the institute ethical clearance committee to conduct biosignals acquisition studies. The volunteers were asked to sit in the supine position and the sensors were placed at the predefined position as described in the previous paragraph. The photograph of a volunteer with the sensor placements has been shown in Fig. 5.

The developed device was able to acquire all the three signals in an efficient manner without any noise or distortion. The device was found to be quite user friendly. The complete setup of the device has been shown in Fig. 6.

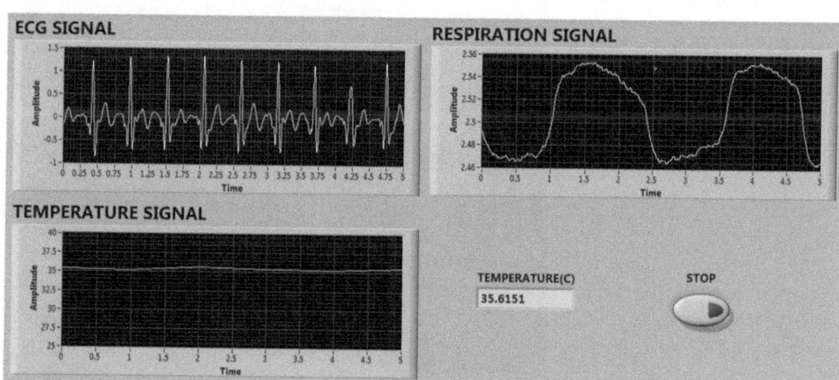

Fig. 4 Picture showing representative output from the sensors

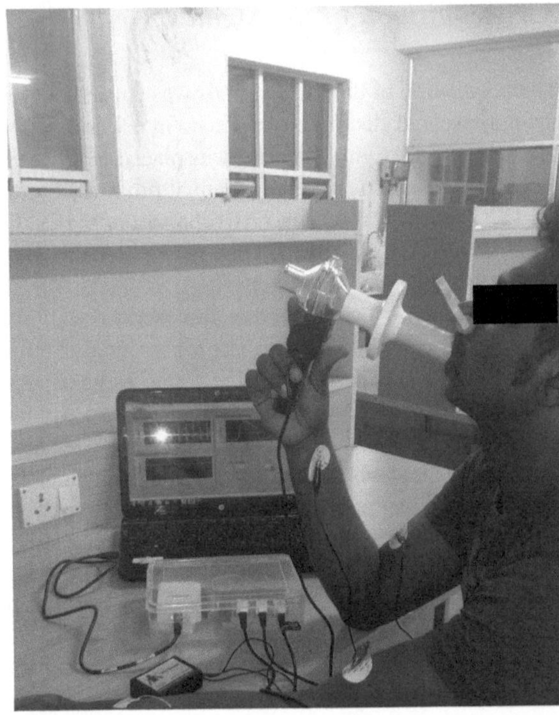

Fig. 5 Photograph of volunteer with sensor placement

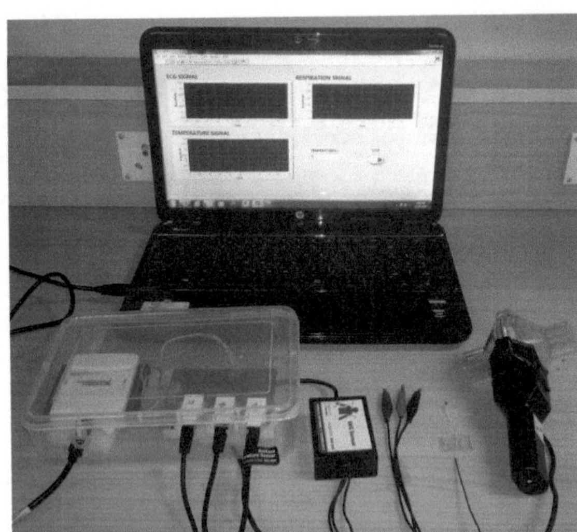

Fig. 6 Complete setup of the device

4 Conclusion

The current study describe the designing of a multichannel biosignals acquisition system using NI USB-6009. The software for controlling the hardware platform was developed in LabVIEW. The biosignals were acquired efficiently and displayed using LabVIEW program. This device can be used in ambulatory monitoring applications for continuous acquisition of ECG, spirogram and body surface temperature.

Acknowledgments Authors extend the acknowledgement for the logistical support provided by National Institute of Technology, Rourkela during the completion of the study.

References

1. Corral-Peñafiel J, Pepin J-L, Barbe F (2013) Ambulatory monitoring in the diagnosis and management of obstructive sleep apnoea syndrome. Eur Respir Rev 22(129):312–324
2. Branzila M, David V Wireless intelligent systems for biosignals monitoring using low cost devices
3. Zhang J-G, Zhao X (2014) Design of the chaotic signal generator based on LABVIEW. Sens Transducers 163:1726–5479
4. Shen L, Chaoran Y (2012) Identification of parameters for a DC-motor by LabVIEW
5. Selvakumar G, et al (2007) Wavelet decomposition for detection and classification of critical ECG arrhythmias. In: proceeding of the 8th WSEAS international conference on mathematics and computers in biology and chemistry pp 80–84
6. Granado Navarro M.Á (2012) Arduino based acquisition system for control applications
7. Sahoo S et al (2014) Wireless transmission of alarm signals from baby incubators to neonatal nursing station. In: automation, control, energy and systems (ACES) 2014 first international conference pp 1–5
8. Biel L et al (2001) ECG analysis: a new approach in human identification. IEEE Trans Instrum Meas 50(3):808–812
9. Kumar U et al (2013) Design of low-cost continuous temperature and water spillage monitoring system. In: IEEE 2013 international conference on information communication and embedded systems (ICICES)
10. Heywood D et al (2009) Integrating the Vernier LabPro™ with Squeak Etoys. In: Society for information technology and teacher education international conference

Arsenic Removal Through Combined Method Using Synthetic Versus Natural Coagulant

Trina Dutta and Sangita Bhattacherjee

Abstract As Arsenic contamination in drinking water has become a matter of severe concern, researchers across the globe have been trying to develop an efficient yet economic method for removal of arsenic from water. Authors of this paper investigated combined treatment methodology namely, coagulation followed by microfiltration (MF) to remove arsenic below permissible level of 10 ppb. Simulated solutions of arsenate and arsenite salts (100 ppb) were prepared. A 95.83 % reduction of arsenic for arsenate solution and 91.99 % reduction of arsenic in case of arsenite solution were achieved in suitable pH range (8–10) when the solution was subjected to coagulation using ferric chloride coagulant, followed by MF using 0.45 μm polyethersulphone membrane. In search of more cost effective and eco-friendly yet viable route, authors explored the effectiveness of crushed Shelled *Moringa oleifera* Seed (SMOS) as a natural coagulant. A 91.01 % decontamination of arsenic for arsenate solution and 70.61 % for arenite solution were achieved in suitable pH range 7–9. This alternative method of coagulation, bio-adsorption (amino acid-arsenic interaction) followed by MF achieved in appreciable arsenic removal efficiency compared to inorganic Ferric Chloride, a synthetic coagulant. The sludge generated in case of ferric chloride was found to be toxic and highly corrosive compared to that obtained with *Moringa oleifera* seeds.

Keywords Arsenic poisoning · Coagulation followed by microfiltration · Ferric chloride versus natural coagulant *Moringa oleifera* · Arsenic removal

T. Dutta (✉)
Chemistry Department, JIS College of Engineering, Kalyani, Nadia 741235, India
e-mail: Trina.d80@gmail.com

S. Bhattacherjee
Chemical Engineering Department, Heritage Institute of Technology, Kolkata 700107, India
e-mail: sangita.bhattacharjee@yahoo.co.in

S. Gupta et al. (eds.), *Advancements of Medical Electronics*,
Lecture Notes in Bioengineering, DOI 10.1007/978-81-322-2256-9_30

1 Introduction

Nowadays, arsenic poisoning has become one of the major environmental concerns in the world as millions of human beings have been exposed to excessive arsenic through contaminated ground water and surface water used for drinking. Arsenic, a metalloid in group VA, is classified as Group 1 carcinogenic substance based on powerful epidemiological evidence [1]. Hence, a continuous investigation is being carried out across the world to develop an efficient yet economical technology for reducing arsenic concentration below maximum contamination level (MCL) of 10 ppb as implemented by US Environment Protection Agency in the year 2001.

There are many possible routes of human exposure to arsenic from both natural and anthropogenic sources. Arsenic occurs as a constituent in more than 200 minerals, although it primarily exists as arsenopyrite and as a constituent in several other sulfide minerals. The introduction of arsenic into drinking water can occur as a result of its natural geological presence in local bedrock. Arsenic-containing bedrock formations of this sort are known in Bangladesh, West Bengal (India), and regions of China, and many cases of endemic contamination by arsenic with serious consequences to human health are known from these areas. Significant natural contamination of surface waters and soil can arise when arsenic-rich geothermal fluids encounter surface waters. When humans are implicated in causing or exacerbating arsenic pollution, the cause can usually be traced to mining or mining-related activities [2].

The acceptable level as defined by WHO for maximum concentrations of arsenic in safe drinking water is 0.01 mg/L. Arsenic contaminated water typically contains arsenous acid (As III) and arsenic acid (As V) or their derivatives. Their names as "acids" is a formality, these species are not aggressive acids but are merely the soluble forms of arsenic near neutral pH. These compounds are extracted from the underlying rocks that surround the aquifer. Arsenic acid tends to exist as the ions $[HAsO_4]^{2-}$ and $[H_2AsO_4]^-$ in neutral water, whereas arsenous acid is not ionized.

Arsenic removal from water is an important subject worldwide, which has recently attracted great attentions. A variety of treatment processes has been developed for arsenic elimination from water, including coagulation (precipitation), adsorption, ion exchange, membrane filtration, electrocoagulation, biological process, iron oxide-coated sand, high gradient magnetic separation and natural iron ores, manganese greensand etc.

Investigators have also used modern technology such as membrane technology which includes reverse osmosis, nanofiltration, ultrafiltration, microfiltration to remove arsenic species from drinking water. Reverse osmosis, can reduce arsenic limit below 10 ppb. But in developing countries like ours considering with low annual income and low electric popularization, highly efficient RO and NF technology seems difficult to be applied due to its high-energy consumption.

Microfiltration (MF), is a low-pressure technique and can remove particles with a molecular weight above 50,000. The pore size of MF membranes is too large to effectively remove dissolved or colloidal arsenic species, however, the MF can remove particulate forms of arsenic from water. Therefore, the arsenic removal efficiency by MF membrane is highly dependent on the size distribution of arsenic bearing particles in the water. Unfortunately, the percentage of particulate arsenic normally is not too high in water. In order to increase the arsenic removal efficiency by MF technique, the technique, which can increase arsenic particle size, such as coagulation [3] and flocculation processes [4], can be used to assist the MF techniques.

In this context, the authors have investigated the performance of a combined method, namely coagulation followed by microfiltration in removal of arsenic from contaminated water. In one study, ferric Chloride salt was selected as synthetic coagulant and another study *Moringa olifera* was used as natural coagulant. Microfiltration in both the cases was used to remove arsenic particulates larger than 0.5 μm. Both the studies were carried out and results were reported.

2 Objective

In case of synthetic coagulant, Ferric Chloride, the coagulation process consisted of the addition of an iron-based coagulant, ferric chloride that would get hydrolysed in water to form $Fe(OH)_3$ precipitate in the pH range of 4−10. Ferric hydroxide floc having a net positive charge on the surface of the precipitate formed, would favourably adsorb negatively charged arsenate by surface complexation. Arsenite being neutral in charge, would also likely to be adsorbed on ferric hydroxide flocs, was as effective and reliable synthetic coagulant due to its dense-discrete floc formation over wide pH range and due to formation of less toxic and compact sludge.

But presence of synthetic coagulant (iron chloride, polyaluminum chloride etc.) in processed water may have several serious drawbacks viz. Alzheimer's disease, health problems, carcinogenic effects if used for drinking purposes. Hence such treatments have not been appreciated [5].

Therefore the necessity of natural coagulant instead of synthetic coagulant was required to be used for arsenic removal. The process of water treatment using natural coagulant is called bio-sorption. Biosorption has gained important credibility during recent years because of its ecofriendly nature, excellent performance, and low cost domestic technique for remediating even heavily metal loaded waste water [6−8]. Recently, biological materials such as algae, bacteria, fungi, yeast [9, 10], aquatic moss Fontinalis antipyretica [11] and agricultural waste products like sunflower stalk [12], sargassum seaweed biomass [13], pinus cone biomass [14], rice husk [15], olive pomace [16], cocoa shells [17], and grape stalk waste [18] have been recognized as cheap natural sorbents for the removal of toxic metals. However, there is still a strong challenge in developing economical and commonly available biosorbents for Arsenic removal.

Moringa oleifera (Saihjan or drumstick) is one of the biosorbent, also known as 'Miracle Tree'. It is a cosmopolitan tropical drought tolerant tree, available throughout the year and is well documented for its various pharmacological importances such as analgesic, antihypertensive activity, and anti-inflammatory effects [19–21]. Further, the coagulating property of the seed powder of this plant has been used for various aspects of water treatment such as turbidity, alkalinity, total dissolved solids, and hardness [5, 22–25]. In Sudan, Northern Nigeria and other developing countries, dried *M. oleifera* seeds are frequently used for removal of turbidity in water as low technology domestic method [26]. However, its biosorption behavior for the removal of toxic metals from water has not been given attention. Moreover, the use of *Moringa oleifera* has an added advantage over the chemical treatment of water because not only it is cheap and eco-friendly but also it is biological and has been reported as edible. Moringa extract contain dimeric cationic proteins or polypeptide (natural organic polymer) with various functional groups, particularly low molecular weight amino acids [27]. These proteinacious amino acids have a variety of structurally related pH-dependent properties of generating either positively or negatively charged sites for attracting the anionic or cationic species of metal ions, respectively [28]. It has molecular weight of 13 kDa and isoelectric points in the range 4–8 [29]. Moreover its efficiency as coagulant is not affected by pH of the solution.

In the light of the above, the major objective of this study was to investigate removal efficiencies of arsenate and arsenite from simulated water by combination of coagulation followed by microfiltration by varying dosages of the natural coagulant-cum-polyelectrolyte *M. oleifera* as well as ferric chloride. The optimum dose has been investigated to reach MCL standard and to make it cost effective as well.

3 Materials and Methods

Sodium arsenate salt ($Na_2HAsO_4 \cdot 7H_2O$), (Molecular Weight = 312) and Sodium arsenite ($NaAsO_2$) (Molecular Weight = 129.9) AR grade procured from Loba Chemie Pvt. Ltd. Ferric Chloride salt, sodium hydroxide pellets were procured from E. Merck, India. The required solutions were prepared with ultrapure deionised water. 1 N Hydrochloric acid and 1 N Sodium Hydroxide solutions were prepared for pH adjustment.

Moringa oleifera Lamarck seeds were collected during summer. Seeds were washed thoroughly with double-distilled water to remove the adhering dirt, dried at 60 °C for 24 h. For shelled seeds (SMOS), the husks enveloping each seed were removed and the kernel was ground to a fine powder using a blender. No other chemical or physical treatments were used prior to experiments.

In the study, polyethersulfone (PES) membrane, a hydrophilic membrane constructed from pure polyethersulfone polymer was used during MF. PES membrane filters are designed to remove particulates during general filtration. The polyethersulfone membrane filters have excellent flow speeds and, connected to it, a high filterable volume. Biologic and pharmaceutical solutions can be filtered, in the wide pH-range of pH 1–14, because of their low protein adsorption.

4 Experimental Procedure

Accurate amount of sodium arsenate was weighed and dissolved in 500 ml of ultrapure water to prepare 100 μg/l of arsenic solution. In 1st iteration 50 ml of this solution was taken in each of 4 different conical flasks. 10, 15, 20, 25 % $FeCl_3$ solutions were prepared 0.10 ml of $FeCl_3$ solution was added to each of the 4 flasks respectively. The pH of the solution was increased and kept in the range of 8−10 by adding 1 N NaOH solution. These were stirred magnetically using a magnetic stirrer for 15 min at 30 °C. Flocs are formed upon increasing the pH. The solution was kept for half an hour to settle down the flocs and was filtered using Whatman filter paper (Grade 41) to remove suspended coarse particles (20−25 μm). For removal of finer particles, the supernatant solution was then microfiltered using a "all glass vacuum filtration unit" (make: Sartorius A.G., Göttingen, Germany), fitted with portable vacuum pump (Sartorius, A.G., Göttingen, Germany), with polyether sulfone (PES) membrane (47 mm diameter, pore size 0.45 μm) being used as filter media. In the 2nd set of study, 50 ml of similar simulated arsenate solution was taken in four different conical flask. 0.1, 0.5, 1 and 1.5 g of powdered *Moringa oleifera* Seed (SMOS) were mixed with the solution respectively and pH value adjusted in the range 7−9. Then the samples were kept on the shaker for 45 min at 110−120 rpm. After settling down big flocks, the soutions were filtered by whatman filter paper (Grade 41) and supernatant is followed by microfiltration (Figs. 1 and 2).

The arsenic concentration of permeate collected from both the processes were then measured by Perkin-Elmer Atomic absorption spectroscopy (AAS). Similar procedure has been followed for arsenite salt also and results were noted.

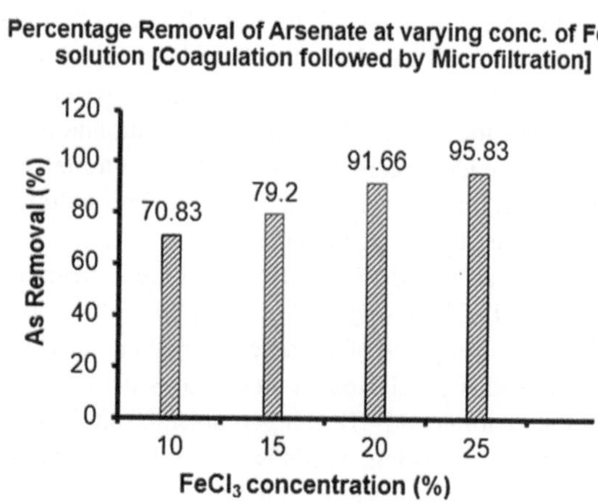

Fig. 1 $FeCl_3$ conc (%) on arsenate removal (%)

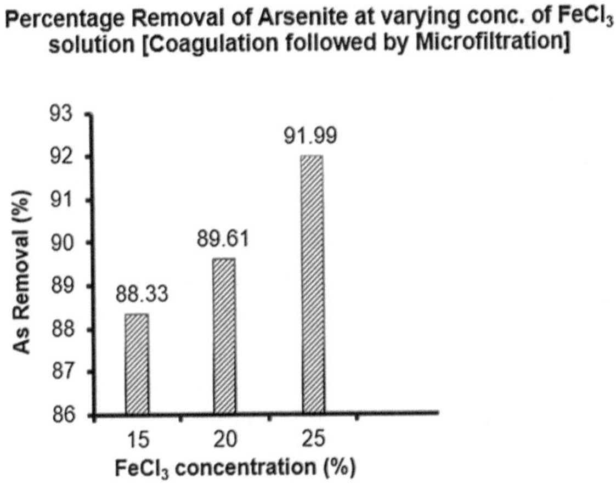

Fig. 2 $FeCl_3$ conc (%) on arsenite removal (%)

5 Result and Discussion

The removal of As(V) and As(III) from simulated water has been investigated by coagulation/adsorption followed by microfiltration process.

In 1st iteration, $FeCl_3$ has been taken as coagulant here because Iron based coagulants are generally more effective than aluminum based coagulants for arsenic removal from water on a weight basis [30]. Ferric chloride is a primary coagulant. Primary coagulants are used to cause particles destabilised and begin to clump together. When ferric chloride is dissolved in water the solution becomes strongly acidic as a result of hydrolysis.

As the pH was adjusted, ferric hydroxide precipitated from the solution. As the pH decreased, the number of positively charged sites on the ferric hydroxide particles increased. The precipitated ferric hydroxide had a net positive charge on the surface and the negatively charged arsenate anions got adsorbed on the surface of ferric hydroxide precipitate by surface complexation models. In this way, the arsenate got removed from the solution. The sedimentation and filtration processes then remove arsenic particulates.

It has been found that both arsenate and arsenite could be effectively removed with 20 % $FeCl_3$ dosage followed by microfiltration.

In the subsequent of treatment, crushed, powdery natural coagulant *Moringa oleifera* seed was selected. As it contains dimeric cationic polypeptide it can easily coagulate with negatively charged arsenic salt. It had been reported that an aqueous solution of the Moringa seeds is a heterogeneous complex mixture, having cationic polypeptides with various functional groups, especially low molecular weight amino acids [27]. Amino acids have been reported as efficient phytochelators that work at even at low concentrations and having the tendency to interact with metal ions and significantly enhance their mobility. The proteinacious amino acids, depending upon the pH, possess both the negatively and positively charged ends and are, thus, capable of generating the appropriate atmosphere for attracting the anionic or cationic species of the metal ions.

Here pH value was maintained between 7 and 9. In this pH range, As(V) exists in monovalent ($H_2AsO_4^-$), and divalent ($HAsO_4^{2-}$) anionic species and As(III) exists in ($H_2AsO_3^-$ and $HAsO_3^{2-}$) anionic species [8]. The positively charged group of amino acids can hold the negatively charged monovalent arsenic species.

The majority of amino acids present in the target biomass have isoelectric points in the range 4.0–8.0 [29]. In this range of pH, over 90 % of the amino acid molecules are in the ionized state. The negatively charged monovalent arsenate or arsenite species may be held by the positively charged group of amino acids. With the increase in pH range, the carboxylic group of the amino acids would progressively be deprotonated as carboxylate ligands, simultaneously protonating the amino group. Such positively charged amino groups facilitate the SMOS-Arsenic binding. SMOS contains proteins and has a net positive charge at pH below 8.5 (pKa > 8.6) [31]. These positively charged proteins are also considered to be active moieties for binding metal ions.

Here the optimum dose has been found as 0.5 g per 50 ml means 10 g/L for both arsenate and arsenite salt. The removal of arsenic resulted from precipitation, co-precipitation and adsorption mechanisms. With further increase in coagulant dosage, arsenic removal had not improved significantly. On the other hand a further increase in coagulant dose causes restabilization of the particles as the charge reversal on the colloids occurs [32]. So it can be concluded that lower dosage of investigated natural coagulants was better than higher ones. This is very important not only for process economy but also for lower organic matter load in processed water because it is known that high organic load might cause microbial growth [33]. Arsenic removal was calculated by the formula

$$\% \text{ Removal} = (C_i - C_f)/C_i \times 100$$

C_i is the initial concentration (μg/L) and C_f is the final arsenic concentration (μg/L) (Figs. 3 and 4).

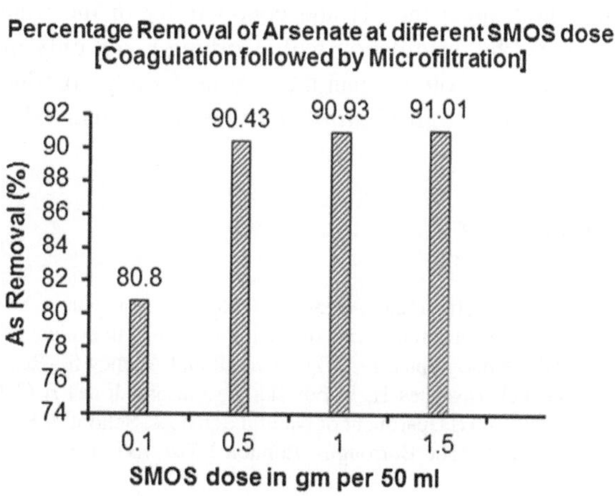

Fig. 3 SMOS dose (gm/50 ml) on arsenate removal (%)

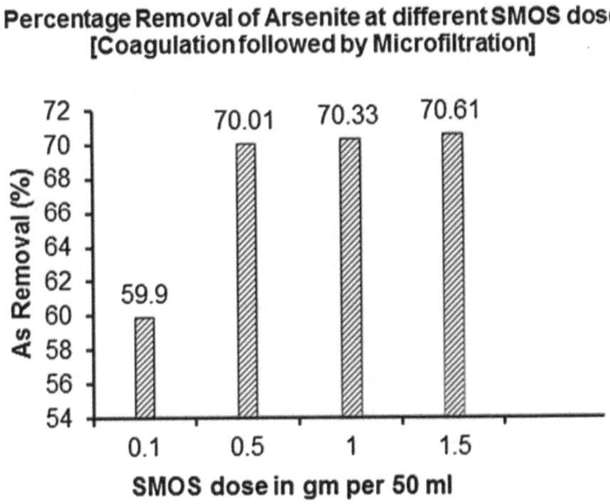

Fig. 4 SMOS dose (gm/50 ml) on arsenite removal (%)

6 Conclusion

From the present study, it can be concluded that arsenic contamination can be brought down below MCL limit of 10 ppb by using natural biosorbent, *Moringa oleifera*. A comparable removal of 91.01 % for arsenate salt and 70.61 % for arsenite salt have been obtained using SMOS compared to synthetic coagulant, $FeCl_3$. More over natural coagulant is cheaper has no toxicity and corrosiveness as in case with synthetic coagulant. Moreover the produced sludge can be used as bio-fertilizers and it becomes an added advantage of this treatment.

Shelled *M. oleifera* seeds (SMOS) provides an exciting opportunity under the domain of Green Processes for domestic, environment friendly, low-cost methods for the decontamination of toxic metals from aqueous systems. It is believed that, in the near future, SMOS could be a potential challenger of synthetic coagulants for the treatment of contaminated drinking water. As a coagulant, Moringa is non-toxic and biodegradable. It is environmentally friendly, and unlike $FeCl_3$, does not significantly affect the pH and conductivity of the water after the treatment. Sludge produced by coagulation with Moringa is not only innocuous but also four to five times less in volume than the chemical sludge produced by $FeCl_3$ coagulation. So, as a coagulant, *Moringa oleifera* can be a potentially viable substitute of $FeCl_3$.

References

1. IARC (International Agency for Research on Cancer) (1987) IARC monographs on the evaluation of carcinogenic risk to humans: overall evaluations of carcinogenicity: an updating of IARC monographs, 1–42(7). International Agency for Research on Cancer Lyon, pp 100–206
2. Garelick H, Jones H, Dybowska A, Valsami-Jones E (2008) Rev Environ Contam Toxicol 197:17−60 (Department of Natural Sciences, School of Health and Social Sciences, Middlesex University, The Burroughs, London NW4 4BT, UK)

3. Ghurey G, Clifford D, Tripp A (2004) Iron coagulation and direct microfiltration to remove arsenic from groundwater. J AWWA 96(4):143–152
4. Han B, Runnells T, Zimbron J, Wick-ramasinghe R (2002) Arsenic removal from drinking water by flocculation and microfiltration. Desalination 145:293–298
5. Okuda T, Baes AU, Nishijima W, Okuda M (2001) Coagulation mechanism of salt solution-extracted active component in Moringa oleifera seeds. Water Res 35:830–834
6. Chandrasekhar K, Kamala CT, Chary NS, Anjaneyulu Y (2001) Removal of heavy metals using a plant biomass with reference to environmental control. Int J Miner Process 68:37–45
7. Basu A, Kumar S, Mukherjee S (2003) Arsenic reduction from environment by water lettuce (Pistia stratiotes L.). Ind J Environ Health 45:143–150
8. Ghimire KN, Inoue K, Makino K, Miyajima T (2002) Adsorptive removal of arsenic using orange juice residue. Sep Sci Technol 37:2785–2799
9. Matuschka B, Strauba G (1993) Biosorption of metals by a waste biomass. J Chem Technol Biotechnol 37:412–417
10. Pagnanelli F, Papini PM, Toro L, Trifoni M, Veglio F (2000) Biosorption of metal ions on Arthrobacter sp.: biomass characterization and biosorption modeling. Environ Sci Technol 34:2773–2778
11. Martins RJE, Pardo R, Boaventura RAR (2004) Cadmium(II) and Zinc(II) adsorption by the aquatic moss Fontinalis antipyretica: effect of temperature, pH and water hardness. Water Res 38:693–699
12. Sun G, Shi W (1998) Sunflower stalks as adsorbents for the removal of metal ions from waste water. Ind Eng Chem Res 37:1324–1328
13. Kratochvil D, Pimentel P, Volesky B (1998) Removal of trivalent and hexavalent chromium by seaweed biosorbent. Environ Sci Technol 32:2693–2698
14. Ucun H, Bayhan YK, Kaha Y, Cakici A, Algur OF (2002) Biosorption of Chromium(VI) from aqueous solution by cone biomass of Pinus sylvestris. Bioresour Technol 85:155–158
15. Khalid N, Rahman A, Ahmad S, Kiani SN, Ahmmad J (1998) Adsorption of cadmium from aqueous solutions on rice husk. Plant Soil 197:71–78
16. Pagnanelli F, Sara M, Veglio F, Luigi T (2003) Heavy metal removal by olive pomace: biosorbent characterization and equilibrium modeling. Chem Eng Sci 58:4709–4717
17. Tyagi RD, Blais JF, Laroulandie J, Meunier N (2003) Cocoa shells for heavy metal removal from acidic solutions. Bioresour Technol 90:255–263
18. Isabel V, Nuria F, Maria M, Nuria M, Jordi P, Joan S (2004) Removal of copper and nickel ions from aqueous solutions by grape stalks wastes. Water Res 38:992–1002
19. Caceres A, Saravia A, Rizzo S, Zabala L, Leon ED, Nave F (1992) Pharmacologic properties of Moringa oleifera: screening for antispasmodic, anti-inflammatory and diuretic activity. J Ethnopharmacol 36:233–237
20. Marugandan S, Srinivasan K, Tandon SK, Hasan HA (2001) Anti-inflammatory and analgesic activity of some medicinal Plants. J Med Arom Plants Sci 22:56–58
21. Dangi SY, Jolly CI, Narayanan S (2002) Anti-hypertensive activity of the total alkaloids from the leaves of Moringa oleifera. Pharm Biol 40:144–148
22. Suleyman A, Muyibi SA, Evison LM (1995) Moringa oleifera seeds for softening hard water. Water Res 29:1099−1105
23. Ndabigengesere A, Narasiah KS (1998) Quality of water treated by coagulation using Moringa oleifera seeds. Water Res 32:781–791
24. Kalogo Y, Verstraete W (2000) Technical feasibility of the treatment of domestic waste water by CEPS–UASB system. Environ Technol 21:55–65
25. Megat J (2001) Moringa oleifera seeds as a flocculants in waste sludge treatment. Int J Environ Stud 58:185–195
26. Ndabigengesere A, Narasiah KS, Talbot BG (1995) Active agents and mechanism of coagulation of turbid water using Moringa oleifera. Water Res 29:703–710
27. Jose TA, Oliveira, Silveria BS, Vasconcelos LM, Cavada BS, Moriera RA (1999) Compositional and nutritional attributes of seeds from the multipurpose tree Moringa oleifera Lam. J Sci Food Agric 79:815−820

28. Costa G, Michant JC, Guckert G (1997) Amino acids exuded form cadmium concentrations. J Plant Nutr 20:883–900
29. Delvin S (2002) Amino Acids and Proteins, 1st edn. IVY, New Delhi
30. Yuan T, Luo Q, Hu J, Ong S, Ng W (2003) A study on arsenic removal from household drinking water. J Environ Sci Health A 38:1731−1744
31. Makker HPS, Becker K (1997) Nutrients and antiquality factors in different morphological parts of Moringa oleifera tree. J Agric Sci (Cambridge) 128:311−322
32. Duan J, Gregory J, (2003) Coagulation by hydrolysing metal salts. Adv Colloid Interface Sci 100−102:475−502
33. Šciban M, Klašnja M, Antov M, Škrbic B (2009) Removal of water turbidity by natural coagulants obtained from chestnut and acorn. Bioresour Technol 100:6639–6643

Development of Novel Architectures for Patient Care Monitoring System and Diagnosis

M.N. Mamatha

Abstract Designing a highly efficient patient care and monitoring system which can handle multiple patients and multiple parametric measurements from every single patient in real time will lead in improvising the data handling capability at Central Nurse Stations (CNS) and Decentralized Nurse Stations (DCNS). The Bio signal Data Acquisition Systems have been designed to suit patients located at CNS and DCNS in a hospital. The RTL design of Bio signal Data Acquisition System was successfully simulated using Model sim. The design was placed and routed using Xilinx ISE 8.2i tool and the bit stream generated is used for downloading into the targeted FPGA. The FPGA used is XC3S400-5ft256, common for both schemes shown. At 50 MHz operation, the system is capable of wireless communication of 32 numbers of Bio signals up to 400,000 bits/s although the maximum frequency of operation of this design is 89 MHz reported by Xilinx ISE tool. This system is truly upgradeable, be it in extending the capabilities to more number of patients or in improving throughput.

Keywords PCM · CNS · DCNS · DAS · FPGA · ZIGBEE

1 Introduction

The availability of prompt and expert medical care can meaningfully improve health care services particularly at understaffed rural or remote areas [1]. Continents facing continuous threats due to spread of infectious diseases, high levels of infant and maternal mortality, low level of life expectancy and deteriorating health care facilities are the greatest beneficiaries of continuous patient care monitoring and assisted with quick diagnosis techniques whenever required. To handle emergency situation, the main requirement is to continuously monitor intensive care parameters

M.N. Mamatha (✉)
Department of Electronics and Instrumentation, B.M.S.C.E, Bangalore, India
e-mail: Mamathamnbms.intn@bmsce.ac.in

S. Gupta et al. (eds.), *Advancements of Medical Electronics*,
Lecture Notes in Bioengineering, DOI 10.1007/978-81-322-2256-9_31

of patients and simultaneously display all captured information to the competent or decision making experts (doctors) anywhere at any time [2–4]. A closer look at the patient monitoring system reveals that it is the doctor alone who is remotely located whereas the emergency site is at a stable point.

The advent of wireless systems has enabled hospitals to customarily measure and analyze the vital signs of surgical, trauma, and other critical care patients throughout their stay in the hospital. Patient Monitoring devices have now been developed for pulse rate, heart rate, temperature, Oxygen Saturation (SpO_2), Central Venous Pressures (CVP), respiration rate, Pulmonary Pressures, EEG, ECG and blood pressure, among other parameters [5, 6]. These systems are able to interface with hospital information technology to provide caregivers with immediate and regular observation of patient status.

Recent advances in embedded computing systems have led to the emergence of wireless sensor networks consisting of small, battery powered devices with limited computation and radio communication capabilities. Sensor networks permit data gathering and computation to be deeply embedded in the physical environment [7, 8]. This technology has the potential to impact the delivery and study of resuscitative care by allowing vital signs to be automatically detected and fully integrated into the patient care used for real time storage, correlation of hospital records, long term observation, etc.

2 Bio Signal Data Acquisition for Multi-patient Care and Monitoring

The present work focuses on implementing a Bio signal Data Acquisition system with an FPGA which can cater to numerous physiological parameters of the patient that are acquired by the sensors or electrodes as shown in Fig. 1. In spite of the system catering to multi patients, it processes the acquired data rapidly in real time since its design is heavily pipelined and massively parallel. This design may be expressed in two different schemes: One as Decentralized Nurse Station without wireless communication and the other as DCNS with Zigbee or LAN. For the latter scheme, we also need to design a Centralized Nurse Station for receiving the acquired data, storing data and Diagnostics. Since this system is exact inverse of DCNS and involved, only DCNS design has been undertaken in the present work.

3 Scheme Proposed for Patient Care Monitoring System

The proposed work comprises the design of Single as well as Multi Patient Care Acquisition Systems involving the following developments.

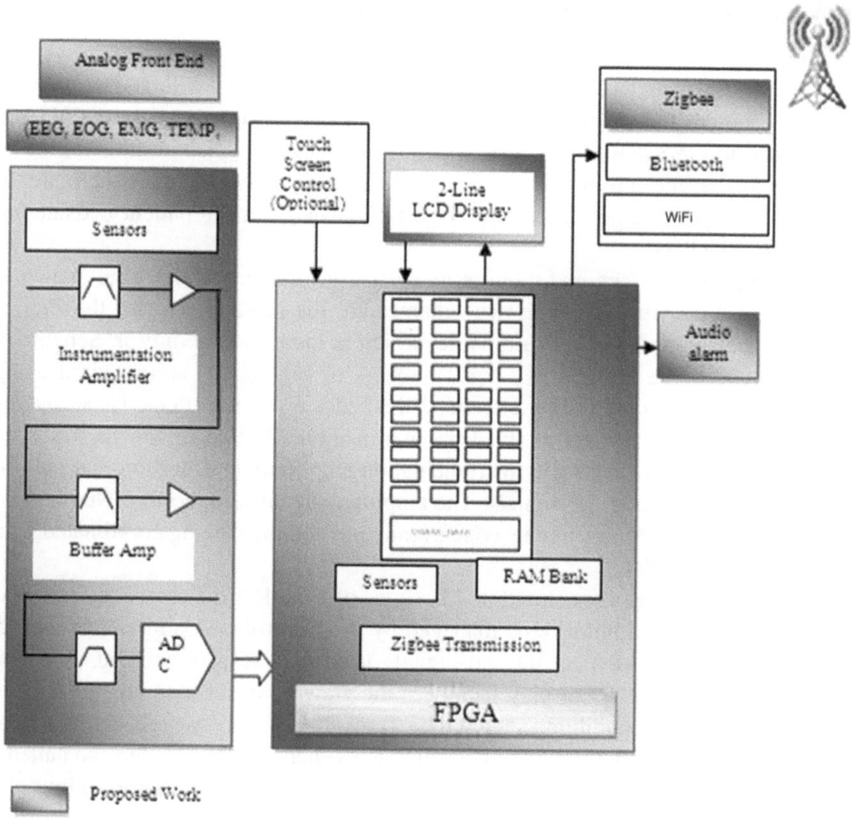

Fig. 1 Block diagram of the proposed bio-signal data acquisition system

1. Development of novel algorithms.
2. Design of architectures for patient care and monitoring for different schemes.
   - (i) A single FPGA which can store a number of physiological parameters from each patient and may be equipped in semi-special ward which houses up to 8 patients in a decentralized nurse station is used. A localized data acquisition is proposed to make it cost effective.
   - (ii) In this scheme, a higher capacity FPGA would be used to acquire bio signals from multi patients and communicate them to a Central Nurse Station in real time. The system shall be endowed with wireless duplex communication ability.

4 Architectures of Schemes Proposed

EEG, EOG, EMG and Temperature signals from multi patients will be continuously acquired, stored and displayed at DCNS. The acquired signals at the DCNS are transmitted to CNS XBee Pro, also popularly known as Zigbee at 400,000 bits/s. This means that the throughput of the system is 40,000 Bytes/s since there are 10 bits to be transmitted per character. The Bio signals from the patient bedside are hard wired to DCNS.

The Multi patient care and monitoring system for Scheme 1 and 2 are shown in Figs. 2 and 3 respectively. The two schemes are the same except for the Zigbee controller design and XBee Pro module, which is incorporated only in Scheme 2.

As shown in the Figs. 1 and 2, 32 numbers of bio signals are acquired using sensors and signal conditioning circuits at the bedside of the patient. These signals are hardwired to amplifiers at the Data Acquisition Module of the DCNS. The signal conditioning circuits use Instrumentation amplifiers in order to get a faithful reproduction of the original signals. The amplified signals are fed to 4 Numbers of ADC 0809, which acquire signals concurrently, all activities being coordinated by a Centralized Controller.

The acquired signals are stored in 4 K Bytes of Dual RAM Bank. The Patient as well as the System Monitoring is managed by the Controller. Using the keyboard, Patient information such as ID and signals used can be programmed in a non volatile storage. An LCD display facilitates the viewing of these inputs in addition to displaying alarm conditions such as the Temperature is very high, reminder for taking medicines etc. A sound alarm accompanies such displays to alert the patients or nursing staff concerned. The features mentioned so far are common to both the schemes.

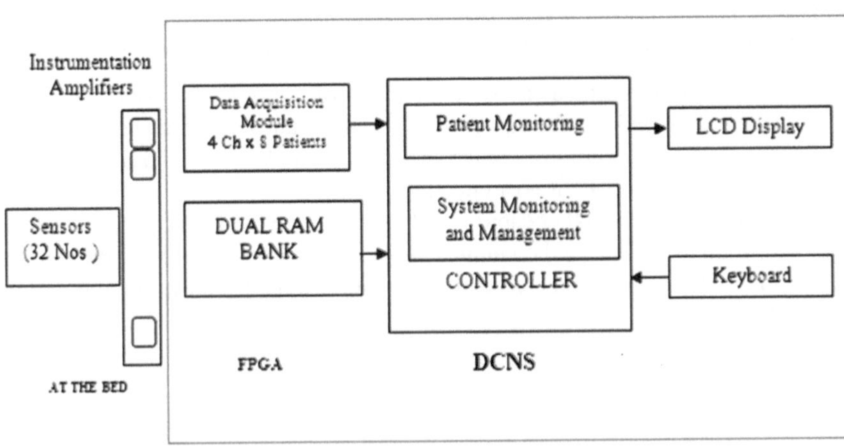

Fig. 2 Multi patient care monitoring system

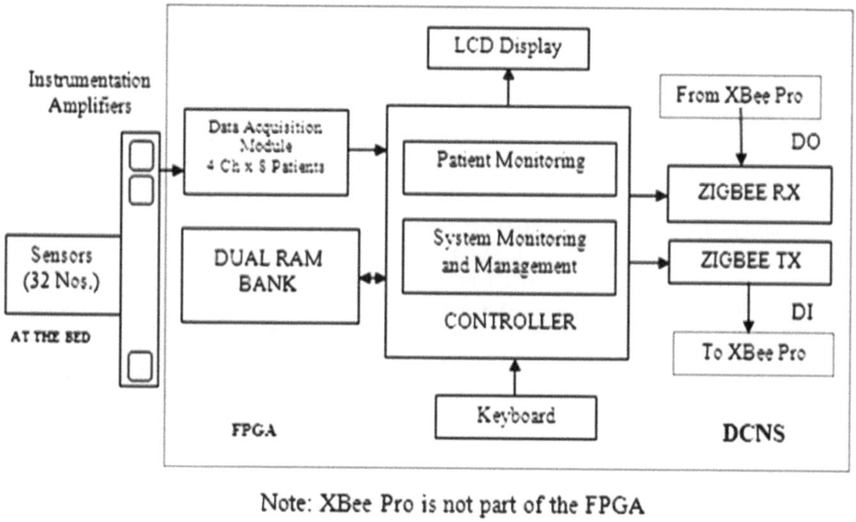

Fig. 3 Multi patient care monitoring transmission

As shown in Fig. 3, a Zigbee Controller is designed in order to communicate the acquired Bio signals to a wireless XBee Pro module. The Zigbee transmitter and receiver are connected to DI and DO pins respectively of the XBee Pro.

5 Proposed Bio Signal Data Acquisition System

The architecture for Scheme 2 only is described in Fig. 4, here since the Scheme 1 is a subset of Scheme 2. The top Architectural design of Bio Signal Data Acquisition System as implemented in this work was presented in order to facilitate the development of the Algorithm. Figures 2 and 3 shows the overall architecture of Bio-DAS. The top design module "bio_das" instantiates three other sub modules "bio_da" which controls 4 Nos. of ADCs, Dual RAM Bank (dualram_bank) and Zigbee Controller (zigbeec).

The detailed Architecture of Bio signal data acquisition system has been realized on a single FPGA in this work.

5.1 Detailed Architecture of Dual RAM Bank

The ADC outputs are stored in intermediate buffers "din1" to "din4" acquired from ADC signals "oadc1" to "oadc4" respectively. These signals are fed to Dual RAM Bank. It contains 32 Nos. of Dual RAMs, which in turn consists of two RAMs,

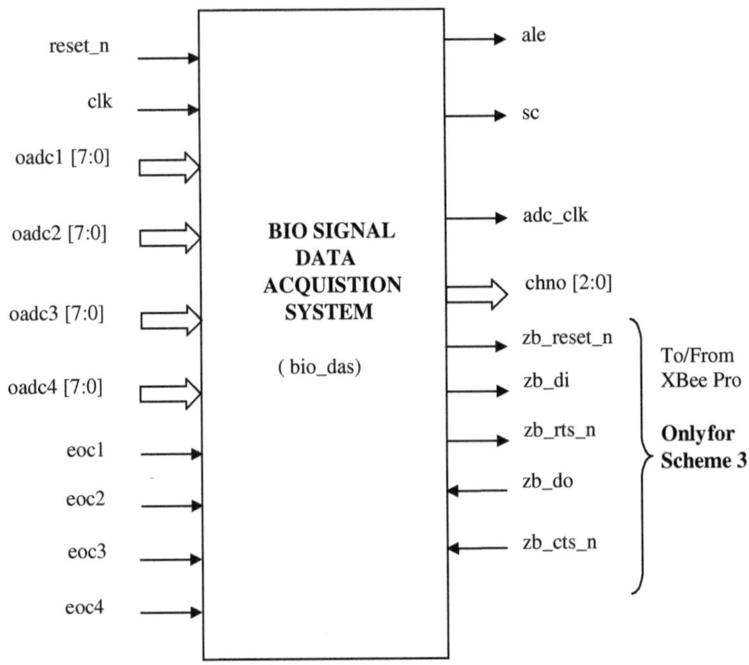

Fig. 4 Architecture of bio signal data acquisition system

RAM1 and RAM2 of size 64 Bytes each. Each location in RAM is of 8 bits size. The control input signal "rnw" configures one of the two RAMs in write only mode and the other in read only mode alternately. A 'high' at "rnw" configures RAM1 in write mode and RAM2 in read mode and vice versa. Once configured, the RAM in write mode gets all the data until "rnw" toggles. The memory size of the RAM in the proposed work is 32 × 2 × 64 Bytes.

The signal "rnw" is connected to RAM 1 directly, whereas its inverted signal is connected to RAM2. The outputs of each RAM "dout1" and "dout2" are multiplexed and issued as the output "dout".

As shown therein, the data bus "din1" to "din4" are input to the Dual RAMs marked for Patients P1/P2, P3/P4, P5/P6 and P7/P8. For each Patient, 4 Nos. of Dual RAMs are allocated so that they may store 4 × 2 × 64 Bytes of samples at the most for the four types of signals: EEG, EOG, EMG and Temperature put together.

The validity of this data input signal ("din_valid1" to "din_valid32") for all the 32 Dual RAMs are asserted individually by the "bio_da" module. It may be noted that four valid signals such as "din_valid1", "din_valid9", "din_valid17" and "din_valid25" are activated simultaneously so that the corresponding samples may be stored concurrently. While reading, "dout_sel1" to "dout_sel32" are activated one after another for every 64 Bytes read.

The XBee Pro module of Rhydolabz is engineered to meet IEEE 802.15.4 standards and support the unique needs of low-cost, low-power, wireless sensor networks. This module requires minimal power and provides reliable delivery of data between devices.

5.2 Detailed Architecture of Zigbee Controller

The XBee Pro Wireless Module interfaces to a host device such as the proposed single FPGA chip DCNS through a logic level, asynchronous serial port. Through its serial port, the module can communicate with any designed logic. The module has a transmitter and a receiver. The interface between the DCNS and the XBee Pro Module is shown in Fig. 5. At the other end of wireless communication, another XBee Pro Module interfaces with a CNS, which design is beyond the scope of the present work.

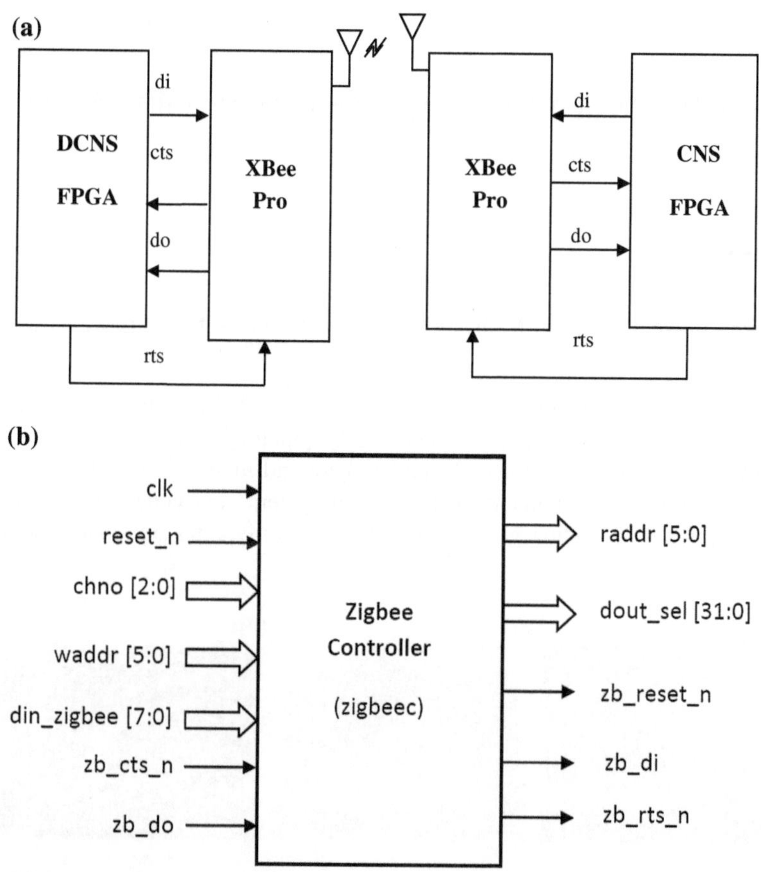

Fig. 5 **a** XBee pro interfaces to DCNS and CNS, **b** XBee, **c** Simulated output

(c)

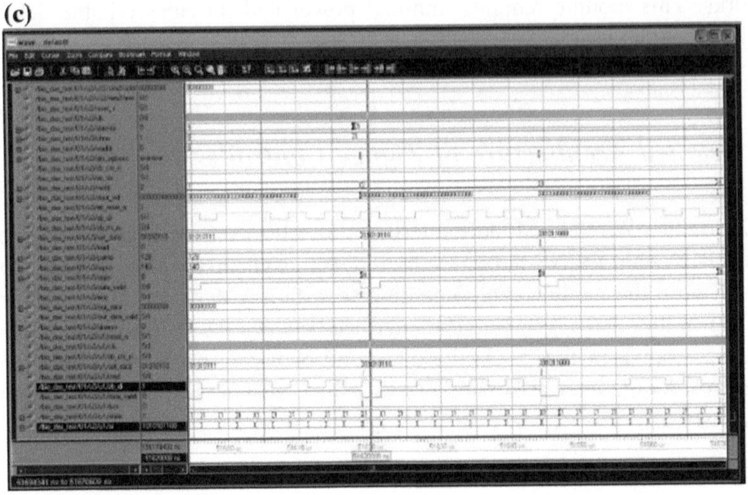

Fig. 5 (continued)

The architecture of **XBee Pro Interfaces to DCNS and CNS** is shown in Fig. 5a–c.

6 Results

The experimental set up of Multi PCM with DAC and ADC cards are as shown in Fig. 6

- Developed fast, novel algorithms suitable for FPGA/ASIC implementation. designed a massively parallel and highly pipelined architecture suitable for FPGA implementation for multi patient care and monitoring system.
- The complete bio signal data acquisition system was coded in verilog conforming to RTL coding guidelines and successfully simulated using modelsim.

(a) (b) (c)

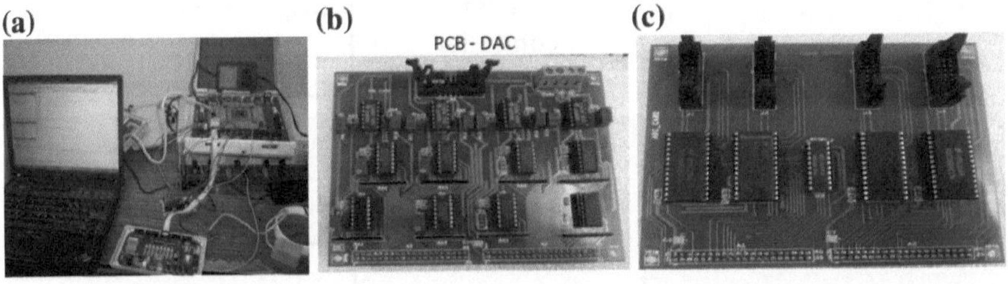

Fig. 6 **a** Hardware set up of scheme 1 and 2 multi PCM system, **b** DAC, **c** ADC FPGA implementation

- The designed system was synthesised, placed and routed with XCS 400-5ft256 FPGA. The real time system is capable of transmitting up to 32 numbers of bio signal data of patients at 400,000 bits/s using a wireless communication channel.

7 Conclusion

- Bio signal DAQ has been designed for multi patient care and monitoring and is suitable for FPGA implementation.
- Acquire four different bio signals such as EEG, EOG, EMG and temperature from eight patients simultaneously using designed sensors.
- Conditioning circuits were also designed to meet the exacting specifications of amplifying low amplitude bio signals.
- Novel algorithms have been developed for acquiring and monitoring the bio signals for decentralized nurse stations. These algorithms were designed as architectures and coded in Verilog.
- The analog signals from bio sensors were digitized by a "bio signal data acquisition" module and were stored in "dual ram bank". These were transmitted serially through a controller module "zigbeec" to a CNS and the signals acquired may be analyzed and diagnosed.

8 Scope for Future Work

Following are some of the suggestions related to this work, which can be undertaken by enthusiastic researchers in near future.

- A DCNS was designed. The system can be further enhanced by designing and integrating a lossless compression encoder before the XBee Pro Wireless module.
- A Centralized Nurse Station can also be designed. It may also incorporate compression decoder in its design.
- The patients' beds may be made adjustable for flat, reclining or sitting posture using motors controlled by patients themselves using eyewinks, Flex etc. or even thought.
- Pressure ulcers or bedsores are one of the most common complications of patients who cannot change positions in a bed all by themselves. By splitting the bed into two halves, provide posture adjustment using motors, driver by patients.

References

1. Gomez H, Camacho J, Yelicich B, Moraes L, Biestro A, Puppo C (2010) Development of a multimodal monitoring platform for medical research. In: Annual international conference of IEEE, engineering in medicine and biology society, Buenos Aires, Argentina, pp 2358–2361
2. Tanha AR, Popovic D (1997) An intelligent monitoring system for patient health care. In: 19th international conference IEEE, engineering in medicine and biology society, Chicago, IL, USA, pp 1086–1087
3. Reilent E, Loobas I, Pahtma R, Kuusik A (2010) Medical and context data acquisition system for patient home monitoring. In: 12th biennial baltic electronics conference, Estonia, pp 269–272
4. Zhan Y, Silvers CT, Randolph AG (2007) Real-time evaluation of patient monitoring algorithms for critical care at the bedside. In: Proceedings of the 29th annual international conference of the IEEE engineering in medicine and biology society cite international, France, pp 2783–2786
5. Takizawa K, Li H-B, Hamaguchi K, Kohno R (2007) Wireless patient monitoring using IEEE 802.15.4a WPAN. In: IEEE, Singapore, pp 235–240
6. Dilmaghani RS, Bobarshad H, Ghavami M, Choobkar S, Wolfe C (2011) Wireless sensor networks for monitoring physiological signals of multiple patients. IEEE Trans Biomed Circuits Syst 5(4):347–356
7. Tafa Z, Stojanović R (2006) Bluetooth–based approach to monitoring biomedical signals. In: Proceedings of the 5th WSEAS international conference on telecommunications and informatics, Istanbul, Turkey, pp 415–420
8. Kogure Y, Matsuoka H, Kinouchi Y, Akutagawa M (2005) The development of a remote patient monitoring system using java-enabled mobile phones. In: Proceedings of the 2005 IEEE engineering in medicine and biology 27th annual conference, Shanghai, China, pp 2157–2260

Review on Biocompatibility of ZnO Nano Particles

Ananya Barman

Abstract ZnO nano-particles have some unique properties like piezoelectric, semiconducting, catalytic properties and antibacterial activities. Thus these particles are widely used in optoelectronics, sensors, transducers, energy conversion and also in medical sciences. Zinc oxide nano-particles have the potential to function as natural selective killers of all highly proliferating cells, whether cancerous or not. The application of zinc oxide nano-particles in cancer therapy looks intriguing and exciting, specific tumor cell targeting will be essential (e.g., by nano-particle functionalization with cell ligands) because these nano-particles are killers of all rapidly proliferating cells, irrespective of their benign or malignant nature. So the cellular level biocompatibility and biosafety of ZnO is studied here. Hela cell line showed a complete biocompatibility to ZnO nanostructures from low to high NW concentrations beyond a couple of production periods. The L929 cell line showed a good reproduction behavior at lower NW concentration, but when the concentration was close to 100 μg/ml, the viability dropped to ~ 50 %. Our study shows the biocompetability and biosafety of ZnO NWs when they are applied in biological applications at normal concentration range.

Keywords Zinc oxide nanoparticles · Dispersion · Selective cytotoxicity · Mesenchymal stem cells

1 Introduction

Nanostructures of zinc oxide (ZnO) have attracted much interest because of their unique piezoelectric, semiconducting, and catalytic properties [1, 2] and a wide range of applications in optoelectronics, sensors, transducers, energy conversion and medical sciences [3–10]. Recently, utilizing the coupled piezoelectric and semiconducting properties, ZnO nanowire (NW) arrays and nanobelts have been applied

A. Barman (✉)
Chemistry Department, JIS College of Engineering, 741235 Kalyani, Nadia, India
e-mail: Ananya_bhu@yahoo.co.in

S. Gupta et al. (eds.), *Advancements of Medical Electronics*,
Lecture Notes in Bioengineering, DOI 10.1007/978-81-322-2256-9_32

for converting mechanical energy into electricity and building piezotronic devices [3, 5, 11].

Again ZnO nano particles have potential biomedical and lifescience applications. zinc oxide nanoparticles offer significant benefits, and are used in several products and systems, including sunscreens, biosensors, food additives, pigments, rubber manufacture, and electronic aterials [1]. In the biomedical field, one of the most important attributes of these nanoparticles is their antibacterial activity, with published reports confirming the efficacy of zinc oxide nanoparticle-based preparations as prophylactic agents against bacterial infections [11, 12].

Few reports had been published on the cytotoxicity of ZnO nano particles on mammalian cells. Other reports suggest that these nanoparticles are nontoxic to cultured human dermal fibroblasts [13] but exhibit toxicity towards neuroblastoma cells [14] and vascular endothelial cells [15] and induce apoptosis in neural stem cells [16]. It has been reported that nanoparticle size influences cell viability. Jones et al. pointed out that zinc oxide particles of 8 nm in size were more toxic than larger zinc oxide particles (50–70 nm) in Staphylococcus aureus [17]. Recently, Hanley et al. observed that there is an inverse relationship between nanoparticle size and cytotoxicity in mammalian cells, as well as nanoparticle size and reactive oxygen species production [18], while Deng et al. showed that zinc oxide nanoparticles manifested dose-dependent, but no size-dependent, toxic effects on neural stem cells [16].

Many of the reported studies on zinc oxide nanoparticle toxicity have significant limitations for two reasons. Firstly, some reports used zinc oxide nanoparticles with different characteristics (size, shape, and purity) and, secondly, the dispersion protocols used have often been unsuitable for biological use. Thus, the first requirement for the biomedical use of zinc oxide nanoparticles is an effective biocompatible protocol for aqueous dispersion.

The potential application of zinc oxide nanoparticles in medicine as a follow-on from the report by Hanley et al. that these nanoparticles induce toxicity in a cell-specific and proliferation-dependent manner. This group demonstrated that zinc oxide nanoparticles exhibit a strong preferential ability to kill rapidly dividing cancerous T cells (28–35x) but not normal cells [19]. A recent report confirmed that these nanoparticles exert a cytotoxic effect on human glioma cells, but not on normal human astrocytes. The mechanisms of toxicity appear to involve the generation of reactive oxygen species, with cancerous cells producing higher inducible levels than normal cells.

2 Materials and Methods

2.1 Preparation of ZnO Nano Particles

Due to its vast areas of application, various synthetic methods have been employed to grow a variety of ZnO nanostructures, including nanoparticles, nanowires, nanorods, nanotubes, nanobelts, and other complex morphologies [20–33].

There are many methods we have, to synthesize ZnO nano particles, like metallurgical process, mechanical, chemical, mechanochemical process, controlled precipitation, sol-gel method and solvothermal and hydrothermal method. Mainly sol-gel method or hydrothermal method present low cost and environment friendly synthetic route, most of the literature for ZnO nano-particles is based on the solution method. In addition, synthesis of ZnO nano-particles in the solution requires a well defined shape and size of ZnO nano-particles.

In this regards, Monge et al. [34] reported room-temperature organometallic synthesis of ZnO nano-particles of controlled shape and size in solution. The principle of this experiment was based on the decomposition of organometallic precursor to the oxidized material in air. It was reported [35] that when a solution of dicyclohexylzinc(II) compound $[Zn(c\text{-}C_6H_{11})_2]$ in tetrahydrofuron (THF) was left standing at room temperature in open air, the solvent evaporated slowly and left a white luminescent residue, which was further characterized by X-Ray diffraction (XRD) and transmission electron microscopy (TEM) and confirmed as agglomerated ZnO nanoparticles with a zincite structure having lack of defined shape and size.

The ZnO NWs were grown by a vapor-liquid-solid (VLS) process in a horizontal tube furnace, as reported previously [36, 37]. Gold nanoparticles were used as the catalyst and the NWs were supported by a polycrystalline alumina substrate.

In some of the reported study 0.2 mmol of zinc nitrate hexahydrate, Zn $(NO_3)_2 \cdot 6H_2O$, and 2 mmol oleic acid (OA) were dissolved in 20 mL triethylene glycol (TREG). The mixture was loaded into a three-neck flask and was stirred at room temperature for 30 min. The original colorless clear solution started to turn milk white. Then, the mixture was heated to 240 °C with continuous stirring for 120 min. A mass of bubbles appeared when the temperature was up to 150 °C, which demonstrated the reaction of ester elimination. After being cooled to room temperature and separated by centrifugation, the nanocrystals were first washed using toluene and then washed using absolute ethanol to remove the unreacted molecules. The final particles were collected by centrifugation and redispersed in deionized water under agitation for further characterization.

Again 1.86 mmol of zinc acetate dihydrate, also Zn $(Ac)_2 \cdot 2H_2O$, was dissolved in 10 mL ethanol in a flask under vigorous stirring and refluxed for 90 min at 68 °C. The obtained Zn $(Ac)_2$ solution was then cooled down to room temperature. 4.11 mmol of KOH was dissolved in 5 mL of ethanol and kept in an ultrasonic bath for 40 min at room temperature. The obtained KOH solution was slowly added to Zn $(Ac)_2$ solution and was stirred at room temperature for 60 min. Then 0.5 mL of deionized water and 0.34 mmol of 3-aminopropyltriethoxysilane (APTES) which was dissolved in 5 mL ethanol were added into the reaction system. After 120 min of constant stirring at room temperature, the precipitate was isolated by centrifugation. The nanocrystals were first washed using toluene and then washed using absolute ethanol to remove the unreacted molecules. The final particles were collected by centrifugation and redispersed in de-ionized water under agitation for further characterization.

After synthesis, the ZnO nano-particles are characterized by DLS, UVand FT-IR Spectra, SEM, TEM, X-RAY Diffraction, EDAX and the optical properties also observed.

3 Biocompatibility of ZnO Nano Particles

3.1 Cell Culture and Interaction with Zno

For the biocompatible study, 5 mg of ZnO NWs were used as the test sample, which were removed from the polycrystalline alumina substrate, then put into a polypropylene centrifuge tube, which was filled with 5 mL sterile phosphate buffer solution (PBS, pH 7.2). The average diameter of the NWs is 1 μm and the average length is 200 μm. The NWs were dispersed by ultrasonication for 10 min. Then samples with concentrations of 1000, 100, 10, and 1 μg/ml, respectively, were prepared (Fig. 1a).

Two cell lines from different origins of tissues were utilized [38]. One was Hela cell line (American type Culture Collection, ATCC, CCL-2, Homo sapiens), a kind of epithelial cell. The other one was L-929 cell line (ATCC, CCL-1, Mus musculus), from subcutaneous connective tissue. The physical interaction of NWs and

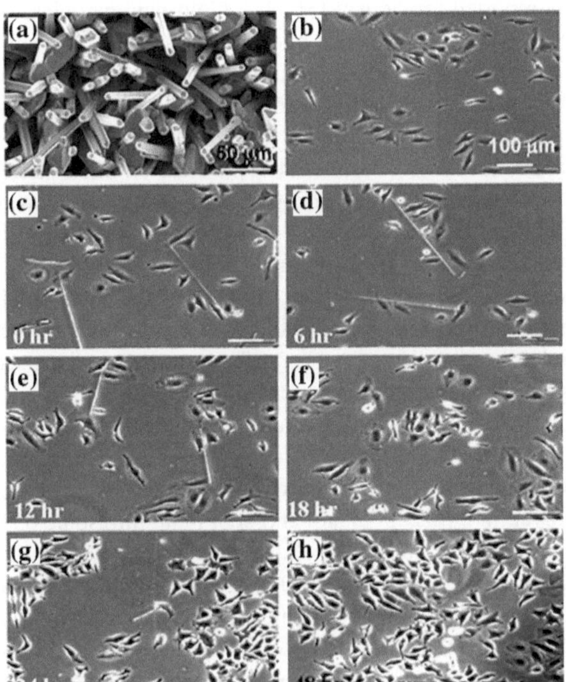

Fig. 1 Effect of ZnO Nws on the growth and reproduction of cells as a function of time. **a** As-grown ZnO nanowires on an alumina substrate. **b** Hela cell have been cultured for 4 h. **c–h** Hela cells cultured with ZnO nanowires in solution after growing for 0, 6, 12, 18, 24 and 48 h respectively. The Cells grew and reproduced with the presence of ZnO nanowires. *Panels* **b–h** were recorded at the same magnification so that the number of cells in each image represents its concentration. The *scale bar* is 100 μm

Fig. 2 SEM images of Hela cells on ZnO NWs arrays. **a** Two Hela cells are growing on the surface of ZnO NWs arrays. **b** Cells are upheld by the NWs. Some ZnO NWs are phagocytosed into the Hela cell (pointed out by the *red arrow*). The diameter and the length of the nanowires are ~100 nm and ~1.5 μm respectively

Hela cells was investigated by inverted microscopy and scanning electron microscopy (SEM). ZnO NWs array was in high density and the average diameter and the length of the nanowires were ~100 nm and ~1.5 μm, respectively. After seeding Hela cells on the ZnO NWs arrays, the substrate was incubated in 37 °C, 5 % CO_2 for 24 h with DMEM cell culture medium. As the cells in the culture medium subsided on the NWs substrate, the cells were upheld by ZnO NWs (Fig. 2a). To maintain the morphology of the cells for SEM, immobilized process will be actualized to the cells adhered on the substrate. SEM images of the samples are shown in Fig. 2a, b. Figure 2a shows two Hela cells began to extend and grow on the surface of ZnO NWs. Hela cells phagocytosed some of the broken ZnO NWs into the cell membrane (Fig. 2b). No external force or inducement was necessary for the phagocytosis, owing to the self-activity of the cell. The results indicate that ZnO nanowires have fine biocompatibility and can be readily phagocytosed by the cells. he Hela cells were still growing and reproducing even though the NWs were directly in contact with them (Fig. 1c–h). Also, we can see the degradation of ZnO NWs in cell culture medium with the time changing (Fig. 1c–h).

Some study were made on human neuroblastoma cell line and bone marrow cells from tibia and femur. These cells are the biological model of proliferating cells. The study demonstrated that zinc oxide nano-particles induce formation of excess reactive oxygen species resulting in alteration and damage of cellular proteins, DNA, and lipids, which can cause cell death [39–42]. We used carboxy-H2DCFDA fluorescence as a reporter of intracellular oxidant production.

It was reported that initial evaluation of the biocompatibility of ZnO nano particles was performed using heamolysis assays. This results further confirm the biocompatibility [43].

4 Results and Discussion

ZnO NWs are biocompatible and biosafe to the two cell lines (Fig. 3). The viabilities of Hela cells cultured with NWs for 12 and 24 h showed no difference. The 48 h cultured cells showed only a slight reduction in viability at a high concentration of 100 μg/ml (Fig. 3a).

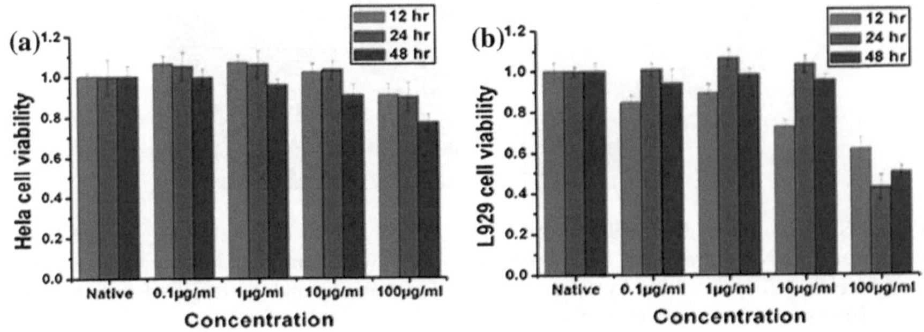

Fig. 3 Cell viability tested by MTT as a function of ZnO NW concentration and time. **a** Cell viability tested by MTT method as a function of ZnO NWs for 12, 24, 48 h. **b** Viability of L929 cell line in MTT test, cultured with different concentrations of ZnO NWs for 12, 24, 48 h

More than 95 % of the Hela cells were alive after the test, and there was no significant difference in viability among the plates of three time groups (SPSS, Paired-sample t test) (Table 1). In the 48 h MTT experiment with the highest NW concentration of 100 μg/ml, the viability of the Hela cells was a little lower than that of the sample without NWs, but the viability was still more than 75 % (Fig. 3a).

Table 1 shows the statistical analysis of Paired-sample t tests using SPSS. If the values of paired-sample t tests are larger than 0.05, it means that there is no significant difference between the compared groups. Most of the values shown are larger than 0.05, indicating adding ZnO NWs did not affect the viability of the cells. For 96-well plates of planting L929 cells, the viabilities showed some variations (Fig. 3b). For the 12 h plate, the viability of L929 cells was lower than those of the other two 24 and 48 h time-sequence plates (Table 1), indicating that the L929 cell was in a frail period and more sensitive to ZnO NWs when it was first cultured for less than 12 h. After the culturing time exceeded 12 h, the viability of the cells remained strong and was better than 95 % even at relatively high NWs concentration. However, the viability of the cells dropped significantly when the NW concentration reached 100 μg/ml. Therefore, the NWs are considered to be completely biocompatible and biosafe at NWs concentration lower than 100 μg/ml.

Table 1 Paired-sample T-test, SPSS

Hela/concentration (μ g/ml)	0.1	1	10	100
12 h Sig. (2 tailed)	0.087	0.103	0.541	0.059
24 h Sig. (2 tailed)	0.471	0.346	0.524	0.060
48 h Sig. (2 tailed)	0.124	0.736	0.131	0.000
L929/concentration (μ g/ml)	0.1	1	10	100
12 h Sig. (2 tailed)	0.039	0.155	0.007	0.001
24 h Sig. (2 tailed)	0.725	0.059	0.287	0.000
48 h Sig. (2 tailed)	0.182	0.545	0.142	0.000

Other zinc oxide nanoparticle dispersions have already been used for cytotoxicity studies. Some published reports indicate that zinc oxide nanoparticles can be toxic to mammalian cells [14–16]. Hanley et al. proposed that the mechanism of zinc oxide nanoparticle toxicity involves the generation of reactive oxygen species. These authors also suggested cell-specific behavior, with cancer cells producing higher inducible levels of reactive oxygen species than their normal counterparts following exposure to zinc oxide nano-particles [44].

Cell proliferation assays performed in a human neuroblastoma cell line demonstrate that the cytotoxicity induced by zinc oxide nano-particles is dose-dependent. No significant changes in cell viability of nano-particle concentrations of 10 μg/mL was observed. At these concentrations, we observed a spurious small increase in viability compared with the control due to nanoparticle absorbance at the working wavelength. Cell viability dropped significantly at a nanoparticle concentration of 20 μg/mL, and was associated with production of reactive oxygen species. Based on published work, we have developed a model of zinc oxide nanoparticle cytotoxicity. Essentially the process involves internalization of the nano-particles via receptor-mediated endocytosis, hydrolysis of the nanoparticles via zinc ions within the lysosomes, and release of zinc ions into the cytosol.

When zinc oxide nano-particles are added to growth medium, the particles are coated by proteins in the medium. Cell surface receptors bind the protein adsorbed onto the nanoparticles, and the nano-particles enter cells via receptor-mediated endocytosis. When nano-particle containing endosomes fuse with lysosomes, the pH drops dramatically, approaching 5. The rate of zinc oxide hydrolysis was seen as a function of pH; whereas under physiological conditions the fraction hydrolyzed is negligible (0.02 %), the hydrolysis of zinc oxide is complete at pH 5.75. The zinc ions induce lysis of the lysosomal membrane, and the ions are released into the cytosol. Cytosolic accumulation of zinc ions triggers pathways which ultimately cause cell death.

Specifically, Xia et al. demonstrated that zinc oxide dissociation disrupts the cellular homeostasis of zinc, leading to lysosomal and mitochondrial damage, and ultimately cell death by inhibiting cellular respiration through interference with cytochrome bc1 in complex III and with α-ketoglutarate dehydrogenase in complex I [39]. Other researchers have under-lined that zinc ion-mediated production of reactive oxygen species promotes two important mechanisms, i.e., cytoplasmic release of calcium ions and interaction with the cytoplasmic membrane, causing loss of membrane integrity and leading to calcium influx through membrane channels [40].

Over the last two decades, mesenchymal stem cell-based therapy has progressed rapidly from preclinical to early clinical Phase I and II studies in a range of human diseases (www.clinicaltrialgov). Osteocytes used in this work were obtained via osteogenic differentiation of mesenchymal stem cells. These two populations, although biologically related, show opposite proliferation rates. We evaluated cell death induced by zinc oxide nano-particles using flow cytometry. The increased fluorescent intensity in the presence of zinc oxide nano-particles was more significant in mesenchymal stem cells than in the differentiated osteogenic lineage.

The different behavior of the two cell types is likely due to the different cellular interaction with zinc oxide nano-particles rather than to any differences in uptake of the nano-particles. The strong interaction between osteocytes and the nano-particles is reflected by the parallel strong increase of side scatter compared with the control cultures, indicative of increased cellular complexity due to nano-particle cell interaction [45].

Based on these findings, we confirm that zinc oxide nano-particles have the potential to function as natural selective killers of all highly proliferating cells, whether cancerous or not. In conclusion, although the application of zinc oxide nanoparticles in cancer therapy looks intriguing and exciting, specific tumor cell targeting will be essential (e.g., by nano-particle functionalization with cell ligands) because these nano-particles are killers of all rapidly proliferating cells, irrespective of their benign or malignant nature.

5 Conclusion

In conclusion all these studies show the biocompatibility and biosafety of ZnO nano particles when they are applied in biological applications at normal concentration range. This is an important conclusion for their applications in in vivo biomedical science and engineering. The threshold of intracellular ZnO NP concentration required to induce cell death in proliferating cells is 0.4 ± 0.02 mM. Finally, flow cytometry analysis revealed that the threshold dose of zinc oxide nano-particles was lethal to proliferating pluripotent mesenchymal stem cells but exhibited negligible cytotoxic effects to osteogenically differentiated mesenchymal stem cells. These results confirm the ZnO NP selective cytotoxic action on rapidly proliferating cells, whether benign or malignant.

In addition, relatively small size, ease of transport within tissues/organs, ability to cross plasma membranes, and potential targeting of biologically active molecules will facilitate biomedical applications of nanoparticles in the field of medicine.

References

1. Lieber CM, Wang ZL (2007) MRS Bull 32:99–104
2. Norton DP, Heo YW, Ivill MP, Pearton SJ, Chisholm MF, Steiner T (2004) Mater Today 7 (6):34–40
3. Wang ZL, Song JH (2006) Science 312:242–246
4. Law M, Sirbuly D, Johnson J, Goldberger J, Saykally R, Yang P (2004) Science 305:1269
5. Wang XD, Song JH, Liu J, Wang ZL (2007) Science 316:102–105
6. Johnson JC, Yan HQ, Schaller RD, Petersen PB, Yang PD, Saykally RJ (2002) Nano Lett 2 (4):279–283
7. Arnold MS, Avouris P, Pan ZW, Wang ZL (2003) J Phys Chem B 107(3):659–663
8. Lao CS, Kuang Q, Wang ZL, Park MC, Deng Y (2007) Appl Phys Lett 90:262107

9. Dorfman A, Kumar N, Hahm S (2006) J Adv Mater 18:2685
10. Liu TY, Liao HC, Lin CC, Hu SH, Chen SY (2006) Langmuir 22(13):5804–5809
11. Qin Y, Wang XD, Wang ZL (2008) Nature 451:809–813
12. Huang ZB, Zheng X, Yan DH et al (2008) Toxicological effect of ZnO nano-particles based on bacteria. Langmuir 24(8):4140–4144
13. Reddy KM, Feris K, Bell J, Wingett DG, Hanley C, Punnoose A (2007) Selective toxicity of zinc oxide nanoparticles to prokaryotic and eukaryotic systems. Appl Phys Lett 90 (21):2139021–2139023
14. Jeng HA, Swanson J (2006) Toxicity of metal oxide nanoparticles in mammalian cells. J Environ Sci Health A Tox Hazard Subst Environ Eng 41(12):2699–2711
15. Gojova A, Guo B, Kota RS, Rutledge JC, Kennedy IM, Barakat AI (2007) Induction of inflammation in vascular endothelial cells by metal oxide nanoparticles: effect of particle composition. Environ Health Perspect 115(3):403–409
16. Deng XY, Luan QX, Chen WT et al (2009) Nanosized zinc oxide particles induce neural stem cell apoptosis. Nanotechnology 20(11):115101
17. Jones N, Ray B, Ranjit KT, Manna AC (2008) Antibacterial activity of ZnO nanoparticle suspensions on a broad spectrum of microorganisms. FEMS Microbiol Lett 279(1):71–76
18. Hanley C, Thurber A, Hanna C, Punnoose A, Zhang J, Wingett DG (2009) The influences of cell type and ZnO nanoparticle size on immune cell cytotoxicity and cytokine induction. Nanoscale Res Lett 4(12):1409–1420
19. Hanley C, Layne J, Punnoose A et al (2008) Preferential killing of cancer cells and activated human T cells using ZnO nanoparticles. Nanotechnology 19(29):295103
20. Song RQ, Xu AW, Deng B, Li Q, Chen GY (2007) Adv Funct Mater 17:296
21. Gao PX, Ding Y, Mai W, Hughes WL, Lao C, Wang ZL (2005) Science 309:1700
22. Hu P, Liu Y, Wang X, Fu L, Zhu D (2003) Chem Commun 50:1304
23. Zhang J, Sun L, Liao C, Yan C (2002) Chem Commun 3:262
24. Tang Q, Zhou W,Shen J, Zhang W, Kong L, Qian Y (2004) Chem Commun 10(6):2004
25. Greene LE, Law M, Goldberger J, Kim F, Johnson JC, Zhang Y, Saykally RJ, Yang P (2003) Angew Chem 42:3031
26. Kong XY, Ding Y, Yang R, Wang ZL (2004) Science 303:1348
27. Qian D, Jiang JZ, Hansen PL (2003) Chem Commun 9(9):1078
28. Zhong X, Knoll W (2005) Chem Commun 9(9):1158
29. Li F, Ding Y, Gao P, Xin X, Wang ZL (2004) Angew Chem 43:5238
30. Ding GQ, Shen WZ, Zheng MJ, Fan DH (2006) Appl Phys Lett 88:103106
31. Shpeizer BG, Bakhmutov VI, Clearfield A (2006) Micro Meso Mater 90:81
32. Wang X, Summers CJ, Wang ZL (2004) Adv Mater 16:1215
33. Polarz S, Orlov AV, Schüth F, Lu AH (2007) Chem Eur J 13:592
34. Monge M, Kahn ML, Maisonnat A, Chaudret B (2003) Angew Chem Int Ed 42:5321
35. Carnes CL, Klabunde KJ (2000) Langmuir 16:3764
36. Wang XD, Summers CJ, Wang ZL (2004) Nano Lett 4:423–426
37. Yang P, Yan H, Mao S, Russo R, Johnson J, Saykally R, Morris N, Pham J, He R, Choi H (2002) Adv Func Mater 12:323
38. Hanks CT, Wataha JC, Sun ZL (1996) Dent Mater 12:186–193
39. Xia T, Kovochich M, Liong M et al (2008) Comparison of the mechanism of toxicity of zinc oxide and cerium oxide nanoparticles based on dissolution and oxidative stress properties. ACS Nano 2(10):2121–2134
40. Huang CC, Aronstam RS, Chen DR, Huang YW (2010) Oxidative stress, calcium homeostasis, and altered gene expression in human lung epithelial cells exposed to ZnO nanoparticles. Toxicol In Vitro 24(1):45–55
41. Florianczyk B, Trojanowski T (2009) Inhibition of respiratory processes by overabundance of zinc in neuronal cells. Folia Neuropathol 47(3):234–239

42. Gazaryan IG, Krasnikov BF, Ashby GA, Thorneley RNF, Kristal BS, Brown AM (2002) Zinc is a potent inhibitor of thiol oxidoreductase activity and stimulates reactive oxygen species production by lipoamide dehydrogenase. J Biol Chem 277(12):10064–10072
43. Dobrovolskaia MA, Clogston JD, Neun BW, Hall JB, Patri AK, McNeil SE (2008) Nano Lett 8:2180–2187
44. Wang ZL (2004) Zinc oxide nanostructures: growth, properties and applications. J Phys Condens Matter 16(25):R829–R858
45. Cai D, Blair D, Dufort FJ et al (2008) Interaction between carbon nano-tubes and mammalian cells: characterization by flow cytometry and application. Nanotechnology 19(34):1–10

Tailoring Characteristic Wavelength Range of Circular Quantum Dots for Detecting Signature of Virus in IR Region

Swapan Bhattacharyya and Arpan Deyasi

Abstract Characteristic wavelength carrying the signature of virus is analytically determined through its match with the radiating wavelength comes out from computation of intersubband transition energies of different circular quantum dots, namely quantum ring and quantum disk. Time-independent Schrödinger equation is solved subject to the applied electric field along the axis, and first and second order Bessel functions are considered for computation of energy subbands. Non-monotonic spacing of quantized energy states have been observed by changing different dimensions of the quantum dots. Three lowest confinement states along with subband energies are plotted with different structural parameters, and also with external field. Comparative study reveals that better tuning of intersubband transition energy can be achieved in quantum ring than quantum disk having similar structural parameters; which reveals the fact that characteristic wavelength from quantum ring can track wider rage of virus signature. Tailoring of wavelength can be revealed by notifying the blueshift/redshift in absorption spectra in the choice of frequency region.

Keywords Characteristic wavelength · Quantum ring · Quantum disk · Electric field · Intersubband transition energy · Signature of virus

S. Bhattacharyya (✉)
Dept of Electronics and Communication Engineering, JIS College of Engineering, Kalyani 741235, India
e-mail: swapanbhattacharyya@ieee.org

A. Deyasi
Dept of Electronics and Communication Engineering, RCC Institute of Information Technology, Kolkata 700015, India
e-mail: deyasi_arpan@yahoo.co.in

S. Gupta et al. (eds.), *Advancements of Medical Electronics*,
Lecture Notes in Bioengineering, DOI 10.1007/978-81-322-2256-9_33

1 Introduction

Low dimensional quantum structures have already attracted the interest of researchers due to its possibility in designing novel electronic [1–3] and optoelectronic [4–6] devices which have been physically realizable due to the advancement of growth techniques [7, 8]. Quantum well, wire and dot are the new devices where carrier motion is restricted along one, two or three dimensions so that it becomes comparable with de- Broglie wavelength, and it leads to the quantization of energy levels. Different geometrical variations of quantum dots are already proposed for computation of eigenstates for specific application purposes [9, 10], among which the ring and disk-shaped dots attract special attention because they are proposed to be utilized in designing Qubit [11], to realize the dream of quantum computers [12].

A ring-shaped quantum dot may be termed as quantum ring, which can also be considered as a "hole" created at the middle of a circular quantum dot; whereas a solid cylindrical-shaped one-dimensional quantum dot may be termed as quantum disk. This particular class of circular quantum dot attracted quantum physicists due to the possibility of experimental observations of the Aharonov-Bohm effect [13].

Kish et al. [14] demonstrated the candidature of heterostrucutre quantum ring as laser with good current confinement and photon confinement. Han et al. [15] experimentally investigated the same with square geometry. Filikhin et al. [16] calculated the quantized energy levels of semiconductor quantum ring using non-linear Schrödinger equation with the consideration of energy-dependence of effective mass. Llorens et al. [17] also calculated electronic states of a quantum ring under lateral electric field and compared it with the results of a quantum disk. Energy level of the disk is also computed by Lamouche and Lépine [18] in absence of any field. Peeters and Schweigert [19] calculated the same in presence of magnetic field. Hassanien et al. [20] calculated the same for parabolic potential profile in presence and absence of electric field. Kikuchi et al. [21] fabricated heterostructure quantum disk for designing novel LED. Susa [22] theoretically established that quantum disk can be utilized in designing distributed feedback laser with very precise control of threshold current density.

Knowledge of optical transition between two subbands can be used in specialized applications e.g., non-invasive medical diagnosis, environment monitoring etc. It can also be utilized to monitor and control food quality. The characteristic wavelength of a few viruses generally lies in the IR range; hence its identification can only be possible with its proper matching with the radiating/absorbing wavelength of similar range from a device. The most important feature of these applications is the characteristic wavelength due to the intersubband transition energy, originated due to the transition form a particular quantum state to another quantum state. This, in turn, depends on quantized states; which are function of structural parameters of the device. Addition of field in precise direction will add flexibility to the purpose as it can help to tune the emitted/detected wavelength in either of blueshift/redshift directions. Hence computation of intersubband transition plays the vital role in determining signature of the virus.

In this paper, characteristic wavelength range of two different circular quantum dots is analytically computed from the result of intersubband transition energy. Eigenstates are obtained by solving time-independent Schrödinger equation. Electric field is applied along the axis of the structure, and radius as well as thickness is varied to observe the effect on lowest three quantization states. Quantum ring and quantum disk are considered for vis-à-vis comparative analysis of eigenstates with variation of structural parameters, and also with electric field up to a moderate range. Dimensions and field are chosen in such a way transition energies lay in IR region, so that signature of virus fall inside that range can suitably be traced out.

2 Mathematical Modeling

2.1 Calculating Eigenstate of Quantum Disk

We consider the schematic structure of a cylindrical quantum disk of radius '*b*' and thickness '*d*' as shown in Fig. 1a under external electric field (*F*) applied perpendicular to the plane of the disk. Using the cylindrical co-ordinate system, the time-independent Schrödinger equation for the electron wavefunction Ψ can be written as

$$-\frac{\hbar^2}{2m^*}\left[\frac{1}{\rho}\frac{\partial}{\partial\rho}\left(\rho\frac{\partial\Psi}{\partial\rho}\right)+\frac{1}{\rho^2}\frac{\partial^2\Psi}{\partial\rho^2}+\frac{\partial^2\Psi}{\partial z^2}\right]-eF\Psi=E\Psi \tag{1}$$

where $\rho \leq b$ and $0 \leq z \leq d$, *F* being the externally applied electric field, *E* is the energy eigenvalue, m^* is the effective mass of electron, *e* the electronic charge and $\hbar$ the Dirac constant. Here we assume parabolic band structure.

The left hand side of Eq. (2) may be conceived as effective Hamiltonian operator ($\hat{H}$) acting on wavefunction ψ. We can break up the Hamiltonian operator ($\hat{H}$) into two parts as

$$\hat{H}=\hat{H}_0+\hat{H}' \tag{2}$$

The above equation satisfies the condition

$$\hat{H}_0\Psi_0=E_0\Psi_0 \tag{3}$$

where

$$\widehat{H}_0=-\frac{\hbar^2}{2m^*}\left[\frac{1}{\rho}\frac{\partial}{\partial\rho}\left(\rho\frac{\partial\Psi}{\partial\rho}\right)+\frac{1}{\rho^2}\frac{\partial^2\Psi}{\partial\rho^2}+\frac{\partial^2\Psi}{\partial z^2}\right] \tag{4}$$

where E_0 and ψ_0 are the unperturbed values of energy eigenvalue and the wavefunction respectively.

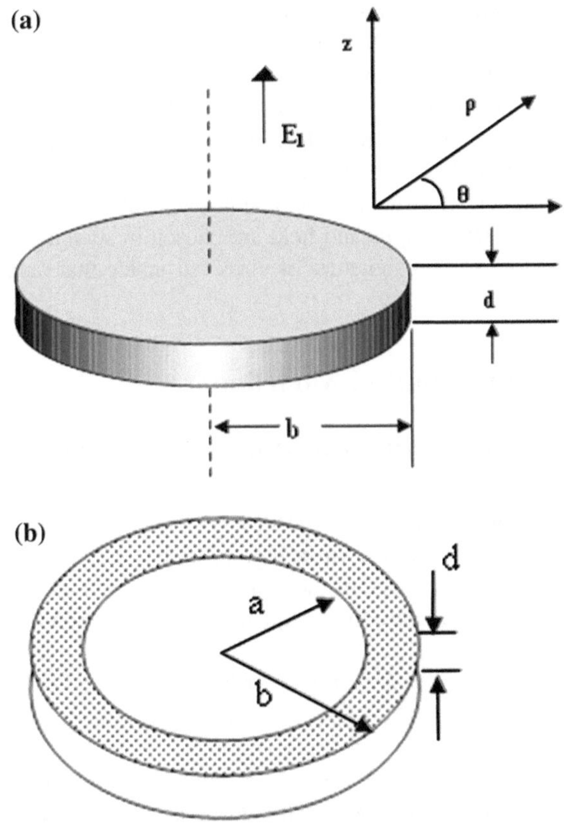

Fig. 1 **a** Schematic structure of a cylindrical quantum disk, **b** Cylindrical quantum ring

Choosing Ψ_0 = R(ρ) Θ (θ)Z(z) and applying the method of separation of variables, the θ-dependent component of the wavefunction Θ has solution given by

$$\Theta \sim \exp(\pm im\theta) \tag{5}$$

for m = 0, 1, 2,…

The ρ-dependent component of the wavefunction, R, can be determined by solving the differential equation

$$\rho^2 \frac{\partial^2 R}{\partial \rho^2} + \frac{\partial R}{\partial \rho} + (\lambda^2 \rho^2 - m^2)R = 0 \tag{6}$$

for 0 ≤ ρ ≤ b, where

$$\lambda^2 = \frac{2\,m^*(E_0 - E_n)}{\hbar^2} \tag{7}$$

E_n is the quantized energy due to motion in the z-direction. The general solution of Eq. (7) is obtained in terms of m-th order Bessel functions $J_m(\lambda)$ and $Y_m(\lambda)$ of first kind and second kind respectively. With the application of boundary conditions, $R = 0$ at $\rho = 0$ ($\rho = b$), it can be seen that only some discrete values of λ are allowed. These discrete values can be obtained by solving the following determinants

$$\begin{vmatrix} J_m(\lambda_1) & Y_m(\lambda_1) \\ J_m(\lambda_2) & Y_m(\lambda_2) \end{vmatrix} = 0 \tag{8}$$

where $\lambda_1(\lambda_2)$ corresponds to the value of λ at $\rho = 0$ ($\rho = b$).

The solution of (9) requires two quantum numbers, m and l (say), which indicates the l-th zero (solution) of the m-th order Bessel function. Thus, the unperturbed energy eigenvalue E_0 can be obtained using Eq. (8) and, the energy eigenvalue is quantized with three quantum numbers (m, l, n).

The exact eigenvalue can be approximately derived first by operating $\hat{H}$ on the wavefunction Ψ_0 and subsequently by using the relation

$$E' = E_0 + H'_{ss} \tag{9}$$

where

$$H'_{ss} = \frac{\int_V \Psi^*_{0s} \hat{H}' \Psi_{0s}\, dV}{\int_V \Psi^*_{0s} \Psi_{0s}\, dV} \tag{10}$$

Ψ^*_{0s} being the complex conjugate of the unperturbed wavefunction Ψ_{0s} of state ‘s’. Using various substitutions and simplifications, the energy eigenvalue of electron in the disk is written as

$$E_{zn} = \frac{1}{2}\left[\frac{\frac{\hbar^2}{2m^*}\left(\frac{n\pi}{d}\right)^2 - \frac{eFd}{2}}{+\sqrt{\frac{\hbar^2}{2m^*}\left(\frac{n\pi}{d}\right)^2\left\{\frac{\hbar^2}{2m^*}\left(\frac{n\pi}{d}\right)^2 - eFd\right\}}}\right] \tag{11}$$

2.2 Calculating Eigenstate of Quantum Ring

We consider the schematic structure of the cylindrical quantum ring of inner radius ‘a’, outer radius ‘b’ and thickness ‘d’ as shown in Fig. 1b. Using the cylindrical coordinate system, the time-independent Schrödinger equation for the electron wavefunction Ψ is same as Eq. (1) where the domain is changed as $a \le \rho \le b$ and $0 \le z \le d$.

The final expression of energy eigenvalue of quantum ring with approximation of not very large F is given by

$$E_{lmn} = \frac{\hbar^2}{2m^*}\lambda_{ml}^2 + \frac{1}{2}\left[\begin{array}{l} \frac{\hbar^2}{2m^*}\left(\frac{n\pi}{d}\right)^2 - \frac{eFd}{2} \\ + \sqrt{\frac{\hbar^2}{2m^*}\left(\frac{n\pi}{d}\right)^2\left\{\frac{\hbar^2}{2m^*}\left(\frac{n\pi}{d}\right)^2 - eFd\right\}} \end{array}\right] \tag{12}$$

3 Results and Discussion

Using Eqs. (11) and (12), the energy eigenvalues of an n-GaAs quantum disk and ring are computed as function of different structural parameters and also of external electric field.

Figure 2 shows the variation of eigenenergies for the lowest three states with outer diameter of the disk in presence and absence of electric field. From the plot, it is observed that by increasing the diameter of the disk, energy monotonically decreases. This is due to the fact that with increase in dimension, quantum confinement decreases, which lowers the eigenenergy. The rate of decrement is significant with lower diameter, whereas the rate reduces with higher size of the disk. Application of the axial electric field further lowers the eigenvalue for same dimension. Similar feature may be observed for quantum ring when plotted with

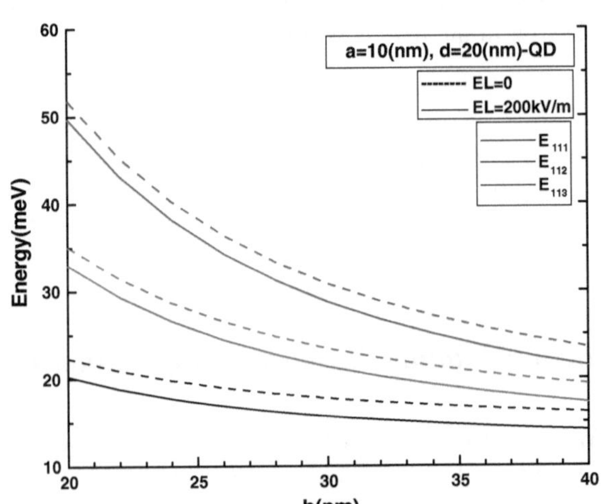

Fig. 2 Lowest three energy states with diameter of quantum disk in presence and absence of electric field

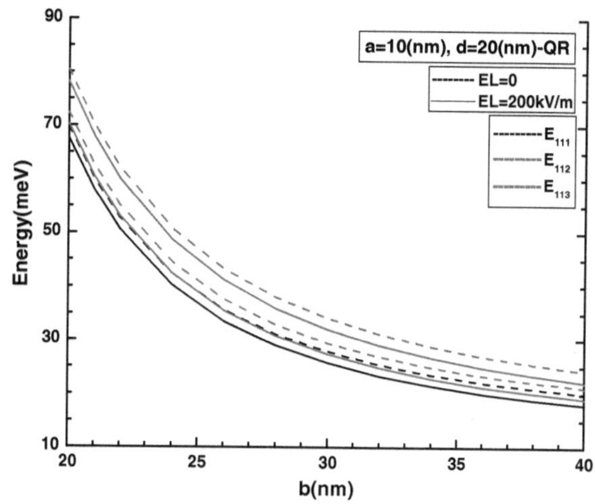

Fig. 3 Lowest three energy states with outer diameter of quantum ring

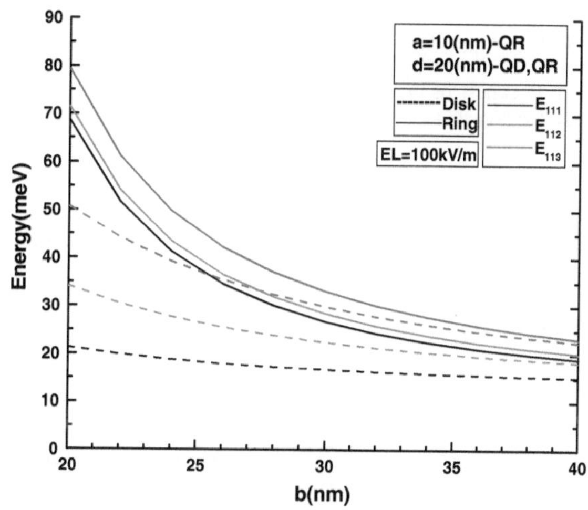

Fig. 4 Comparative study of lowest three energy states with outer diameter of the ring and disk in presence of electric field

thickness, as shown in Fig. 3. The notable feature observed from the plot is that the curvature of graph for any eigenstate remains same in absence of field compared to the plot generated in presence of field.

Figure 4 shows the comparative analysis for lowest three eigenstates with outer diameter for quantum disk and ring having similar dimensions in presence of axial electric field. It may be seen form the result that higher magnitude of eigenstates can be obtained form quantum ring than that of the disk. The reason behind the difference in magnitude for any eigenstate is that the ring possesses air dielectric in the core region which has lower dielectric constant than the solid GaAs material present in the disk. By virtue of the heterostructure formed between air and semiconductor, quantum confinement becomes higher for the ring than that of the disk. The

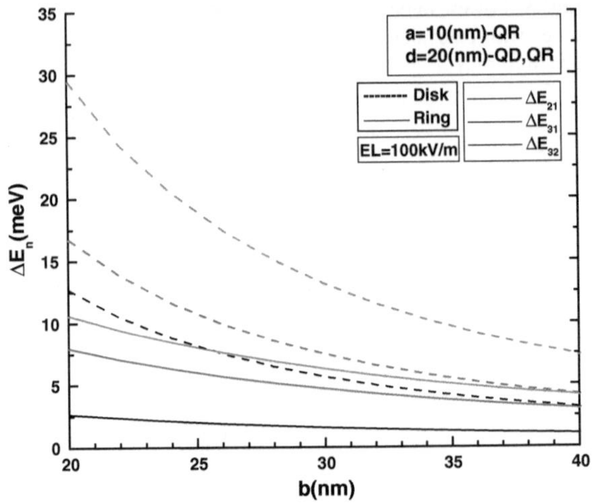

Fig. 5 Comparative study of intersubband transition energy with outer diameter of the ring and disk in presence of electric field

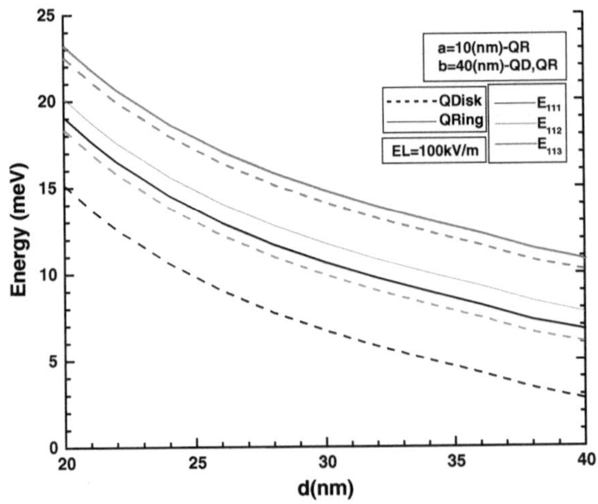

Fig. 6 Comparative study of lowest three energy states with thickness of the ring and disk for constant electric field

difference is significant for lower diameter, but the gap decreases with higher dimension. Corresponding intersubband transition energy is plotted in Fig. 5.

From the plot, it is observed that transition energy variation is significant for quantum ring than that of quantum disk, which speaks about the fact that quantum ring can be used as a better optoelectronic transmitter than that of disk with similar shape and size in presence of external electric field.

Similar comparative analysis is made between ring and disk with thickness of the structure, shown in Fig. 6. From the plot, it is observed that with increase of thickness, energy monotonically decreases. The rate is nonlinear for very lower thickness value, but it becomes almost linear thereafter. The physics behind the nature of energy profile is identical to the previous explanation, and so is not

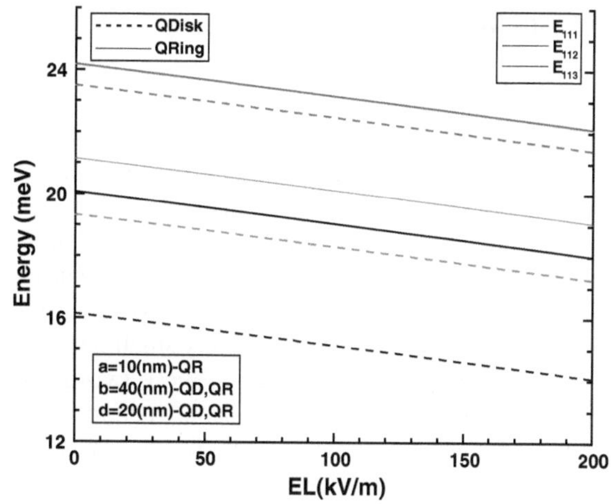

Fig. 7 Comparative study of lowest three energy states with electric field in the n-GaAs quantum ring and disk

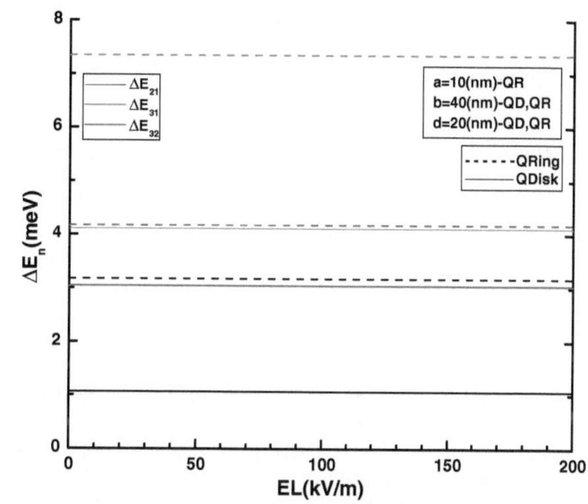

Fig. 8 Comparative study of intersubband transition energy with electric field in the n-GaAs quantum ring and disk

mentioned again to avoid the repetition. Hence intersubband transition energies calculated for the lowest three states become almost constant throughout the range of interest which indicates the fact that tuning is not possible in this case.

Longitudinal electric field adds further dimension for tuning of energy for both the structures. By increasing the strength of field, it may be seen that energy decreases monotonically, as shown in Fig. 7. The rate of reduction is almost linear in low-to-medium field range. Thus eigenstates of circular quantum dots can be externally tuned by applying suitable electric field. Variation of intersubband energy is plotted in Fig. 8. From the figure, it may be concluded that transition energy slowly increases with increase of electric field, which helps electrical tuning of optoelectronic properties for lasing applications.

4 Conclusion

Characteristic wavelength range of quantum ring and quantum disk are determined by analytically computing intersubband transition energies. Different structural parameters are varied along with low-to-moderate axial electric field to calculate the range. Comparative studies are made considering lowest three eigenstates of the two quantum dots of similar size subject to similar external conditions, which reveals the fact that quantum ring provides higher magnitude of eigenenergies than quantum disk. Thus intersubband transition energy can be better varied in case of quantum ring than the disk over a very small range of dimensional change. Effect of electrical tuning is quite similar on these two quantum dots within our choice of interest. Results help to design nanostructure optical detector in required wavelength range for detecting signature of virus which emits characteristic wavelength within the region of choice. Appropriate tuning of dimensions of the structure and axial electric field helps to find the exact signatory wavelength of the virus.

References

1. Zhuang L, Guo L, Chou SY (1998) Silicon single-electron quantum-dot transistor switch operating at room temperature. Appl Phys Lett 72:1205–1207
2. Pigorsch C, Wegscheider W, Klix W, Stenzel R (1997) 3D-Simulation of novel quantum wire transistor. Phys Status Solidi B 204:346–349
3. Urban D, Braun M, König J (2007) Theory of a magnetically controlled quantum-dot spin transistor. Phys Rev B 76:125306
4. Cui D, Jian X, Sheng-Yong X, Paradee G, Lewis BA, Gerhold MD (2006) Infrared photodiode based on colloidal PbSe nanocrystal quantum dots. IEEE Trans Nanotechnol 5:362–367
5. Gvozdic DM, Schlachetzki A (2005) Modulation response of V-groove quantum-wire lasers. IEEE J Quantum Electron 41:842–847
6. Kunz M et al (2005) High-speed quantum dot lasers and amplifiers for optical data communication. Appl Phys A 80:1179–1182
7. Somaschini C, Bietti S, Sanguinetti S, Koguchi N, Fedorov A (2010) Self-assembled GaAs/AlGaAs coupled quantum ring-disk structures by droplet epitaxy. Nanotechnology 21:125601
8. Kuramochi E, Temmyo J, Kamada H, Tamamura T (1999) Spatial ordering of self-organized InGaAs/AlGaAas quantum disks on GaAs (311)B substrates. J Electron Mater 28:445–451
9. Li Y, Voskoboynikov O, Lee CP, Sze SM (2001) Electron energy state dependence on the shape and size of semiconductor quantum dots. J Appl Phys 90:6416–6420
10. Li Y, Voskoboynikov O, Lee CP, Sze SM (2001) Computer simulation of electron energy levels for different shape InAs/GaAs semiconductor quantum dots. Comput Phys Commun 141:66–72
11. Dong QR, Li SS, Niu ZCh, Feng SL, Zheng HZ (2004) Electronic structure of self-assembled InAs quantum disks in an axial magnetic field and two-electron quantum-disk qubit. J Appl Phys 96:3277–3281
12. Loss D, DiVincenzo DP (1998) Quantum computation with quantum dots. Phys Rev A 57:120–126
13. Mühle A, Wegscheider W, Haug RJ (2007) Coupling in concentric double quantum rings. Appl Phys Lett 91:133116

14. Kish FA, Caracci SJ, Maranowski SA, Holonyak N, Smith SC, Burnham RD (1992) Planar native-oxide $Al_xGa_{1-x}As$-GaAs quantum well heterostructure ring laser diodes. Appl Phys Lett 60:1582–1584
15. Han H, Forbes DV, Coleman JJ (1995) InGaAs-AlGaAs-GaAs strained-layer quantum-well heterostructure square ring lasers. IEEE J Quantum Electron 31:1994–1997
16. Filikhin I, Vlahovic B, Deyneka E (2006) Modeling of InAs/GaAs self-assembled heterostructures: quantum dot to quantum ring transformation. J Vac Sci Technol A: Vac Surf Films 24:1249–1251
17. Llorens JM, Trallero-Giner C, García-Cristóbal A, Cantarero A (2002) Energy levels of a quantum ring in a lateral electric field. Microelectron J 33:355–359
18. Lamouche G, Lépine Y (1995) Ground state of a quantum disk by the effective-index method. Phys Rev B 51:1950–1953
19. Peeters FM, Schweigert VA (1996) Two-electron quantum disks. Phys Rev B 53:1468–1474
20. Hassanien HH, Abdelmoly SS, Elmeshad N (2006) Exact solution of finite parabolic potential disk-like quantum dot with and without electric field. FIZIKA-A 15:209–218
21. Kikuchi A, Kawai M, Tada M, Kishino K (2004) InGaN/GaN Multiple quantum disk nanocolumn light-emitting diodes grown on <111> Si substrate. Jpn J Appl Phys 43:L1524–L1526
22. Susa N (1998) Feasibility study on the application of the quantum disk to the gain-coupled distributed feedback laser. IEEE J Quantum Electron 34:1317–1324

Methodology for a Low-Cost Vision-Based Rehabilitation System for Stroke Patients

Arpita Ray Sarkar, Goutam Sanyal and Somajyoti Majumder

Abstract Stroke is a life threatening phenomenon throughout the world caused due to the blockage (by clots) or bursting of arteries. As a result permanent or semi-permanent neurological damage may occur that requires proper rehabilitation to overcome the deficiency of communication or communication disorder of the stroke patient with the outer world. This causes delay in overall recovery and affects the general hygiene of the patient. Computer vision based interaction using gazes may be helpful for such cases. In all those methodologies as a mandatory step eye tracking has to be performed. Present work uses a low cost web camera for eye tracking using Haar feature-based cascade function in comparison with the costlier eye tracking systems available in the market. This method easily detects the eye balls from the video online with less computational load. Several experiments have been carried out to evaluate the performance in different background, lighting conditions and quality of images.

Keywords Computer vision · Object detection · Face recognition · Patient rehabilitation

A.R. Sarkar (✉) · G. Sanyal
Department of Computer Science and Engineering, National Institute of Technology, Durgapur, India
e-mail: sarkartrishna@gmail.com

G. Sanyal
e-mail: nitsanyal@gmail.com

S. Majumder
Surface Robotics Lab, CSIR—Central Mechanical Engineering Research Institute, Durgapur, India
e-mail: sjm@cmeri.res.in

S. Gupta et al. (eds.), *Advancements of Medical Electronics*,
Lecture Notes in Bioengineering, DOI 10.1007/978-81-322-2256-9_34

1 Introduction

The term stroke is well known to the world, especially to the aged people, as it is the second common cause of death throughout the world [1]. Stroke is caused due to bursting or blockage of arteries by clots [2]. Stroke leads to malfunction of the brain as the arteries that carry oxygen and nutrients are damaged. As a result, either death or permanent neurological damage will occur based on the severity and location of the damage. Research has proved that different parts of the brain are responsible for different sensory and motor functions. Damage to a particular location will cause a disability in respective function. Patient may lead almost normal life after undergoing some therapies, if the stroke is not severe. But constant support from the family and friends and intensive rehabilitation by the healthcare professionals are required, if the stroke is severe and the disability is permanent. Stroke affects physical, cognitive and emotional functioning. The followings are the most common after-effects of stroke found in the patients [2].

- Vascular Dementia: Loss of thinking ability
- Aphasia: A communication disorder
- Memory: Short-term or long-term memory loss
- Depression: Biological, behavioural or social factors cause such depression
- Pseudobulbar Affect (PBA): A medical condition which causes sudden crying or laughing

All the above disabilities in terms lead to cause lack of communication or communication—disorder between the patient and outer world including family members, attending healthcare professionals and doctor. Lack or limited communication delays the recovery of the patient as well as affects the general hygiene of the patient.

This situation can be overcome by introducing an automatic rehabilitation device using computer vision-based system which will provide 24 h constant support. Such system may be operated by the patient either through gazes or gestures, if possible or through only nodding head or moving the iris. The input system for such devices can be customized based on the type and degree of disability of the patient.

2 State-of-the-Art

Each year approximately 20 million people are affected by stroke and out of them only 15 million people survive [3]. These lucky 15 million people may need a communication system for rehabilitation. Several research organizations, academic institutions, professional companies are engaged in developing low-cost rehabilitation systems for stroke patients. Most of these systems use mainly vision-based methodology. Some other devices use robotic systems or tactile sensing. Successful attempts have been made to use smart phones, tablets and other similar gadgets to implement those rehabilitation systems. Jack et al. has developed a desktop PC based virtual reality system for rehabilitating hand function for the stroke patients [4]. The proposed system uses a Cyber Glove and a Rutgers Master II-ND (RMII) force feedback gloves as input devices. An efficient gesture based system for impaired dexterity has been developed by Ushaw et al. using a tablet [5]. Sucar et al. has presented the development of a cheaper vision-based system for intensive movement training [6]. Reinkensmeyer et al. have proposed a web-based rehabilitation system that eliminates the always-present therapist and it consists of web-based library of status test, several therapy games as well as progress charts [7].

As already mentioned robotic therapy is also popular for rehabilitation for chronic stroke patients [8–10]. MANUS is such a specially designed robot developed by MIT being used for rehabilitation of stroke patients [8]. Another portable, low-cost robotic system has been developed by Huq et al. for post-stroke upper limb rehabilitation [9, 10]. Most of these robots are commonly used in place of a human therapist. They use easy and interactive GUI for both manual and automatic control. Some of them have force feedback options for better usage.

University of Michigan has successfully developed different support systems using iPhone and iPad to overcome various types of communication challenges for the people with aphasia [11]. These support systems include text-to-speech converter, talking picture dictionary, phonemic cues, speech-to-text converter, video calls, talking dictionary system etc. and these can be easily installed on smart phones and tablets. Microsoft Kinect sensor is well known for its use in gaming and equipped with a depth sensor and camera. Research team from Roke Manor Research Ltd in association with Southampton University has developed a gesture recognition software using Microsoft Kinect sensor for supervision of patients with stroke by a physiotherapy clinic over the internet [12].

As the first step, most of the above mentioned devices/systems use suitable eye detection and tracking technology. Some researchers are using commercially available eye tracker system for the purpose and developing own (software) modules to upgrade the performance [13]. A few of the popular commercial eye trackers are Mirametrix S2 eye tracker [14], EyeTech VT2 [15], Gaze Point GP3 desktop eye tracker [16], Grinbath Eye Guide Mobile Tracker System [17], SMI eye tracking glasses [18], Tobii eye tracking glasses [19]. These systems include both hardware and software as a single package. The approximate costs (excluding shipping and handling charges, taxes) of the popular eye tracking systems are presented in the Table 1.

Every year, it has been reported 90–222 cases of stroke per 100,000 in India [1]. Such costlier systems for rehabilitation purpose are difficult to afford by the common people. If a low-cost system can be designed and developed with the available cheapest components/technologies, a large number people will be benefitted. At first stage, an efficient eye tracking system has been developed that will help the stroke patient to indicate his/her requirements through the icons displayed on a

Table 1 Table styles

Popular commercial eye tracking systems	Price
Mirametrix S2 eye tracker	$4,995
EyeTech VT2	$5,000
Gaze point GP3 eye tracker with gazepoint analysis	$1,395
Grinbath complete eye guide® mobile tracker system	$8,999
SMI eye tracking glasses 2.0	$30,800
Tobii eye tracking glasses	$15,800

wall-mounted monitor/TV/display placed in front of the patient to his/her family members/attendant through gazes/gestures. Initially a low resolution camera of US$ 8 was used and subsequently a high resolution HD camera of approximate US$ 50 [INR 3000] and a desktop PC of approximate US$ 650 [INR 40000] have been used for implementation.

3 Concept Model

The proposed system will be used for rehabilitation purpose of the stroke patients. It will consist of a display unit/monitor or presently available high end television placed in front of the patient at suitable location considering the lighting condition and clear view of the face of the user. Several large size icons will be displayed on the screen. These icons may include 'Call the nurse', 'Self-guide for Physical Exercises', 'Read Books', 'Listen Music', 'Watch Movies', 'Ask for Water', 'Ask

Fig. 1 The concept model of the proposed system with the patient lying halfway on the bed along with the display and camera unit placed on the wall at the front

for Food' etc. The patient needs to locate the desired icon using gazes and if necessary follow the onscreen instructions. A high definition camera will be installed at a fixed location to track the eye balls from a specified distance. Night vision camera is preferred to other normal cameras for viewing even in the dark. Otherwise proper illumination has to be arranged especially during the night. A processing unit will be employed for capturing, storing and processing of images. Windows based processors are preferred due to use of Microsoft Visual Studio 2013 and Open CV software. Initially a desktop PC has been recommended. But with the advancement of technologies small processor or low-cost single board computer or any other windows based gadgets can be used. As shown in Fig. 1 the patient may completely lie down on the bed or may lie down halfway on the bed and the system will work from a distance. The proposed concept model has been presented in Fig. 1.

4 Adopted Methodology

The objective of the present work is to detect the eyeball of the user and track it for activating certain commands displayed on the computer screen placed in front of a stroke patient. To achieve the goal a few steps have to be followed properly. Firstly the face has to be detected amidst of cluttered background from the captured image. Then the eye regions are extracted from the face region. Next step is to detect the eyeball centre from the eye region using eye centre localization. Calibration is another important step to be followed to identify the exact position of the eye centre on the camera window with reference to the interface screen. This will help to select the icons to activate specific commands and proceed accordingly. The generalised system architecture of the proposed system has been shown in Fig. 2.

Till date detection of face region and detection of eye region from the detected face region have been achieved. This method is preferred to detection of eye regions directly as most of the available cheap cameras (webcam) are not of higher resolution. The major disadvantages are that the captured images are blurry and the

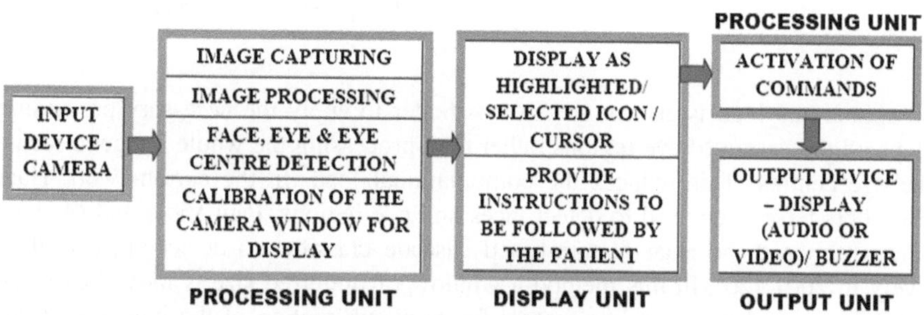

Fig. 2 Generalized system architecture for proposed system

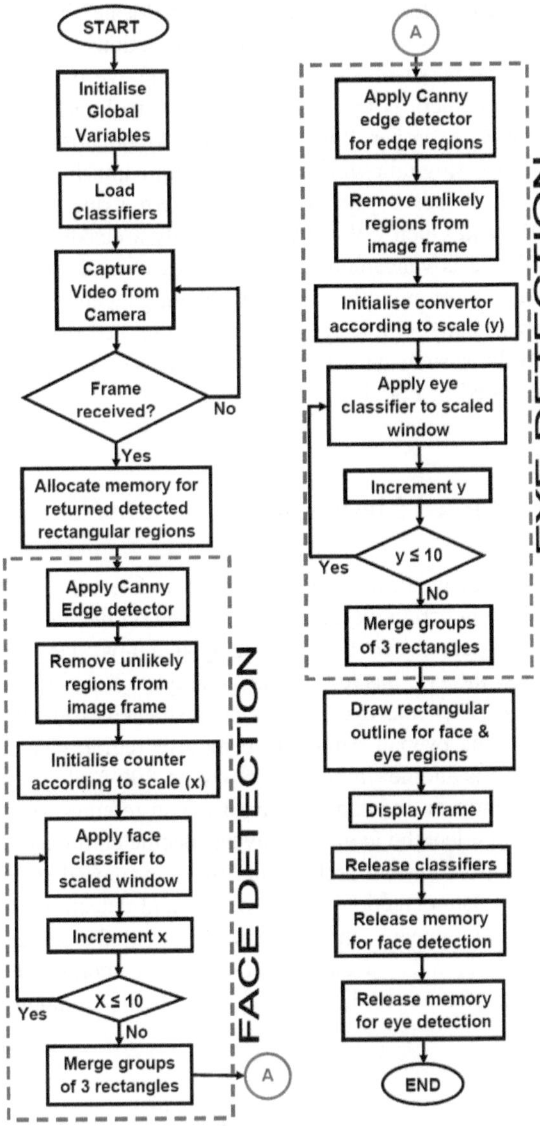

Fig. 3 The flow diagram of the proposed methodology for face detection and eye detection for stroke patients

computational load is also more. So it is better to apply the necessary processing steps on the extracted eye regions rather than processing the whole image to locate the eye centres. This reduces the computational load. In the current work Haar classifiers have been used to detect faces and eye regions. Paul Viola and Michael Jones proposed the Haar feature-based cascade classifier to detect objects effectively in 2001 [20]. In this method a window of preferred size is moved over the input image and the Haar-like feature for each sub-section of the image is calculated. This value is compared to a predefined threshold that differentiates objects from non-objects. This is a machine learning technique where the function is trained using positive (contains object) and negative (contains non-object) images.

Microsoft Visual Studio 2013 with Open CV libraries has been used as the programming platform. Main advantages of Open CV are that it is easily available and it contains many pre-trained classifiers for detection of face, eye, smile etc. The captured video by the camera is loaded frame by frame and the desired xml files are loaded and applied to each of these frames. The three consecutive detections are merged as one. Pruning is done to exclude the region of less interest i.e. where chance of appearance of face is very less. The smallest window to detect face can be defined based on the need or application. Similar approach is followed to detect eyes. The detected region is then marked by a rectangular box.

The final step is to detect eye centre from the extracted images of eye regions. Several methods are available. A few of them are colour based filtering, shape based filtering, use of Hough circles. Another novel method is to train classifiers for eyeballs like that of faces and eye regions. Presently colour based filtering has been used to detect eye centres. The schematic diagram of the above methods has been presented in Fig. 3.

5 Experiments

Image capturing and processing are the two fundamental steps associated with the present work. The quality of the image plays a vital role in this case. There are several factors those affect the quality. A few of them are type of the camera (CCD or CMOS), resolution of the camera, illumination of the environment, type and position of light, distance between the camera and object etc. These factors should be taken care of to capture a good quality image leading to successful detection.

The experiments have been carried out in normal room (dimension 12 Ft × 12 Ft) as the application of the proposed system will be limited to indoor only i.e. in home or hospital. The isometric view of the experimentation room has been presented above in Fig. 4. There are two doors, one window and one fluorescent light in the room. Though the application of the system will be for the patients either completely lying or lying halfway on the bed, here for the ease of initial experiments user sits on the chair with the window at the back. The display unit along with the camera is placed on a table at a distance of 2 Ft in front of the user. Only a fluorescent light has been used during the experiments in night.

Two different types of USB cameras have been used. Initially a low cost CMOS camera with 300 K pixels resolution (Make: Frontech, Model: JIL 2244) [21] has been used and that has been replaced later by a 720 pixel HD camera (Make: Logitech, Model: HD C525) [22] with zoom and auto focus. The quality of images captured by these two cameras are presented in Fig. 5a, b. Simultaneously a lux-meter has been employed on a plane co-planner with the face to measure the average illumination of the surroundings. The data have also been recorded for reference.

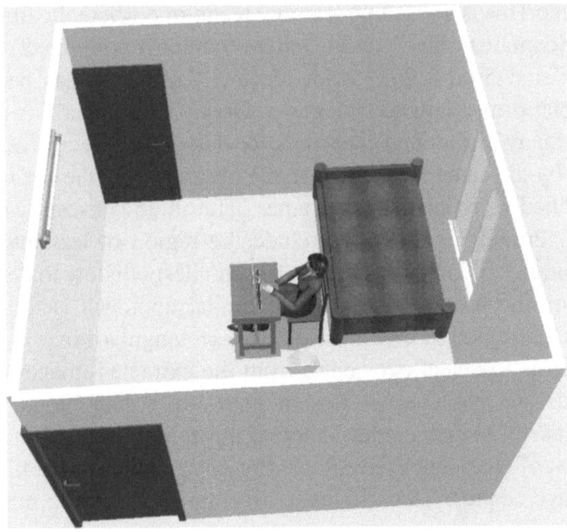

Fig. 4 Isometric CAD view of the experimental system showing that the user sitting in the middle of a room with the display and camera unit in front

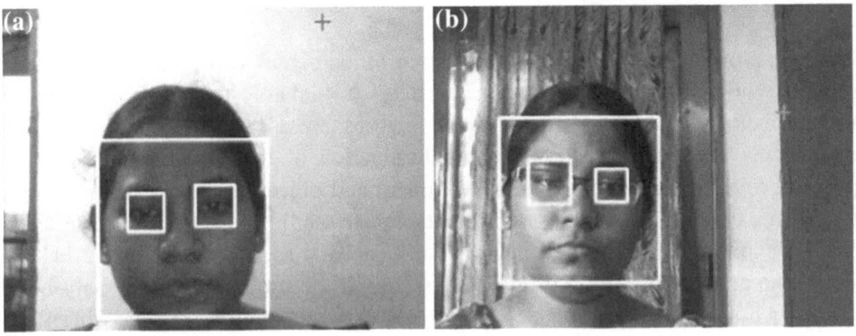

Fig. 5 The online face and eye detection as well as marking from the frame captured by **a** the low resolution camera without spectacles, **b** the high resolution camera of the user with spectacles at same location in different orientation

Continuous videos have been recorded using both the cameras in both day and night with different position of the user and lighting conditions. The programme also constantly detects and marks faces and eyes automatically online. As the user moves the eyeballs, the eye centres are detected and marked online. Both the cases with spectacles and without spectacles have been considered during the experiments. The saved videos and associated data have been analysed and presented in the tables below. The relevant processed data for the low resolution camera with and without spectacles have been given in Tables 2 and 3 presents the processed data for high resolution camera.

Table 2 Video data from the low resolution camera

Total frames	Correct detections	Semi-correct detection
Without spectacles		
1,157	1,110	1,149
1,113	1,025	1,110
1,158	1,143	1,155
1,043	688	1,026
With spectacles		
1,043	473	819
1,158	830	1,064
1,158	609	1,030
1,149	851	1,017

Table 3 Video data from the high resolution camera

Total frames	Correct detections	Semi-correct detection
Without spectacles		
1,161	1,036	1,146
1,042	941	997
1,139	1,080	1,138
1,157	1,152	1,153
With spectacles		
862	715	856
810	717	802
1,158	961	1,143
832	611	821

6 Results and Discussion

The efficiency of detection is calculated for the above mentioned data with reference to the total number of detected frames. Thus the efficiencies of detection for the low resolution camera with and without spectacles have been shown in Fig. 6a, b respectively. Similarly Fig. 7a, b present the efficiencies of detection for the high resolution camera with and without spectacles respectively.

Here the following guideline has been followed to consider the correct detection of eyeballs. The main aim is to detect correctly the eyeballs of the opened eyes. So when the eyes are closed there should be no detection or in other sense if the eyes are closed and no detection is there, it has been considered as correct detection. However detection has also been assumed or manipulated by tracking the single eye and knowing the distance between the centres of the two eyeballs for common normal people. This may fail for squint-eyed people.

The efficiency of correct detection has been denoted by the solid red line and the dotted blue lines represent the semi-correct detection that includes the detection of single eye, double eyes and closed eyes. The performance of the low resolution

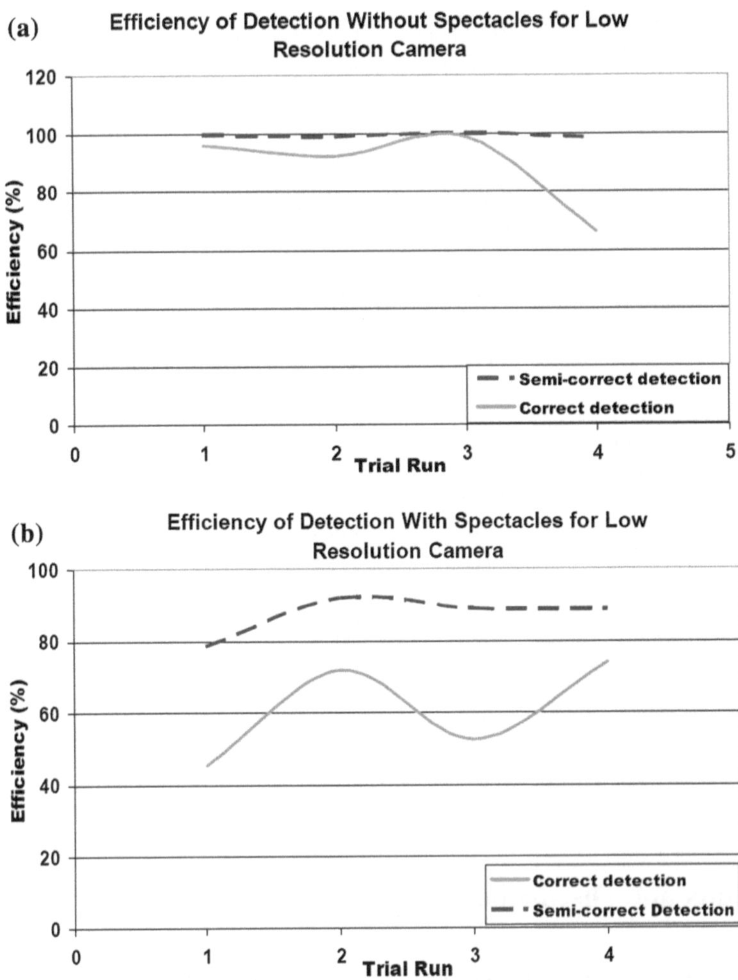

Fig. 6 Efficiency of detection for low resolution camera. **a** Without spectacles. **b** With spectacles

camera without spectacles is found better than that of with spectacles. All the values except one are lying above the range of 80 %. The efficiencies of detection for the low resolution camera with spectacles lie between 40 and 75 % which seems to be very poor for the intended application. This might happened due to improper illumination during the experimentation in the night time.

The efficiencies of detection for the high resolution camera without spectacles always lie above 80 %. This has been achieved due to no glare of the spectacles and experiments are carried out in uniform day light. Also the efficiencies of detection for the high resolution camera with spectacles have been found to be present above 80 %. There has been no definite patterns followed in the above graphs and hence no inference can be drawn. This may be due to the fact that each video is different

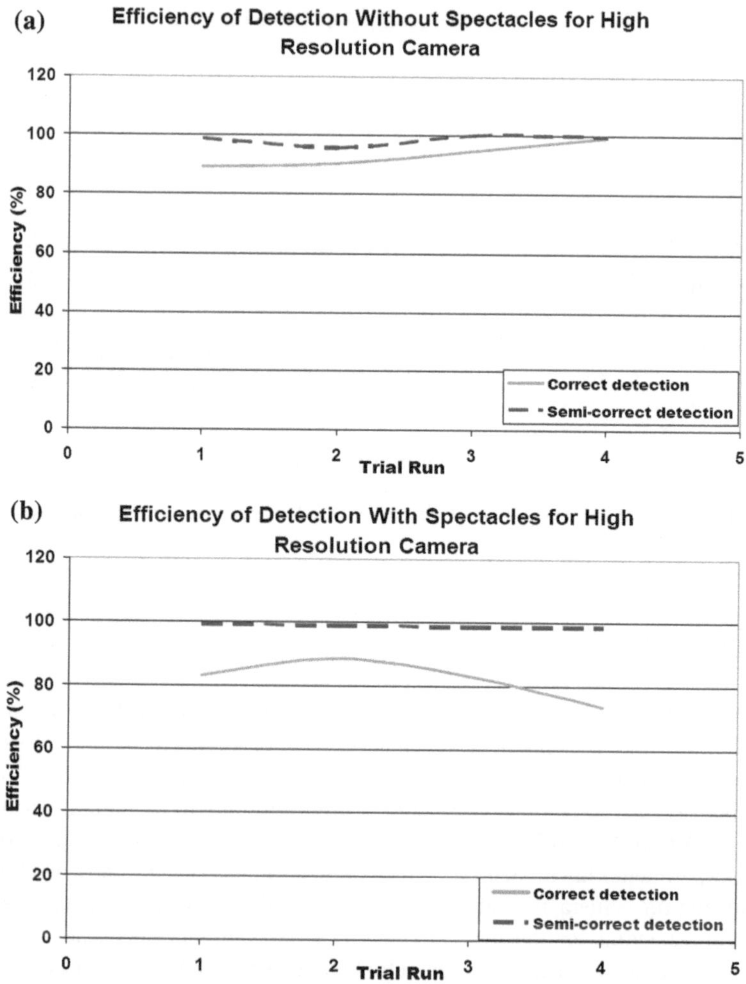

Fig. 7 Efficiency of detection for low resolution camera. **a** Without spectacles. **b** With spectacles

from the other with respect to the orientations of the user, lighting conditions, background etc. However the illumination of the surroundings has a relationship with the efficiency of detection as shown in Fig. 8. The efficiency of detection will increase with the increase in surrounding illumination.

To provide active online support to the stroke patients 100 % efficiency is desirable. There are several reasons for hindrances. The most identified issues are low illumination, low resolution, glaring from the spectacles, incidence of light on the camera, cluttered background etc. These difficulties can be overcome by using high resolution camera, placement of the camera in the direction of light (not opposite), using proper and uniform lighting, clean and clear spectacles (glasses) of the user.

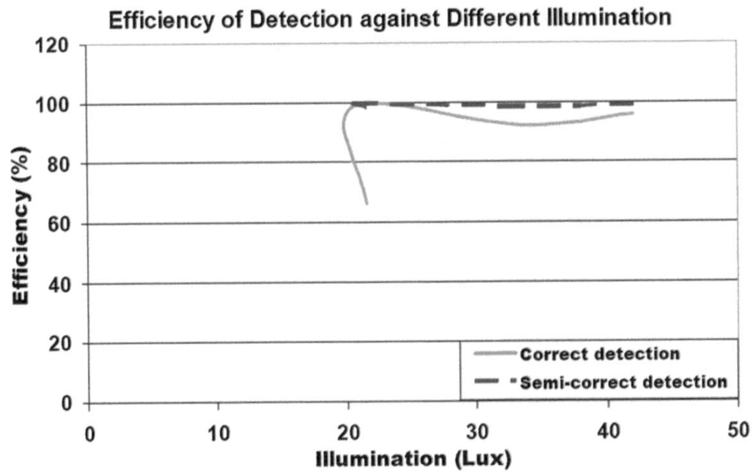

Fig. 8 Efficiency of detection without spectacles for low resolution camera against different environmental illuminations

7 Conclusion

The objective of the present work is to develop a low cost eye tracking system to be used by the stroke patients for rehabilitation. The proposed methodology uses the well known machine learning technique, Haar Classifiers. Haar classifiers provide better performance than other eye tracking algorithm due to its training capability. Microsoft Visual Studio 2013 along with Open CV has been used to carry out the work. Experiments have been performed in a normal living room using low and high resolution camera in both day and night with and without spectacles. Efficiency of detection for the high resolution camera without spectacles has better performance than all others. Several factors are responsible for not achieving 100 % efficiency. Work is in progress to achieve 100 % efficient, robust, rugged system desirable for the rehabilitation of stroke patients [23]. As the future scope, movement of the eyeballs in the camera window is being mapped on the wall-hang monitor or any other suitable display.

Acknowledgments Authors are grateful to the Head and other faculty members of Department of CSE, NIT-Durgapur and SR Lab, CSIR—CMERI Durgapur for their continuous help and tired less support. Authors also thank Dr. D.N. Ray for his continuous suggestions and advices, without which this work would not have completed.

References

1. Taylor FC, Suresh Kumar K (2012) Stroke in India fact sheet (updated 2012)
2. National Stroke Association (2013) Explaining stroke. http://www.stroke.org/site/PageServer?pagename=explainingstroke. Accessed 19 July 2013
3. Dalal P, Bhattacharjee M, Vairale J, Bhat P (2007) UN millennium development goals: can we halt the stroke epidemic in India? Ann Indian Acad Neurol 10(3):130–136
4. Jack D, Boian R, Merians AS, Tremaine M, Burdea GC, Adamovich SV, Recce M, Poizner H (2001) Virtual reality-enhanced stroke rehabilitation. IEEE Trans Neural Syst Rehabil Eng 9 (3):308–318
5. Ushaw G, Ziogas E, Eyre J, Morgan G (2013) An efficient application of gesture recognition from a 2D camera for rehabilitation of patients with impaired dexterity. School of Computing Science Technical Report Series. http://www.cs.ncl.ac.uk/publications/trs/papers/1368.pdf. Accessed 22 Jan 2013
6. Sucar L, Luis R, Leder R, Hernandez J, Sanchez I (2010) Gesture therapy: a vision-based system for upper extremity stroke rehabilitation. In: Proceedings of IEEE engineering medical biology society, 2010, pp 107–111
7. Reinkensmeyer DJ, Pang CT, Nessler JA, Painter CC (2002) Web-based tele rehabilitation for the upper extremity after stroke. IEEE Trans Neural Syst Rehabil Eng 10(2):102–108
8. Fasoli SE, Krebs HI, Stein J, Frontera WR, Hogan N (2003) Effects of robotic therapy on motor impairment and recovery in chronic stroke. Arch Phys Med Rehabil 84:477–482
9. Huq R, Wang R, Lu E, Lacheray H, Mihailidis A (2013) Development of a fuzzy logic based intelligent system for autonomous guidance of poststroke rehabilitation exercise. In: Proceedings of 13th international conference on rehabilitation robotics, 24–26 June 2013, WA
10. Huq R, Lu E, Wang R, Mihailidis A (2012) Development of a portable robot and graphical user interface for haptic rehabilitation exercise. In: Proceedings of 4th IEEE/RAS-EMBS international conference on biomedical robotics and biomechatronics, June 2012, Italy
11. Block M, Mercado L (2013) Talking tech: technology expands communication opportunities for people with aphasia, everyday survival. Springer, Heidelberg, pp 18–19
12. Roke Manor Research Ltd (2013) Microsoft kinect gesture recognition software for stroke patients, Inside Technology, Issue 9. http://www.ttp.com. Accessed 07 June 2013
13. Arrington Research (2013) Eye tracker prices. http://www.arringtonresearch.com/prices.html. Accessed 20 Dec 2013
14. Mirametrix, S2 Eye Tracker (2014) http://mirametrix.com/products/eye-tracker. Accessed 25 Feb 2014
15. iMotions A/S Denmark, Quotation no. 884836000000801041, 21 February 2014
16. Gazepoint Products (2014) http://gazept.com/products. Accessed 15 Feb 2014
17. EyeGuide Mobile Tracking Price (2014) http://www.grinbath.com/content/eyeguider_mobile_tracker_pricing. Accessed 13 June 2014
18. Aerobe Medicare Pvt. Ltd., New Delhi, Quotation no. nil, 07 April 2014
19. Tobii Glasses 2 (2014) http://www.tobii.com/en/eye-tracking-research/global/landingpages/tobii-glasses-2/our-offering. Accessed 25 May 2014
20. Viola P, Jones M (2001) Rapid object detection using a boosted cascade of simple features. In: Proceedings of IEEE conference computer vision and pattern recognition, 2001
21. Product—Frontech (2012) E-brochure. http://www.frontechonline.com/product.php. Accessed 25 May 2012
22. Logitech HD Webcam C525 (2013) http://www.logitech.com/en-us/product/hd-webcam-c525. Accessed 12 Oct 2013
23. Sarkar AR, Sanyal G, Majumder S (2013) Hand gesture recognition systems: a survey. Int J Comput Appl 71(15):25–37

References

Coacervation—A Method for Drug Delivery

Lakshmi Priya Dutta and Mahuya Das

Abstract The present review outlines recent advances in coacervate based research, their historical background and area of diversification. Methods of their preparation, encapsulation, theoretical overview, coacervation induced nano particle formation, applications in various fields have been covered. Chemically modified coacervates used in drug delivery research are discussed critically to evaluate the usefulness of these system in delivering bioactive molecules. From literature survey, it is realized that coacervate based research in drug delivery as well as in proto cellular biology have increased rapidly. Hence the present review is timely.

Keywords Coacervate · Coacervation · Nanomedicine · Biogenesis

1 Introduction

One of the most unsolved, difficult and interesting mystery surrounding the science lies behind the mechanism of life formation via the interaction of bio macro molecules. In relation with this Pasteur have done one experiment on biogenesis [1, 2] which revealed only that the life can not arise spontaneously under conditions that exist on earth today and the pre-biotic environment must be different billions of years ago. In 1929, J.B.S. Haldane presented a brief, paper in The Rationalist Annual that reflects the modern concept on protocell theory according to which life arose on earth as membrane bound system [1–12]. According to them- There is no fundamental difference between a living organism and lifeless matter. The complex combination of manifestations and properties so characteristic of life must have

L.P. Dutta
Department of Nanoscience and Technology, JISCE, Kalyani, Nadia, West Bengal, India

M. Das (✉)
Department of Chemistry, Department of Nanoscience and Technology,
JISCE, Kalyani, Nadia, West Bengal, India
e-mail: d_mahuya@yahoo.com

S. Gupta et al. (eds.), *Advancements of Medical Electronics*,
Lecture Notes in Bioengineering, DOI 10.1007/978-81-322-2256-9_35

arisen in the process of the evolution of matter [4–6]. Oparin have proposed the solution leading to the idea of coacervate which is termed as microspheres of assorted organic bio molecules that are associated by weak interactions [4–10]. All the newer research throughout the world to solve the mystery has been limited towards the evolution of macro molecules like DNA, RNA, protein [1, 3, 4] from the past decades. On the basis of these ideas the interaction between these types of biomacromolecules has broadened the area of investigation starting from the origin of life to the recent advancement in drug delivery [4] including encapsulation.

2 Concept Behind Coacervation

The coacervates are spherical aggregation of organic macro molecules making up a inclusion which is held by hydrophobic forces [1, 3–8]. The boundaries of the coacervates allow the selective adsorption of simple organic molecules from the surrounding medium [13]. The drop that forms the coacervate particle is often a liquid oil and the drops will have a shell of the hydro colloid mixture around them and under correct conditions, the colloids will set to form a solid or semipermeable shell [14, 15]. Coacervate formation is both entropy and enthalpy driven [4, 5] and so a minimum temperature have to be maintained. Coacervate are basically the cell like encapsulator model [12, 16, 17] to host the cellular components and bio molecules. Though Lipid vesicles are considered the prototypical protocell as they can form functional microscopic spherical assemblies due to their amphiphilic nature, but there are alternative models based on colloid phase separation that lead to the emergence of compartments. In many such encapsulation processes like food flavoring [14, 15, 18–20], drug delivery etc where it is desirable to encapsulate as well as compartmentalize a particle, enzymatic materials or drop, or polymeric stimuli responsive drug moiety to protect the interior contents of that particle, to provide controlled release [21] or to experiment the taste of the interior phase and thus coacervation are employed.

3 Mechanism

The coacervation process involves the formation of microscopic droplets of the coacervate phase in the stirred liquid phase resulting usually in unstable colloidal dispersion showing tendency towards coalescence. The droplet size can be preserved and maintained by chemical crosslinking or physical gelation of the polymer in the coacervate phase. The coacervation mechanism is generally subdivided into two groups: simple coacervation and complex coacervation. The simple coacervation (segregative phase separation) involves partial desolvation of the polymer caused either by changing the solvent quality for the polymer in the solution or by changing the solution temperature, the pH in the case of weak PEs, or by adding a

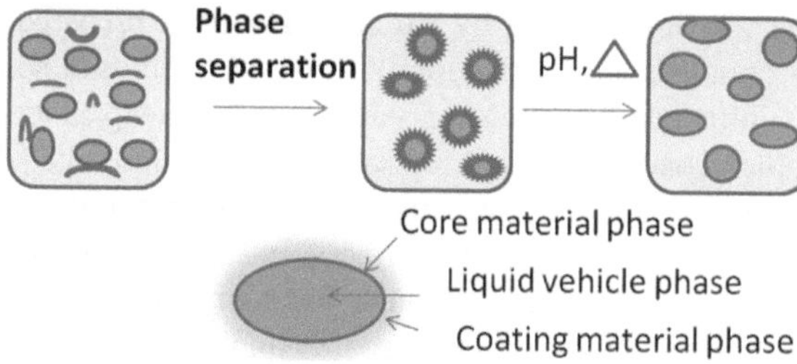

Fig. 1 Coacervation: multi phase distribution phenomenon

nonsolvent for the polymer to the solution. In medical technology, simple coacervation is often used for entrapping drugs into microcapsules.

When solutions of two colloids are mixed at a specific concentration, phase separation [14, 15, 18–20] between the colloidal system have been termed as complex coacervation. Complex coacervation typically possesses very low interfacial free energy [4, 5] in aqueous solution and this property enable the coacervate to engulf a variety of particles in solution. Hence the complex coacervation refers to the associative phase separation that occurs between two oppositely charged polyelectrolytes (PE), leading to the formation of polyelectrolyte complexes, or between a PE and di- and trivalent counterions resulting in so-called "inotropic" hydrogels. In rare cases, complex coacervation may occurs between polymers via hydrogen bonds leading to the formation of hydrogen-bonded complexes.

Fabrication methods of coacervation involves formation of stable micro and nanoparticles under mild conditions without using organic solvents, surfactants, or steric stabilizers. The chemical phases present in coacervate system are as follows (i) a liquid manufacturing vehicle phase (ii) a core material phase and (iii) a coating material phase. To form the three phases, the core material is dispersed in a solution of the coating material; the solvent for the polymer is the liquid manufacturing vehicle phase. The polymer coating material phase can be formed by changing the temperature of the polymer solution or by adding a salt or incompatible polymer [15, 18–20] (Fig. 1).

4 Coacervates Induced Nano Particle Formation

The remarkable achievements attained in the fields of nanoscience and nanotechnology have led to applications in the most diverse fields of medicine. **Nanomedicine** is the application of nanotechnology (the engineering of tiny particles) to the prevention and treatment of disease. In principle, nanomedicine focuses on two main fields: (a) development of new diagnostic systems like nanobiosensors, nanochips and (b) Development of highly efficient delivery vehicles to incorporate the drug induced nano particles within human body. Complex coacervation induced nano particle formation was investigated by Hu et al. Hu et al. prepared chitosan–PAA nanoparticles by template polymerization of PAA in a chitosan solution at 70 °C using $K_2S_2O_8$ as initiator [22]. Chitosan nanoparticles encapsulating siRNA were prepared by complex coacervation of chitosan and polyguluronate [23]. Because of their ease of preparation and extensive modification opportunities, complex coacervate core micelles (C3Ms) are used as a good alternative for diffusional nanoprobes. Carboxymethylcellulose and alginate have also been complexed via the coacervation technique with chitosan to prepare nanoparticles. There are numerous reports on the preparation of chitosan–DNA nanoparticles and chitosan–heparin nanoparticles via phase induced coacervation technique. Thus complex coacervation is a very efficient and useful procedure for obtaining nanoparticles. The speculative subfield of nano technology have emerged a new dimension towards incorporation of nanomedicines within micro capsules as well as formation of nano particles via the microsphere based techniques.

5 Factor Affecting Coacervation

The coacervation procedure is the stimuli responsive system that largely depends on various factors like concentration of the added macro ions, pH, temperature and ionic strength. The concentration of added micro ions are varied with the total polyion mixing concentration increasing with an increase in the polyion concentration up to a maximum at a polyion concentration of approximately 1 % (w/v) and then decreasing. The micro ion concentration at which maximum coacervation occurs changes with poly ion concentration as does the value of this maximum. Kaibara et al. saw no temperature dependence for pH_c and pH_φ for PDADMAC and BSA. On the other hand, because coacervation of PE/micelle system can be entropy-driven, by release of counterions, temperature can induce coacervation.

6 Theoretical Approach

In accordance with the view of oparin-Haldane though coacervation theory is typically based on protocell formation within the coast of archaic water basin some hypothesis (coacervate in coacervate) also made on the basis of coacervate as "culture media". The approach concerns coacervates as suitable culture media from which the first polynucleotide have ever manufactured and around the replicating DNA molecule secondary DNA molecules accumulated which developed gradually to the first prokaryotic cells. A theoretical approach on complex coacervation was first demonstrated by overbeek and voorn on the basis of experimental studies of de Jong [24] and his coworkers. The voorn-overbeek theory practically based on the configurational status of coacervates in multiple phases [15, 25] though they have neglected the Huggins interaction or solute-solvent interaction. A second theory was parallely proposed by Veis [18]; "the dilute phase aggregation model" that correlates the electrostatic attraction phenomenon with simple thermodynamics according to which entropy driven aggregation drives the complex coacervation. There are several basic differences between these two theories. Voorn-overbeek theory have neglected the thermodynamic approach along with solute-solvent interactions and they have utilized the gelatin-acacia system as a major one whereas Veis-Aranyi group deals with gelatin-gelatin system as a major one. However later the voorn-overbeek model was modified by Veis by including the Huggins parameter whereas replacing the debye-huckel parameter [18].

7 Application

Coacervates have been taken over in various field of application of medical technology with their astonishing versatility. The nature of coacervation as described is fully dependent on weak electrostatic interaction and entropy dependent phase separation mainly. Over the course of the years, the research on Coacervates have been shifted from prebiotics to its application in modern medicinal chemistry. In this regard, Scientists have taken different types of oppositely charged polyelectrolites to monitor their self assembling behavior.

The polymer coacervation is a process of liquid–liquid phase separation of a polymer solution into a polymer-rich phase (coacervate phase) and a polymer-lean phase (equilibrium phase) which is generally distinct from precipitation or coagulation—a phase separation process in colloidally unstable systems resulting in the formation of compact aggregates (coagula). Coacervation involves formation of stable micro and nanoparticles or spheres under mild conditions without using organic solvents, surfactants, or steric stabilizers. The fabrication being a reversible process, finds a potential application as vehicles for triggered drug delivery.

The complex coacervation is an increasingly popular approach for the fabrication of particulate drug delivery systems. Stable water-soluble colloidal particles are often formed by oppositely charged PEs mixed together under certain conditions in a ratio by which a nonstoichiometric electrostatic complex is formed in the fabricated coacervate. The existing studies on complex coacervate particles is actually based on the early works of Tsuchida's and Kabanov's groups [19].

In the vast majority of studies in relation to application of complex coacervate, chitosan, a naturally occurring weak polycation, and its derivatives are used as one of the components. Besides biocompatibility, biodegradability, and low toxicity, the popularity of chitosan is explained by its mucoadhesive properties, making it useful for transmucosal drug delivery as well as its ability to open tight junctions between epithelial cells, enhancing the transport of macromolecules across epithelia [19]. Colloids based on electrostatic chitosan–DNA, chitosan–protein, and chitosan–polysaccharide complexes along with chitosan hydrogels crosslinked with polyion tripolyphosphate have drawn much interest due to the formation of strong interaction between high positive charge on the backbone of chitosan and those negatively charged biomacromolecules with them. Because of the electrostatic nature of these complexes, they are intrinsically pH- and ionic strength responsive and are modern tool for drug delivery.

Complex coacervation phase separation behavior via protein-polysaccharide interaction like gelatin gum acacia system was first reported by Oparin [1]. Followed by him different types of protein-polysaccharide system have been taken into account to study the coacervation. As depicted chitosan-gelatin, chitosan-gum acacia, chitosan-DNA system are granted as the major ones for the experimental purpose. Chitosan the deacetylated product of chitin, chemically known as β (1,4) 2-amino-2-deoxy-D-glucose is a hydrophilic polysaccharide renders special cationic property in solution chemistry [20]. Due to it's excellent biodegradable and biocompatible property, this naturally occurring biopolymer act a major role towards medicinal chemistry. Gelatin, gum acacia [14, 15, 18–20, 26] and DNA are anionic biomolecules that interacts with positively charged moieties via supramolecular interactions [14, 15, 18–21, 26, 27]. Formation of lysozyme-heparin complex coacervates [21] at a pH near the physiological pH (pH-7.2) largely resembles the proto cellular type coordination. Among different scenarios, inter molecular complexation between oppositely charged macro molecules like lanthanide based [27] heavy metal, ATP-Polylysine [28–31], polynucleotide-ATP [32] coacervates have emerged a new field of research on microsphere based drug encapsulation as well as the self sustaining nature have emerged a new paradigm in the area of origin of life. Spontaneous self assembling behavior of fatty acid entrapped membrane on the surface of preformed coacervate micro droplets has also been investigated. These types of membrane bounded coacervate microdroplets have a broad range of sequestration property as compared to uncoated coacervate micro droplets. Partitioning of different types of dyes, nano particles, enzymes and coenzymes within the membrane bound [22, 23, 32, 33] as well uncoated coacervate micro droplet inferior have been well established. Work in progress is focused on incorporating more versatile types of nano- and micro phases into complex coacervates to further improve encapsulation properties. Complex coacervation was proposed to play a role in the formation of the underwater bioadhesive of the Sandcastle worm (*Phragmatopoma californica*) [22].

8 Conclusion

Superamphiphiles or supramolecular chemistry based on coacervation have emerged a new scope of research towards various field of science. Today the comprehensive technology of coacervation enables it to satisfy the need as encapsulator. However the most important aspect of coacervate research is their diversity towards application in pharmaceutical industry as well as in food and polymer industry. Moreover electrochemical moieties can be attached through the procedure via noncovalent interactions greatly reducing the need for organic chemical synthesis. Since the coacervate moieties are expected to possesses stimuli responsiveness, they can able to compartmentalize. Thus condensation of wide range of poly electrolytes in aqueous media produces microdroplets that are capable of sequestering wide range of compounds, dyes, drug molecules and nano particles. Thus coacervation is an extremely facile method for non covalent construction of stimuli responsive moieties.

References

1. Oparin AI (1953) The origin of life, 2nd edn. Dover Publications, New York
2. Mansy SS et al (2008) Template-directed synthesis of a genetic polymer in a model protocell. Nature 454:122–125
3. Rasmussen S et al (eds) (2009) Protocells: bridging nonliving and living matter. MIT Press, Cambridge
4. Luisi PL (2006) The emergence of life. Cambridge University Press, Cambridge
5. Hargreaves WR, Deamer DW (1978) Liposomes from ionic, single-chain amphiphiles. Biochemistry 17:3759–3768
6. Szostak JW, Bartel DP, Luisi PL (2001) Synthesizing life. Nature 409:387–390
7. Meierhenrich UJ, Filippi JJ, Meinert C, Vierling P, Dworkin JP (2010) On the origin of primitive cells: from nutrient intake to elongation of encapsulated nucleotides. Angew Chem Int Ed 49:3738–3750
8. Dzieciol AJ, Mann S (2012) Designs for life: protocell models in the laboratory. Chem Soc Rev 41:79–85
9. Deamer DW, Dworkin JP (2005) Chemistry and physics of primitive membranes. Top Curr Chem 259:1–27
10. Apel CL, Deamer DW, Mautner MN (2002) Self-assembled vesicles of monocarboxylic acids and alcohols: conditions for stability and for the encapsulation of biopolymers. Biochim Biophys Acta 1559:1–9
11. Oberholzer T, Wick R, Luisi PL, Biebricher CK (1995) Enzymatic RNA replication in self-reproducing vesicles: an approach to a minimal cell. Biochem Biophys Res Commun 207:250–257
12. Chen IA, Szostak JW (2004) Membrane growth can generate a transmembrane pH gradient in fatty acid vesicles. Proc Natl Acad Sci USA 101:7965–7970
13. Hyman AA, Simons K (2012) Cell biology. Beyond oil and water—phase transitions in cells. Science 337:1047–1049
14. Burgess DJ (1960) Complex coacervates of gelatine. J Phys Chem 64 1203–1210
15. Overbeek JTG, Voorn MJ (1957) Phase separation in polyelectrolyte solutions. Theory of complex coacervation. J Cell Comp Physiol 49(Supp I):7
16. Zhu TF, Adamala K, Zhang N, Szostak JW (2012) Photochemically driven redox chemistry induces protocell membrane pearling and division. Proc Natl Acad Sci USA 109:9828–9832
17. Adamala K, Szostak JW (2013) Competition between model protocells driven by an encapsulated catalyst. Nat Chem 5:495–501
18. Veis A (1961) Phase separation in polyelectrolyte solutions. II. Interaction effects. J Phys Chem 65:1798–1803
19. Motornov M, Roiter Y, Tokarev I, Minko S (2010) Prog Polym Sci 35:174–211
20. Sionkowska A, Wisniewski M, Skopinska J, Kennedy CJ, Wess TJ (2004) The photochemical stability of collagen-chitosan blends. Biomaterials 162:545–554
21. Schmitt C, Sanchez C, Thomas F, Hardy J (1999) Complex coacervation between h-lactoglobulin and acacia gum in aqueous media. Food Hydrocoll 13:483–496
22. Stewart RJ, Wang CS, Shao H (2011) Complex coacervates as a foundation for synthetic underwater adhesives. Adv Colloid Interface Sci 167:85–93

23. Hu Y, Jiang X, Ding Y, Ge H, Yuan Y, Yang C (2002) Biomaterials 23:3193–3201
24. Bungenberg de Jong HG, Kruyt HR (1929) Coacervation (partial miscibility in colloid systems). Proc K Ned Akad Wet 32:849–856
25. Overbeek JTG, Voorn MJ (1957) Phase separation in polyelectrolyte solutions. Theory of complex coacervation. J Cell Comp Physiol 49(1):7–26
26. Burgess DJ, Carless JE (1984) Microelectrophoretic studies of gelatin and acacia for the prediction of complex coacervation. J Colloid Interface Sci 98(1):1–8
27. Wang J, Velders AH, Gianolio E, Aime S, Vergeldt FJ, Van As H, Yan Y, Drechsler M, de Keizer A, Cohen Stuarta MA, van der Guchta J (2013) Controlled mixing of lanthanide(III) ions in coacervate core micelles. Chem Commun 3736
28. Seyrek E, Dubin PL, Tribet C, Gamble EA (2003) Ionic strength dependence of protein polyelectrolyte interactions. Biomacromolecules 273–282
29. Poon W, Pusey P, Lekkerkerker H (1996) Colloids in suspense. Phys World 55:3762
30. Mattison KW, Brittain IJ, Dubin PL (1995) Protein–polyelectrolyte phase boundaries. Biotechnol Prog 11:632–637
31. Weinbreck HS, Rollema RH (2004) Tromp, diffusivity of whey protein and gum arabic in their coacervates. Langmuir 20:6389–6395
32. Koga S, Williams DS, Perriman AW, Mann S (2011) Peptide–nucleotide microdroplets as a step towards a membrane-free protocell model. Nat Chem 3(9):720
33. Lee DW, Yun K-S, Ban H-S, Choe W, Lee SK, Lee KY (2009) J Control Rel 139:146–152
34. Fulton AB (1982) How crowded is the cytoplasm? Cell 30:345–347
35. Bakker MAE, Galema SA, Visser A (1999) Microcapsules of gelatin and carboxy methyl cellulose. European Patent Application EP 0 937 496 A2, Unilever NV, NL; Unilever PLC, GB (Bangs WE, Reineccius GA 1981)
36. Tiebackx FWZ (1911) Gleichzeitige Ausflockung zweier Kolloide. Chem Ind Kolloide 8:198–201

A Simulation Study of Nanoscale Ultrathin-Body InAsSb-on-Insulator MOSFETs

Swagata Bhattacherjee and Subhasri Dutta

Abstract In this paper, we report, a simulation study of nanoscale ultra thin body InAsSb channel n-MOSFETs. Our work is based on numerical simulation using ATLAS, a 2-D device simulator. Accuracy of the model has been verified by comparing simulation results with the reported experimental data. The proposed model has been employed to calculate the drain current, transconductance of InAsSb channel MOSFETs for different gate and drain voltages and also to compute Short Channel Effects.

Keywords Analog circuit applications · InAsSb DG MOSFETs · Transconductance · Threshold voltage roll-off · Subthreshold slope

1 Introduction

Silicon is on the verge of its fundamental limits due to the aggressive scaling of the modern complementary metal oxide semiconductor field effect transistor (MOSFET) [1]. Existing technology trends indicate that more non-Si elements are being added to the Si transistor for enhancing its scalability and performance for the future CMOS technology. Recently high mobility channel materials has attracted extensive attention due to its improved carrier mobility [2]. Among all known semiconductors, $InAs_xSb_{1-x}$ has one of the highest electron mobilities [2–9] and saturation velocities. However, due to its smaller band gap and larger permittivity,

S. Bhattacherjee (✉)
Department of Physics, JIS College of Engineering, Block A, Phase III, Kalyani, India
e-mail: swagata.bhattacherjee@gmail.com

S. Dutta
Department of Nano Science and Technology, JIS College of Engineering, Block A, Phase III, Kalyani, India
e-mail: abiswas5@rediffmail.com

S. Gupta et al. (eds.), *Advancements of Medical Electronics*,
Lecture Notes in Bioengineering, DOI 10.1007/978-81-322-2256-9_36

InAsSb channel MOSFETs suffer from high value of leakage current and reduced electrostatic integrity in short channel devices.

There have been a few approaches available in the literature [4, 5] in order to evaluate the current voltage characteristic of InAsSb channel devices pertaining to DG MOSFETs. In [4], an experimental work on the fabrication of ultrathin-body InAsSb-on-insulator n-type field effect transistors (FETs) with ultrahigh electron mobilities has been reported. Another group [4, 5] worked with depletion and enhancement-mode InAsSb MOSFETs by integrating a composite high-κ gate stack considering ultrathin $InAs_{0.7}Sb_{0.3}$ quantum well structure. But till date no work has been delt with short channel effects of InAsSb channel MOSFETS.

In the present paper first time we have studied the simulation work on short channel effects of InAsSb channel MOSFETs. Also the performance of nanoscale InAsSb channel MOSFETs over the entire region of operation have been studied. Further our predicted results have been compared with reported experimental data [8] to ensure the validity of our proposed model.

1.1 Model Development

The cross section of the n-type InAsSb channel MOSFET considered in our study is shown in Fig. 1. The details of the process flow for fabrication of such a device are reported in [8]. The device comprises a 500-nm n-type ultrathin $InAs_{0.7}Sb_{0.3}$ layers of different thicknesses (T_{InAsSb} = 7 and 17 nm) with a doping concentration of 1×10^{17} cm^{-3}. The gate insulator consists of a 10-nm thick ZrO_2 dielectric layer, while Ni is used for formation of ohmic source and drain contacts. For top-gate FETs, a 10-nm-thick ZrO_2 while for bottom gate a 50 nm SiO_2 gate dielectric is used. In the investigations we have employed 2-D numerical device simulator

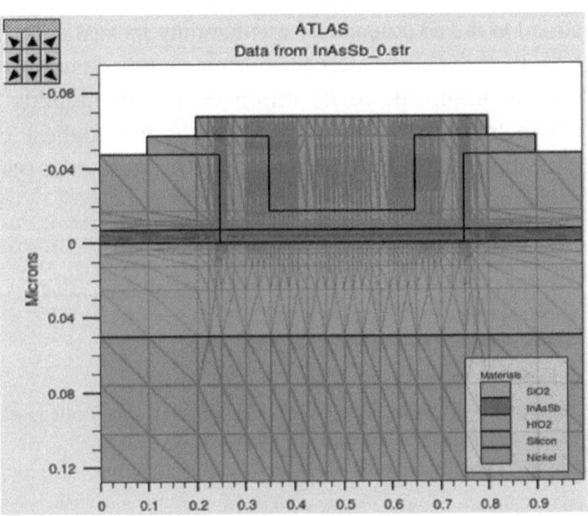

Fig. 1 Schematic diagram of a $InAs_{0.7}Sb_{0.3}$ channel MOSFET

ATLAS to simulate the InAsSb MOSFET. Two-dimensional device simulations are done for the InAsSb-channel MOSFET, as shown in Fig. 1. As the channel material is an alloy, its dielectric constant and bandgap are computed using linear interpolation among the corresponding reported parameters of the constituents [9]. The computed values of dielectric constant and bandgap of 17.7 and 0.174 eV, respectively, for $InAs_{0.7}Sb_{0.3}$ are used in our simulation. An intrinsic carrier concentration of 1.8×10^{16} cm^{-3} for InAsSb is used in our study. Experimental findings reveal that the InAsSb/high-*k* interface contains appreciable amount of interface trap charge density in the range of 10^{13} eV^{-1} cm^{-2} [8] which is incorporated in our simulation. Physical models that were taken into account in the simulation included the electric field dependent carrier mobility and the velocity saturation. The measured value of electron mobility in InAsSb channel n MOSFETs are used to obtain empirically the field dependent relationship for mobility [9]. This functional variation of mobility with respect to the effective field has been incorporated as a library file in the simulator by employing the C-interpreter tool. Various electrical parameters like electron concentration, surface potential, electric field, drain current, etc. have been obtained from the device simulation of InAsSb channel MOSFETs. Figure 1 shows Schematic diagram of an InAsSb channel n-MOSFET displaying different regions such as source, drain and gate along with mesh points within the grid as obtained from ATLAS. All dimensions are in μm.

2 Results and Discussions

Figure 2 shows the surface potential variation of the structure referred in Fig. 1. A comparison of the simulated device characteristics with the experimental results as reported in [8] for the InAsSb MOSFET is shown in Fig. 3 for three different values

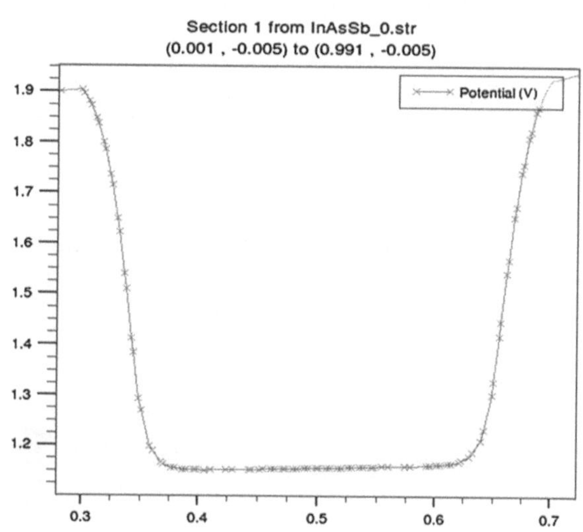

Fig. 2 Variation of surface potential with channel length. Other parameters are l = 500 nm, t_{InAsSb} = 7 nm, $D_{it} = 1 \times 10^{13}$ eV^{-1} cm^{-2}, t_{eq} = 1.5 nm, V_{gs} = −0.2 V and V_{ds} = 50 mV

Fig. 3 Dependence of drain current with drain bias for three different gate biases. Other parameters are $l = 500$ nm, $t_{InAsSb} = 7$ nm, $D_{it} = 1 \times 10^{13}\ eV^{-1}\ cm^{-2}$ and $t_{eq} = 1.5$ nm. *Solid symbols* represent simulation data and *open symbols* for experimental results

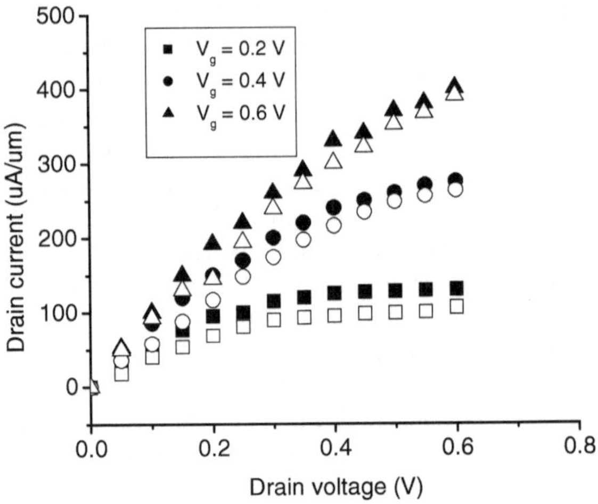

of the drain bias. It is followed from the graph that the results obtained from our model matches well with the experimental data which ensures validity of our model. Figure 4 shows the variation of drain current in both the linear and log scales with gate to source voltage for InAsSb and Si channel devices considering two different channel lengths. For lower channel length the drain current is higher as expected but it is observed that instead of having higher values of ON current in InAsSb channel device, a large amount of OFF current relative to the Si is also present. High OFF current and high value of subthreshold slope for InAsSb channel MOSFETs (161 mV/dec for 90 nm channel length) makes the device inferior for low power applications. Figure 5 depicts the transconductance curves of InAsSb control devices for a wide range of gate bias voltages to capture the subthreshold, linear and saturation regions of operation. As follows from the curve that transconductance shows a higher value for InAsSb channel device. But in case of

Fig. 4 Variation of drain current with gate bias for two different channel lengths. *Open* and *closed symbols* are for Si and InAsSb channel devices respectively. Other parameters are $t_{InAsSb} = 7$ nm, $V_{ds} = 0.05$ V and $t_{eq} = 1.5$ nm

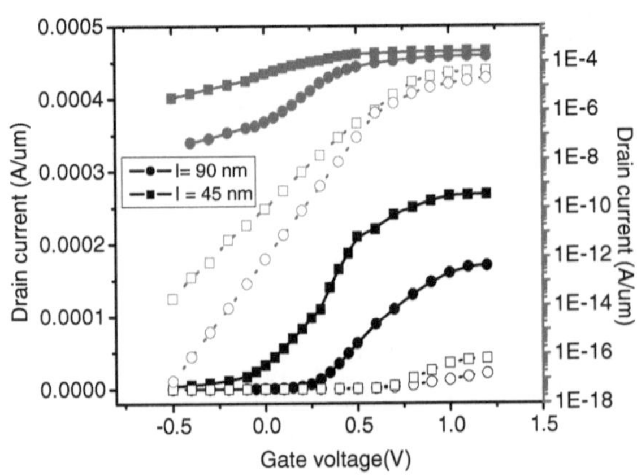

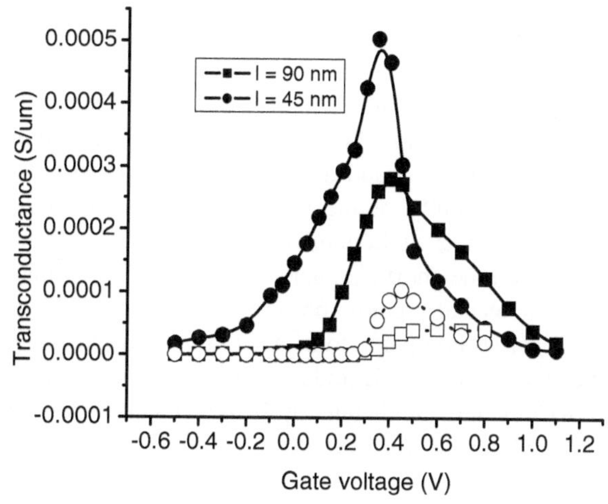

Fig. 5 Plot of transconductance with gate bias for different channel lengths. Other parameters are $t_{InAsSb} = 7$ nm, $V_{ds} = 0.05$ V and $t_{eq} = 1.5$ nm. *Open* and *closed symbols* are for Si and InAsSb channel devices, respectively

InAsSb channel device, transconductance reduces very fast with increasing gate bias while Si channel device can retain it for a longer time.

Figure 6 depicts variation of threshold voltage roll-off with channel length for InAsSb channel deices and its Si counterparts. From the graph it is found that short channel effects is higher for InAsSb channel devices due to its high dielectric constant and beyond 65 nm it is become dominating. So 65 nm is the scaling limit for the current structure.

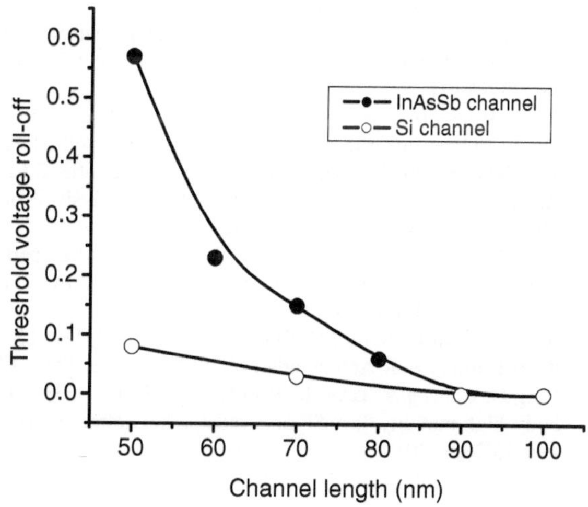

Fig. 6 Variation of threshold voltage roll-off with channel length. Other parameters are $t_{InAsSb} = 7$ nm, $V_{ds} = 0.05$ V and $t_{eq} = 1.5$ nm. *Open* and *closed symbols* are for Si and InAsSb channel devices, respectively

3 Conclusion

In this paper first time we have investigated simulation study of short $InAs_xSb_{1-x}$ channel MOSFETs. Predicted simulation results have been verified with reported experimental datas and a good match has been found among the two which confirms the validity of the proposed model. Our investigations show that InAsSb channel devices yield improved results for ON current and transconductance. But at the same time short channel effects is also a dominating factor for InAsSb channel devices specially below 65 nm channel length. InAsSb has a much smaller direct band gap as compared to Si which gives rise to a higher leakage current. Its higher dielectric constant degrades electrostatic integrity in short channel devices. Thus replacement of Si channel by InAsSb channel devices requires a few critical issues to be addressed in InAsSb channel MOS technology.

References

1. Kim SH, Yokoyama M, Taoka N, Nakane R, Yasuda T, Ichikawa O, Fukuhara N, Hata M, Takenaka M, Takagi S (2011) Enhancement technologies and physical understanding of electron mobility in III–V n-MOSFETs with strain and MOS interface buffer engineering. In: IEEE IEDM Tech Dig 13.4.1
2. Bennett BR, Magno R, Boos JB, Kruppa W, Ancona MG (2005) Antimonide-based compound semiconductors for electronic devices: a review. Solid State Electron 49:1875–1895
3. Ali A et al (2010) Advanced composite high-k gate stack for mixed anion arsenide antimonide quantum well transistor. In: IEEE IEDM Tech Dig 6.3.1
4. Ali A, Madan H, Misra R, Agrawal A, Schiffer P, Boos JB, Bennett BR, Datta S (2011) Experimental determination of quantum and centroid capacitance in arsenide–antimonide quantum-well MOSFETs incorporating nonparabolicity effect. IEEE Trans Electron Devices 58:1397–1403
5. Ali A, Madan H, Agrawal A, Ramirez I, Misra R, Boos JB, Bennett BR, Lindemuth J, Datta S (2011) Enhancement-mode antimonide quantum-well MOSFETs with high electron mobility and gigahertz small-signal switching performance. IEEE Electron Device Lett 32:1689–1691
6. Pal HS, Low T, Lundstrom MS (2008) NEGF analysis of InGaAs Schottky barrier double gate MOSFETs. In: IEEE IEDM Tech Dig 1
7. Xuan Y, Wu YQ, Shen T, Yang T, Ye PD (2007) High performance submicron inversion-type enhancement-mode InGaAs MOSFETs with ALD Al_2O_3, HfO_2 and HfAlO as gate dielectrics. In: IEEE IEDM Tech Dig 637
8. Ali A, Madan H, Barth MJ, Boos JB, Bennett BR, Datta S (2013) Effect of interface states on the performance of Antimonide nMOSFETs. IEEE Trans Electron Devices 34:360–362
9. Fang H, Chuang S, Takei K, Kim HS, Plis E, Liu C-H, Krishna S, Javey A (2012) Ultrathin-body high-mobility InAsSb-on insulator field-effect transistors. IEEE Trans Electron Device Lett 33:2012–2014

Author Index

S. Gupta et al. (eds.), *Advancements of Medical Electronics*,
Lecture Notes in Bioengineering, DOI 10.1007/978-81-322-2256-9